T0140257

Invariant Integrals in Physics

Genady P. Cherepanov

Invariant Integrals in Physics

Genady P. Cherepanov
Miami, FL, USA

ISBN 978-3-030-28339-1 ISBN 978-3-030-28337-7 (eBook)
https://doi.org/10.1007/978-3-030-28337-7

This Springer imprint is published by the registered company Springer Nature Switzerland AG
The registered company address is: Gewerbestrasse 11, 6330 Cham, Switzerland

Preface

In this book, basic laws of physics are derived from the law of energy conservation using invariant or path-independent integrals. This new approach allowed the present author to also discover some new laws generalizing Archimedes' buoyancy principle, Coulomb's Law of rolling, Newton's Law of gravity, Coulomb's Law for electric charges, and some others.

The book presents these findings; it is meant to be used in all walks of life and at school—wherever physics is necessary. For those who seek new opportunities for discovering new laws, it may be of a major interest. Some minimal mathematical knowledge, including the divergence theorem, elementary tensor skills, and analytical functions, would be helpful. However, the promise of discoveries much exceed these technical difficulties.

I express a deep gratitude to my son Yury G. Cherepanov who has a MS in math and a MA in business; he was an invaluable help in the preparation of this book.

Miami, USA
June 2019

Genady P. Cherepanov
Honorary Life Member of the New York
Academy of Sciences elected on December 8, 1976
together with Linus C. Pauling

Introduction

The value of an invariant integral does not depend on its path, surface, or volume of integration which can thus be deformed without changing this value. Invariant integrals express the laws of conservation of energy, mass, or momentum. Some path-independent contour integrals first appeared still in the eighteenth century in the works of Euler and Bernoulli and later played the main role in the mathematical theory of complex variables created by Cauchy, Riemann, and others. This theory found many applications to two-dimensional problems of fluid dynamics and elasticity. Recently, it appeared that invariant integrals in any dimensions can be used as the unique tool for deriving classical laws of physics and even for discovering some new laws.

The term "physics" comes from the ancient Greek "Φυσϊκη" which is the name of the most known book by Aristotle (384–322 BC), approximately translated as "nature." Today, physics is the main science that deals with energy, force, and matter of nature. Archimedes, Galileo Galilei, Isaac Newton, Gottfried Leibniz, Robert Hook, Leonhard Euler, Charles Coulomb, Georg Riemann, James Maxwell, Hendrik Lorentz, Jules Poincare, Albert Einstein, Nikolai Joukowski, John Eshelby, George Irwin, and others left their names on the main physical laws of nature. Some laws of these men are mentioned in Chap. 1, although they are set forth sometimes very differently from the traditional approach. In the rest of the book, there are no equations that would not belong to the present author, including the term "invariant integral" which was introduced about fifty years ago.

In this book, all physical laws are calculated from some invariant integrals which express the conservation of energy, mass, or momentum. This new approach allows us to unify the laws of theoretical physics, to simplify their derivation, and to discover some novel or more universal laws. For example, Newton's Law of gravity is generalized in order to take into account cosmic forces of repulsion and describe the growth and shape of the universe. Archimedes' principle of buoyancy is modified to take into account the surface tension of liquids. Even Coulomb's Laws describing the interaction of electric charges and the rolling friction are substantially repaired and generalized. For example, it appeared that relativistic electrons can interact in a weird way, namely, by attracting one another and coalescing into dense packs.

Moreover, invariant integrals provide an alternative to differential equations of the mathematical physics because they suggest a straightforward approach to the solution of boundary value problems. However, the corresponding numerical procedures are outside the framework of this book where one can find only exact analytical solutions. Physics today is still imperfect and sometimes self-contradicting, though. We say "vacuum" is a sort of nothing but it is in vacuum where some quite material and well-measured things like electromagnetic waves exist. Gravitation is an even more mysterious property of matter. Opposite to the general theory of relativity but based on numerous probes of the WMAP and Planck satellite missions over many years, the universe is flat which justifies the new approach to cosmology treated in this book.

The book can be a text for the special course "Invariant Integrals in Physics." It can also serve as a complementary textbook for graduate students specializing in physics and its applications.

Contents

List of Figures

Chapter 1
The Laws of Classical Physics

Abstract This chapter serves to introduce and discuss the method of invariant integrals using some well-known laws of physics. This method is applied to derive and calculate the buoyancy principle (Archimedes), the force of inertia and motion laws (Newton and Galileo), Einstein's equation connecting mass and energy, the law of gravity (Newton), the lift force of wings and the theory of flight (Kutta and Joukowski), the driving force of dislocations and foreign atoms in elastic materials (Peach, Koehler, and Eshelby), and Coulomb's Law of the interaction force of electric charges. This chapter is for everybody interested in physics.

The chapter reviews some basic laws of classical physics and derives them from invariant integrals.

1.1 Archimedes' Buoyancy Principle

Archimedes (287–212 BC) living in Sicily can be called the Grandfather of Science because he invented screw engine, spirals, buoyancy law, and calculus (by integrating area under parabola) among many other discoveries. Navigation, shipbuilding and aerostation are based on his laws. It was an epoch of bloody struggle between the arising Rome and powerful Carthage, with Sicily being between the jaws of both; the war tremendously stimulated Archimedes' inventions.

Let us consider the statics of a heavy fluid like water in a lake, sea, or anywhere else. The state of the fluid is determined by pressure which is a linear function of depth. Suppose a body of volume V is submerged in the fluid so that every point of the surface is under pressure of the fluid. The components of the resultant force the fluid acts upon the body are:

$$F_i = \int_S p n_i \mathrm{d}S \quad (i = 1, 2, 3) \tag{1.1.1}$$

G. P. Cherepanov, *Invariant Integrals in Physics*,
https://doi.org/10.1007/978-3-030-28337-7_1

Here p is the pressure of the fluid on any closed surface S embracing the body; and n_i are the components of the unit vector normal to this surface. This force called the buoyancy force is equal to the weight of the body if it rests.

Now, let us remove the body and fill up its volume by the same fluid so that the pressure field outside volume V remains the same. Applying the divergence theorem to the integral of Eq. (1.1.1) provides

$$F_i = \int_V p_{,i} dV = \rho V g_i \quad (i = 1, 2, 3) \tag{1.1.2}$$

Here $p_{,i} = \partial p/\partial x_i = \rho g_i$; ρ is the fluid density; g_i are the components of the gravity acceleration vector; and x_1, x_2 and x_3 are the Cartesian coordinates.

If the x_3-axis is chosen along the gravity force, then, from Eq. (1.1.2), we have

$$F_3 = -\rho g V, \quad F_1 = F_2 = 0. \tag{1.1.3}$$

Here g is the value of the gravity acceleration directed oppositely to the x_3- axis so that $p_{,1} = p_{,2} = 0$, and $p_{,3} = -\rho g$.

Further, let us assume that the original body of volume V does not produce any perturbations of the fluid pressure. Then, based on Eqs. (1.1.1)–(1.1.3), we conclude that *the buoyancy force upon the original body is equal to the weight of the fluid in the volume of this body*. It is the famous Archimedes' principle. The buoyancy force does not depend on what is inside the body. It does not depend on the shape/surface of any body of the same volume. Therefore, the integral of Eq. (1.1.1) can be called invariant or independent of its surface of integration if the surface covers the same volume.

And so, this principle is based on the assumption that the body produces no perturbations in the pressure field outside itself. If this assumption holds, we can also use this principle in the case of partial submergence of the body of volume V. As an example, consider the problem of iceberg floating in the sea. In this case, using Archimedes' buoyancy principle, it is easy to find that

$$V_w = \frac{W - gV\rho_a}{g(\rho_w - \rho_a)}, \quad V_a = -\frac{W - gV\rho_w}{g(\rho_w - \rho_a)}. \tag{1.1.4}$$

Here W and V are the total weight and volume of the iceberg, respectively; $\rho_w$and ρ_a are the densities of water and air; and V_w and V_a are the volumes of the iceberg submerged in water and air.

If the density of ice is equal to ρ_i, then from Eq. (1.1.4), it follows that

$$V_w/V_a = (\rho_i - \rho_a)/(\rho_w - \rho_i). \tag{1.1.5}$$

For example, $V_w/V_a = 10$ if the density of water is ten percent more than of ice.

In Chap. 2, we derive a more general law when the body submerged in a fluid changes its local pressure due to surface tension. Also, we consider the buoyancy force of submarines on the bottom, the floating of continents, as well as the buoyancy force in fluidized beds and in extremely viscous media.

1.2 The Law of Motion

Galileo Galilei (1564–1642) called the Father of Physics established the main principle of motion which says that *a body moves along a straight line at a constant speed if no forces act upon the body*. From this principle, it follows that any acceleration of its rectilinear motion or any curvature of its path is caused by a force. Galileo discovered the laws of falling bodies and the parabolic motion of projectiles; however, he became famous for his stubborn support of the Copernican theory in astronomy. It was the Enlightenment epoch, or the Age of Reason, in the history of Europe and the world.

Let us consider the rectilinear motion of an undeformable body of volume V and surface S at velocity v. We introduce the notions of the force of inertia and of the specific kinetic energy of the body using the following two postulates [3]:

(i) *The force of inertia of a body is equal to the balance of specific kinetic energy on the body surface*

$$F_i = \int_S K(\bar{v}) n_i dS \qquad (i = 1, 2, 3); \tag{1.2.1}$$

(ii) *The differential of specific kinetic energy equals the flow of the differential of flow of specific kinetic energy*

$$\mathrm{d}K = \bar{v}\mathrm{d}(\bar{v}K). \tag{1.2.2}$$

Here $K = K(\bar{v})$ is the specific kinetic energy of the body per unit volume; F_i are the components of the force of inertia; S is a closed surface embracing the body; n_i are the components of the outer unit vector which is normal to surface S; and $\bar{v} = v/c$ is the dimensionless velocity of the body where c is a special speed which physical meaning is found below in this section.

The integral in Eq. (1.2.1) is similar to the invariant integral of Eq. (1.1.1), with the specific kinetic energy playing the role of pressure on the body surface in Archimedes' principle. The integral in Eq. (1.2.1) is also called invariant because it does not depend on the shape of integration surface S embracing the body volume. The postulate of

Eq. (1.2.2) provides the definition of the notion of the specific kinetic energy of the body.

Let us re-write the condition equation in Eq. (1.2.2) as follows:

$$\frac{1}{v}\frac{\mathrm{d}K}{\mathrm{d}v} = \frac{1}{c^2}\frac{\mathrm{d}(vK)}{\mathrm{d}v} \quad \text{or} \quad \frac{\mathrm{d}K}{\mathrm{d}v} = \frac{vK}{c^2 - v^2}. \tag{1.2.3}$$

The solution of Eq. (1.2.3) is:

$$K = \frac{E}{\sqrt{1 - \frac{v^2}{c^2}}}. \tag{1.2.4}$$

Here E is a constant.

From Eq. (1.2.4), it follows that any velocity of matter cannot exceed the value of c. Hence, we conclude that c is the speed of light in vacuum because it is the maximum known velocity of propagation of any physical field.

Based on Eq. (1.2.4), function $K(v)$ can be represented by the following series

$$K = E\left(1 + \frac{1}{2}\frac{v^2}{c^2} + \frac{3}{8}\frac{v^4}{c^4} + \frac{5}{16}\frac{v^6}{c^6} + \ldots\right). \tag{1.2.5}$$

Now, let us require that our definition of specific kinetic energy should coincide with the classic definition of kinetic energy for small v/c as $K_0 = \rho_0 v^2/2$, where ρ_0 is the specific mass, or density, of the body at small velocities.

From here and from Eq. (1.2.5), it follows that

$$E = \rho_0 c^2. \tag{1.2.6}$$

From this theory and Eqs. (1.2.5) and (1.2.6), it follows also that any mass m_0 carries, in and by itself, a "silent" energy $E = m_0 c^2$ independent of its velocity. This is Einstein's equation, the most famous and important law of physics. The value of E, the energy of mass, is called the nuclear energy. Some ways of liberation of this energy led to the invention of nuclear weapons and power stations that have already transformed the life on the Earth and may as well determine the future life of the man. Einstein's equation has established a new physical entity, "the mass–energy." Equations (1.2.4)–(1.2.6) are usually derived from the special theory of relativity elaborated by Lorentz, Einstein, and Poincare in the Minkowski "space–time."

For the rectilinear motion of a body along the x_1-axis, using the divergence theorem and Eqs. (1.2.1)–(1.2.6), we get:

$$K_{,1} = \frac{\mathrm{d}K}{\mathrm{d}x_1} = \frac{\mathrm{d}K}{\mathrm{d}v}\frac{\mathrm{d}v}{\mathrm{d}x_1} = \frac{\mathrm{d}K}{\mathrm{d}v}\frac{\mathrm{d}v}{\mathrm{d}t}\frac{1}{v} = \frac{1}{c^2}\frac{\mathrm{d}}{\mathrm{d}t}(vK); \tag{1.2.7}$$

$$F_1 = \int\limits_S K n_1 \mathrm{d}S = \int\limits_V K_{,1} \mathrm{d}V = \int\limits_V \frac{1}{v}\frac{\mathrm{d}K}{\mathrm{d}v}\frac{\mathrm{d}v}{\mathrm{d}t}\mathrm{d}V = \frac{\mathrm{d}}{\mathrm{d}t}\left(\frac{m_0 v}{\sqrt{1-\frac{v^2}{c^2}}}\right). \tag{1.2.8}$$

Here t is time; and $m_0 = \rho_0 V$ and $F_2 = F_3 = 0$. When $v = \text{const}$, then $F_1 = 0$ in accordance with the principle of Galileo Galilei.

And so, from Eq. (1.2.8), it follows that the law of the rectilinear motion has the following shape:

$$\frac{\mathrm{d}}{\mathrm{d}t}\left(\frac{m_0 v}{\sqrt{1-\frac{v^2}{c^2}}}\right) = F_1 \quad \text{for any} \quad v < c; \tag{1.2.9}$$

$$\frac{\mathrm{d}}{\mathrm{d}t}(m_0 v) = F_1 \quad \text{for small } v \ll c. \tag{1.2.10}$$

The law of Eq. (1.2.10) suggested by Isaac Newton (1642–1727) is the famous Newton's Law of motion, and force F_1 is called the force of inertia. In his celebrated treatise *Principia*, Newton treated this approximate equation as the eternal, universal, God-granted law, Lord's proposition in his terms. Indeed, it worked well for the motion of both planets and small fluffs in Torricelli's vacuum tube—and it stood unshakable for two centuries.

Still earlier, Gottfried Leibniz (1646–1716) suggested the law $\mathrm{d}K_0/\mathrm{d}t = vF_1$ which is equivalent to Newton's Law (1.2.10); Leibniz called the kinetic energy $K_0 = \frac{1}{2}m_0 v^2$ the "live force." However, as a result of some bitter controversy powerful Sir Isaac put Leibniz down in obscurity and most of Leibniz discoveries became known much later. Even the grave of Leibniz buried by his servant alone was unknown and occasionally found many years later after his death.

For arbitrary motion along any curvilinear path, the law of motion is written as:

$$\frac{\mathrm{d}}{\mathrm{d}t}\left(\frac{m_0 \vec{v}}{\sqrt{1-\frac{v^2}{c^2}}}\right) = \vec{F} \quad \left(v^2 = |\vec{v}|^2 < c^2\right), \tag{1.2.11}$$

$$\frac{\mathrm{d}}{\mathrm{d}t}(m_0 \vec{v}) = \vec{F} \quad \left(v^2 = |\vec{v}|^2 \ll c^2\right). \tag{1.2.12}$$

Here $\vec{v}$ and $\vec{F}$ are the vectors of velocity and force. As a reminder, these laws of motion, Eqs. (1.2.11) and (1.2.12), are here derived from the invariant integral of Eq. (1.2.1). Equations (1.2.11) are the basic equations of the relativistic mechanics.

Particularly, from Eq. (1.2.12), it follows also that in the case of the uniform, circular motion of a mass around a center the mass is subject to the centrifugal force of inertia $m_0 v^2 r^{-1}$ (where r is the radius of rotation).

It should be noted that all of these equations including Eqs. (1.2.9)–(1.2.12) are written here using Leibniz's calculus. Newton never recognized it—he formulated the laws in his system of "fluxions," which has never been used and has been since forgotten.

In Chap. 10, some unusual properties of relativistic electron beams are studied by using Eqs. (1.2.11).

1.3 The Gravitation

Based on the works of Nicolaus Copernicus (1473–1543) and Tycho Brahe (1546–1601) in astronomy, Brahe's student Johannes Kepler (1571–1630) discovered that planets move around the Sun in one plane along some elliptical, close to circular, orbits so that the square of period of each revolution is directly proportional to the cube of the mean radius of the orbit. In 1672, Robert Hooke (1635–1703) who was the curator of the just established Royal Society in London proposed the inverse-square law of gravitation as an evident consequence of Kepler's laws based on the balance between the gravity and centripetal force of circular motion. In 1679, he assigned Isaac Newton to prove this law for elliptical orbits of planets.

It is noteworthy that Hooke and Newton were never-married Anglican monks who from 1689 to 1703 served to King William III, the former Prince of Orange of Holland, who occupied England in 1689. At that time, Holland was the mighty world power with many colonies in North America. For some time, Newton served as a spy for this king and saved king's treasury from counterfeiters. Despite several years spent in Bedlam, he became the Master of the Royal Mint for providing this important service to the Crown; he kept this office until his death. As the President of the Royal Society after Hooke died, Newton was second to none by his power.

In 1687, Newton published in *Principia* the proof of the inverse-square law of gravitation using Kepler's laws for elliptical orbits of planets. In 1725, after more than twenty years of the bitter priority dispute since Hooke's death, the meeting of the Royal Society declared its President Sir Isaac the author of the gravitation law:

$$F = G\frac{mM}{R^2}. \tag{1.3.1}$$

Here F is the gravity of masses m and M; R is the distance between them; and G is the gravitational constant. It is only in 1805 that Henry Cavendish (1731–1810) determined its value ($G \approx 6.67 \times 10^{-11}$ Nm^2 kg^{-2}).

Below we provide another approach to the theory of gravitation, in which Newton's Law (1.3.1) will be derived. Let us introduce the invariant integral of the gravitational field as follows [1–4]:

$$F_k = \frac{1}{4\pi G}\int\limits_S \left(\frac{1}{2}\varphi_{,i}\varphi_{,i}n_k - \varphi_{,i}\varphi_{,k}n_i\right)\mathrm{d}S \quad (i, k = 1, 2, 3) \tag{1.3.2}$$

Here $\varphi = \varphi(x_1, x_2, x_3)$ is the potential of the gravitational field; the comma in $\varphi_{,1}$, $\varphi_{,2}$ and $\varphi_{,3}$ means the derivatives over x_1, x_2 and x_3; and S is any closed surface of integration. Here and everywhere below the repeated indices mean summation. As we show further, F_k are the components of the resultant force of gravitation of all gravitational masses inside S. In particular, $F_k = 0$ if there are no gravitational masses inside S which means vacuum inside S.

Using the divergence theorem, we transform the integral in Eq. (1.3.2) as follows:

$$\begin{aligned}\int\limits_S \left(\frac{1}{2}\varphi_{,i}\varphi_{,i}n_k - \varphi_{,i}\varphi_{,k}n_i\right)\mathrm{d}S &= \int\limits_V \left\{\frac{1}{2}\left(\varphi_{,i}\varphi_{,i}\right)_{,k} - \left(\varphi_{,i}\varphi_{,k}\right)_{,i}\right\}\mathrm{d}V \\ &= \int\limits_V \left(\varphi_{,ik}\varphi_{,i} - \varphi_{,i}\varphi_{,ki} - \varphi_{,ii}\varphi_{,k}\right)\mathrm{d}V = -\int\limits_V \varphi_{,ii}\varphi_{,k}\mathrm{d}V.\end{aligned} \tag{1.3.3}$$

Here V is the volume inside S.

From Eqs. (1.3.2) and (1.3.3), it follows that in vacuum potential φ meets Laplace's equation

$$\varphi_{,ii} = 0. \tag{1.3.4}$$

In spherical coordinates r, θ, and ψ, Laplace's equation is written as follows:

$$\varphi_{,ii} = \frac{\partial^2\varphi}{\partial r^2} + \frac{2}{r}\frac{\partial\varphi}{\partial r} + \frac{1}{r^2\sin^2\theta}\frac{\partial^2\varphi}{\partial\psi^2} + \frac{1}{r^2}\frac{\partial^2\varphi}{\partial\theta^2} + \frac{1}{r^2}\cot\theta\frac{\partial\varphi}{\partial\theta} = 0. \tag{1.3.5}$$

Let a point source of the gravitational field be at the origin of coordinates. The simplest singular solution of Laplace's equation can be written as

$$\varphi = -G\frac{M}{r} - g_k x_k \quad \left(r^2 = x_k x_k\right) \tag{1.3.6}$$

Here x_1, x_2, and x_3 are the Cartesian rectangular coordinates, and M and g_k are some constants.

By differentiating Eq. (1.3.6) on x_k once, twice, and many times, we can get the whole set of solutions of Laplace's equation singular at the origin of coordinates.

Let this point source of Eq. (1.3.6) be inside the integration surface S. From Eq. (1.3.6), we have

$$\varphi_{,k} = -g_k + GMx_k/r^3. \tag{1.3.7}$$

And so, g_k is the intensity or tension of the unperturbed field of gravitation at the origin of coordinates.

Now, substitute the value of $\varphi_{,k}$ in Eq. (1.3.1) by that of Eq. (1.3.7) and calculate the value of force F_1 using as S the surface of the narrow parallelepiped with faces along $x_1 = \pm L$; $x_2 = \pm L$; and $x_3 = \pm\delta$, where $\delta/L \to 0$ and $L \to 0$:

$$F_1 = \frac{1}{2\pi G} \lim \int_{-L}^{+L}\int_{-L}^{+L} \varphi_{,1}\varphi_{,3} \mathrm{d}x_1 \mathrm{d}x_2$$
$$= \frac{1}{2\pi G} \lim \int_{-L}^{+L}\int_{-L}^{+L} \left(g_1 \frac{GM}{r^3}\delta + g_3 \frac{GM}{r^3} x_1 \right) \mathrm{d}x_1 \mathrm{d}x_2$$
$$= \frac{g_1 M}{2\pi} \lim \int_{-L}^{+L}\int_{-L}^{+L} \frac{\delta \mathrm{d}x_1 \mathrm{d}x_2}{\left(x_1^2 + x_2^2 + \delta^2\right)^{3/2}} = \frac{g_1 M}{2\pi} \int_{-\infty}^{+\infty}\int_{-\infty}^{+\infty} \frac{\mathrm{d}t_1 \mathrm{d}t_2}{\left(1 + t_1^2 + t_2^2\right)^{3/2}}. \quad (1.3.8)$$

In cylindrical coordinates R and ϑ so that $R^2 = t_1^2 + t_2^2$ and $\mathrm{d}t_1 \mathrm{d}t_2 = R\mathrm{d}R\mathrm{d}\vartheta$, we get

$$F_1 = \frac{g_1 M}{2\pi} \int_0^{2\pi} d\vartheta \int_0^{\infty} \frac{R\mathrm{d}R}{\left(1 + R^2\right)^{3/2}} = \frac{1}{2} g_1 M \int_1^{\infty} \frac{\mathrm{d}s}{s^{3/2}} = g_1 M \quad (1.3.9)$$

And so, in the general case, we have

$$F_k = M g_k \quad (i = 1, 2, 3) \quad (1.3.10)$$

To clarify the physical meaning of M, let us calculate the interaction force F_1 of a source M at point (0, 0, 0) upon another source m at point $(R, 0, 0)$ in the infinite vacuum space. According to Eq. (1.3.6), the source M produces the gravitational field of intensity $\varphi_{,1} = -g_1 = \mathrm{GM}/R^2$ at point $(R, 0, 0)$. Hence, based on Eq. (1.3.9), the force acting upon the source m is equal to

$$F_1 = -\mathrm{Gm}MR^{-2} \quad (1.3.11)$$

This is Newton's Law of Eq. (1.3.1); hence, m and M are the masses of the relative sources.

Suppose a heavy point mass m falls freely down to the Earth surface under the action of gravity so that the drag of air can be ignored. In this case, based on Eqs. (1.3.11) and (1.2.12), the free fall acceleration of the mass because of inertial force is equal to

$$g = \mathrm{GM}/R^2. \quad (1.3.12)$$

Here M and R are the mass and radius of the Earth. This verifiable relationship for the chosen value of the gravitational constant proves the identity of the gravitational and inertial mass.

And so, the theory of the Newtonian gravitational field is done here using the invariant integral of Eq. (1.3.2). From the viewpoint of the energy conservation law, the first term in it is the specific potential energy of the field and the second term is the specific work of the field tensions. From the viewpoint of the momentum conservation law, these terms can be treated as the specific momentum and stress tensor of the field.

In Chap. 11, Newton's Law is generalized by account of the cosmic forces of repulsion of the matter which allows one to look into the history of the universe, to calculate its age, and to solve the problems of Dark Energy, Dark Matter, and some other problems of cosmology.

1.4 The Flight

Let us study the motion of bodies in the air which, at low-subsonic velocities, can be considered as an incompressible, inviscid fluid.

Non-viscous fluids. For the stationary flow of an incompressible, inviscid fluid around a body, it is natural to introduce the invariant integral as follows [1–4]:

$$F_k = \int_S (pn_k + \rho v_i n_i v_k)\mathrm{d}S. \quad (i, k = 1, 2, 3) \tag{1.4.1}$$

Here p and v_i are the pressure and velocity components of the fluid; $\rho = \text{const}$ is the fluid density; S is a closed surface around the body; F_k are the components of the resultant force acting upon the body; and n_i are the components of the unit normal vector.

The first term in Eq. (1.4.1) coincides with that in Eq. (1.1.1) of Archimedes' principle for the static problem, and the second term takes account of the motion of the fluid. The sum of both terms provides the flow of the specific fluid momentum through surface S.

By applying the divergence theorem to the integral in Eq. (1.4.1), we get

$$F_k = \int_V (p_{,k} + \rho v_i v_{k,i} + \rho v_{i,i} v_k)\mathrm{d}V \quad (i, k = 1, 2, 3) \tag{1.4.2}$$

Since $v_{i,i} = 0$ for incompressible fluids and volume V inside surface S is arbitrary, from Eq. (1.4.2), it follows that

$$p_{,k} + \rho v_i v_{k,i} = 0 \quad \text{if} \quad F_k = 0 \quad (i, k = 1, 2, 3). \tag{1.4.3}$$

These are Euler's equations and d Alembert's paradox, according to which the drag force on a body moving in an inviscid fluid is equal to zero. And so, the force upon the body, lift, and/or drag can result only from some flow singularities where Euler's equations are not valid. Typically, these are vortices produced by the moving body in the fluid.

From Eqs. (1.4.3), it follows that for irrotational flows (when $v_{i,k} = v_{k,i}$), we get

$$2p + \rho v_i v_i = \text{const};\ v_i = \varphi_{,i} \ \text{ and } \ \varphi_{,ii} = 0 \quad (i = 1, 2, 3) \tag{1.4.4}$$

Here φ is the flow potential. The first equation in Eqs. (1.4.4) is the famous Bernoulli's equation.

Leonhard Euler (1709–1784), a Swiss-Russian genius, can be considered the Father of Hydrodynamics. He was born in Switzerland but took Russia's citizenship where he worked most of his life. He is believed to be the most prolific scientist of all times. It took more than hundred years after his death that the Russian Academy of Sciences could have published most of his works. Euler derived the basic equations of hydrodynamics long before its practical applications. He also solved the basic differential equations and put the ground of the theory of functions of a complex variable. It is interesting that Euler like the ancient philosopher Plato, but distinct from Galileo and Newton, believed in the science of pure reason—the only beauty and logic of mind were of interest for him. Paradoxically, his works were later used in practice more than anyone else's.

Because of broad industrial applications of various blades, wings, and aerofoils, the hydrodynamics of two-dimensional flows when $v_3 = 0$ is of special interest. For these flows, from Eqs. (1.4.4), we get

$$v_1 - iv_2 = \frac{\mathrm{d}w}{\mathrm{d}z}, \quad z = x_1 + ix_2, \quad 2p + \rho\left|\frac{\mathrm{d}w}{\mathrm{d}z}\right|^2 = \text{const.} \tag{1.4.5}$$

Here the complex potential $w(z)$ is an analytical function of the complex variable z to be found.

Let us calculate the force upon an arbitrary cylinder of contour C in the cross section placed in the uniform stream $v_1 = U$, $v_2 = 0$. This force per unit length of the cylinder is given by Eq. (1.4.1) where $v_i n_i = 0$ on C so that we get

$$F_1 - iF_2 = i\oint_C p\mathrm{d}\bar{z} = -\frac{1}{2}i\rho\oint_C \left|\frac{\mathrm{d}w}{\mathrm{d}z}\right|^2 \mathrm{d}\bar{z}. \tag{1.4.6}$$

Here we used Eq. (1.4.5) and the following identity $(n_1 - in_2)\mathrm{d}S = i\mathrm{d}\bar{z}$. Since $dw = d\bar{w}$ on C, we have

$$\left|\frac{\mathrm{d}w}{\mathrm{d}z}\right|^2 \mathrm{d}\bar{z} = \frac{\mathrm{d}w}{\mathrm{d}z}\frac{\mathrm{d}\bar{w}}{\mathrm{d}\bar{z}}\mathrm{d}\bar{z} = \left(\frac{\mathrm{d}w}{\mathrm{d}z}\right)^2 \mathrm{d}z. \tag{1.4.7}$$

From Eqs. (1.4.6) and (1.4.7), we conclude that

$$F_1 - iF_2 = -\frac{1}{2}i\rho \oint\limits_C \left(\frac{\mathrm{d}w}{\mathrm{d}z}\right)^2 \mathrm{d}z. \tag{1.4.8}$$

This is the well-known theorem of Blasius.

Since function $(\mathrm{d}w/\mathrm{d}z)^2$ is analytical, we can arbitrarily deform contour C in the flow domain with no change of the value of force $F_1 - iF_2$. Suppose for large z, the function $\mathrm{d}w/\mathrm{d}z$ behaves as follows:

$$\frac{\mathrm{d}w}{\mathrm{d}z} = U + \frac{Q - i\Gamma}{2\pi z}. \tag{1.4.9}$$

Here $v_1 = U$ is the uniform flow velocity.

The second term describes the sum of the following two flows:

$$v_r = \frac{Q}{2\pi r},\ v_\theta = 0; \tag{1.4.10}$$

$$v_r = 0,\ v_\theta = \frac{\Gamma}{2\pi r}. \tag{1.4.11}$$

Here r and θ are the polar coordinates, and v_r and v_θ are the corresponding components of the fluid velocity.

From Eqs. (1.4.10) and (1.4.11), it follows that quantity Q is the volume source of fluid per unit length and time, and Γ is the vortex circulation of the same dimension, both characterizing the linear singularity of the flow at $x_1 = x_2 = 0$.

And so, the solution of Eqs. (1.4.9)–(1.4.11) describes the flow in the whole unbounded space produced by the uniform flow and the vortex source at the coordinate origin. This flow takes place very far from an arbitrary body streamlined by fluid, if the body can produce this vortex and same fluid source.

Using the theorem of Blasius, Eq. (1.4.8), let us calculate the force of the uniform flow $v_1 = U$, $v_2 = 0$ upon the body of any contour C, which produces vortex and source at infinity described by Eq. (1.4.9). Since integral of Eq. (1.4.8) is invariant, we can deform contour C into the circle of very large radius, $z = R\exp(i\theta)$, where $R \to \infty$, and use Eq. (1.4.9). As a result of the simple calculation, we find

$$F_1 - iF_2 = \rho U(Q - i\Gamma). \tag{1.4.12}$$

This is the force upon profile C produced by the fluid stream. The value of Q has sense in the case of the boundary-layer suction used in some jets, but for most practical applications, it can be put zero. Then, the formula of Eq. (1.4.12) takes the classic shape

$$F_1 + iF_2 = i\rho\Gamma U \quad (F_1 = 0,\ F_2 = \rho\Gamma U) \tag{1.4.13}$$

And so, the lift is equal to the product of air density, flight speed, and circulation. The lift law of Eq. (1.4.13) was derived by Kutta, Joukowski, and Chaplygin in the early 1900s.

The lift of the aerofoil. The law of Eq. (1.4.13) seemed to have opened a way to aviation. However, circulation Γ is a very capricious, shape-dependent property of wing profiles. It was genius of Joukowski who designed special profiles with a cusp serving as the trailing edge of the aerofoil. The value of circulation Γ is easily projected in terms of this Joukowski profile shape, with the velocity at the cusp being finite.

Let us calculate the value of circulation Γ for very thin flat wings with chord a having the infinite span in the x_3-direction to provide the two-dimensional flow. The plane of the wing makes angle of attack α to the stream velocity. Let us designate the characteristic points of the flow domain and profile as follows:

$$A:\ z = \frac{1}{2}a\exp(\pi - \alpha)i;\ B:\ z = \frac{1}{2}a\exp(-i\alpha);$$
$$C_{+}\text{ and }C_{-}:\ z = \pm i0;\ 0 < \alpha < \frac{\pi}{2}$$

The rectilinear cut connecting points A and B in the complex plane z represents the wing profile. Point B is called the trailing edge of the flow. The stagnation point O of the flow where its velocity is zero is located on the frontal side of the wing (Fig. 1.1).

At infinity, the air stream is described by the complex potential of Eq. (1.4.9) where the values of Q and Γ should be found. To solve this problem, Joukowski offered that the flow velocity should be finite at the trailing point B of the profile and, hence, is directed under angle $-\alpha$ to the direction of the uniform flow. This guess appeared to be valid for a broad range of cusped profiles known as Joukowski aerofoils widely used in the construction of wings.

Let us provide the conformal mapping of the flow domain of the $z-$ plane onto the exterior of the unit circle of the parametric complex variable ξ as follows:

$$z = \frac{a}{4}\mathrm{e}^{-i\alpha}\left(\xi + \frac{1}{\xi}\right). \tag{1.4.14}$$

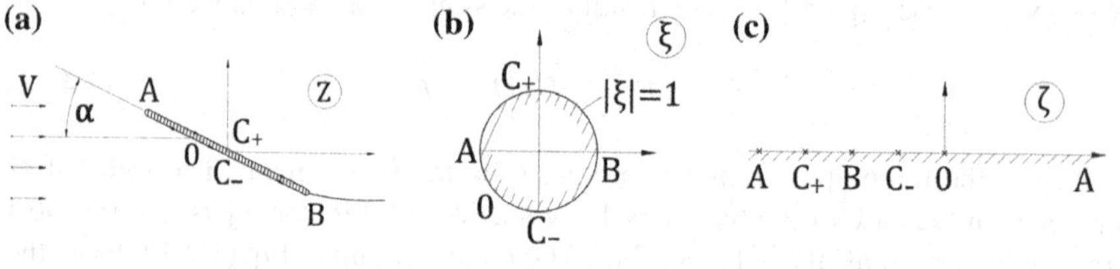

(a) V α A Z 0 C+ C- B

(b) ξ C+ |ξ|=1 A B 0 C-

(c) ζ A C+ B C- 0 A

Fig. 1.1 A thin plate streamlined by uniform flow. **a** The physical z-plane. **b** The parametric ξ-plane. **c** The parametric ζ-plane

The corresponding points A, C_+, B, C_-, O and A of the unit circle on the $\xi-$ plane passed clockwise are:

$$A := -1;\ C_+ : \ \xi = +i;\ B : \ \xi = 1;\ C_- : \ \xi = -i;\ O : \ \xi = \exp(\pi + \delta)i.$$

Here the value of δ has to be found.

The boundary condition equations on the $\xi-$ plane are:

$$\text{OA: } \arg\frac{dw}{dz} = \pi + \alpha; \quad AC_+BC_-O : \ \arg\frac{dw}{dz} = \alpha. \tag{1.4.15}$$

Here arg means the argument of the corresponding function.

Now, let us provide the conformal mapping of the exterior of the unit circle of the ξ-plane onto the upper half-plane of the ζ-plane as follows:

$$\zeta = -D\left(1 + e^{-i\delta}\right)\frac{\xi + e^{i\delta}}{\xi + 1}. \tag{1.4.16}$$

Here D is a positive real constant ($O : \ \zeta = 0;\ A : \ \zeta = \pm\infty$).

From Eq. (1.4.15), we get the following boundary value problem for the analytical function $\ln\dfrac{dw}{dz}$ on the upper half-plane of the $\zeta-$ plane:

$$\begin{aligned} \operatorname{Im}\ln\frac{dw}{dz} &= -\pi + \alpha \quad \text{when} \quad Re\zeta > 0, \operatorname{Im}\zeta = 0; \\ \operatorname{Im}\ln\frac{dw}{dz} &= \alpha \quad \text{when} \quad Re\zeta < 0, \operatorname{Im}\zeta = 0. \end{aligned} \tag{1.4.17}$$

The solution to this problem is evident:

$$ln\frac{dw}{dz} = \ln\zeta + i(-\pi + \alpha) + C; \tag{1.4.18}$$

$$C = \ln U - i(-\pi + \alpha) - \ln\left[-D\left(1 + e^{-i\delta}\right)\right]. \tag{1.4.19}$$

Here the second equation follows from the condition at infinity, Eqs. (1.4.9), (1.4.14), and (1.4.16), when $z \to \infty$ and $\xi \to \infty$. The imaginary part of Eq. (1.4.19) should be equal to zero; from this, it follows that $\delta = 2\alpha$.

By taking into account Eqs. (1.4.14)–(1.4.19), we get the final solution as follows:

$$\frac{dw}{dz} = U\frac{\xi + e^{2i\alpha}}{\xi + 1}, \quad ze^{i\alpha} = \frac{a}{4}\left(\xi + \frac{1}{\xi}\right), \quad (\delta = 2\alpha). \tag{1.4.20}$$

From here, using Eq. (1.4.9) at infinity, we get

$$\Gamma = -\pi a U \sin\alpha, \quad Q = 0. \tag{1.4.21}$$

Finally, from Eqs. (1.4.13) and (1.4.21), we find

$$F_2 = \pi a \rho U^2 \sin\alpha, \quad (F_1 = 0). \tag{1.4.22}$$

This is the lift per unit length of the infinite span of the wing.

For wings of finite spans, this solution is applicable in every wing cross section if $d\Gamma/dx_3 \ll 1$ so that the lift F of the wing based on Eq. (1.4.13) is very close to

$$F = \rho U \int \Gamma(x_3) dx_3. \tag{1.4.23}$$

For example, for swept wings with sweep angle ε when $a = \pm\varepsilon(s - x_3)$, we get the following equation for the lift

$$F = \pi\rho S U^2 \sin\alpha, \quad (\Gamma = -\pi S U \sin\alpha). \tag{1.4.24}$$

Here $S = \varepsilon s^2$ is the area of the triangular wing, and s is its span.

For small angles of attack, we have $F = \alpha\pi\rho U^2 S$ which was derived by Keldysh and Sedov in 1939; see also Landau and Lifshitz [11, 12].

The simple result of Eq. (1.4.24) is a very good approximation for thin swept wings of any cross-sectional shape, if its rear end is the trailing edge of the air flow, that is, as soon as Joukowski pattern works. For too high angles of attack, it is evidently violated and Eq. (1.4.24) becomes invalid. The critical angle of stall can be estimated using the theory of boundary layer. The above theory is valid only for small Mach numbers $M = v/c \ll 1$ (where c is the speed of sound in air).

We add two remarks about electromagnetic analogy and inductive drag.

Electromagnetic analogy. The Kutta–Joukowski equation $F = \rho V \Gamma$ for the lift force is analogous to Ampere's Law $\vec{F} = \vec{j} \times \vec{B}$ where $\vec{j} = I\,\overrightarrow{dl}$ for the force $\vec{F}$ exerted by the magnetic field $\vec{B}$ on the electric current element $I\overrightarrow{dl}$. Moreover, the Kutta–Joukowski equation should be written in the form of the following vector product: $\vec{F} = -\rho\vec{V} \times \vec{\Gamma}$. Here $\vec{V}$ is the velocity vector of the wing, and $\vec{\Gamma}$ is the circulation vector describing the vortex at infinity which produces the lift force (the vectors $-\vec{V}$, $\overrightarrow{\Gamma}$ and $\vec{F}$ form the right-hand triad similarly to that of $\vec{i}$, $\vec{j}$ and $\vec{k}$).

Inductive drag. For wings of finite span, it is necessary to keep in mind the effect of secondary vortices arising in planes normal to the flight direction that creates the force of inductive drag on the wing (the Prandtl effect).

In Chap. 5, the theory of flight based on the invariant integrals will be treated for the subsonic flows (when Mach numbers are less than one).

1.5 The Theory of Elasticity

Let us consider small deformations of the static elastic medium. In this case, the equilibrium equations and the law of energy conservation can be written in the form of the following invariant integrals [3, 5, 6]

$$F_i = \int_S \sigma_{ij} n_j \mathrm{d}S \quad (i, j = 1, 2, 3); \tag{1.5.1}$$

$$\Gamma_k = \int_S \left(U n_k - \sigma_{ij} n_j u_{i,k}\right) \mathrm{d}S \quad (i, j, k = 1, 2, 3). \tag{1.5.2}$$

Here U is the elastic potential per unit volume; u_i and σ_{ij} are the components of the displacement vector and stress tensor; and n_j are the components of the outer unit vector normal to the closed surface S of integration.

The physical meaning of F_i and Γ_k is different. Vector (F_1, F_2, F_3) represents the resultant vector of external forces acting upon the body inside S. The values of Γ_1, Γ_2 and Γ_3 are equal to the change of the elastic energy of the system inside S resulted from the motion of singular points, lines, or surfaces of the elastic field inside S (per unit length along the x_1, x_2 and x_3-axes).

Singular points of the elastic field are concentrated forces and moments of forces, small foreign inclusions, tiny pores and cavities, and the like. Singular lines of the elastic field are the fronts of cracks, dislocations, linear concentrated forces, and so on. Singular surfaces are typically interfaces of different elastic materials or planar pile-ups of dislocations in the elastic field.

Both F_k and Γ_k have the dimension of force. However, while the F_k are real forces, the values of Γ_k usually called moving or driving forces are, as a matter of fact, some specific losses of elastic energy of the body occurring only in the case if the corresponding singularities move. To distinguish them from real forces, we will call them the Γ-forces, with their dimension being like $N \cdot m$ per meter.

Invariant integrals of Eq. (1.5.2) were introduced by the great British scientist J. D. Eshelby, and independently, as a particular case of more general invariant integrals, by this author. It is interesting that Young's modulus E was introduced still by young Euler, but it and the equations of the theory of elasticity formulated later were named after Young, Poisson, Cauchy, Navier, and others.

Let us apply the divergence theorem to Eqs. (1.5.1) and (1.5.2). As a result, we get

$$F_i = \int_V \sigma_{ij,j} \mathrm{d}V \quad (i, j = 1, 2, 3); \tag{1.5.3}$$

$$\Gamma_k = \int_V \left[U_{,k} - \left(\sigma_{ij} u_{i,k}\right)_{,j}\right] \mathrm{d}V \quad (i, j, k = 1, 2, 3). \tag{1.5.4}$$

Here V is the volume inside S.

Since V is arbitrary, from Eqs. (1.5.3) and (1.5.4), it follows that:

$$\sigma_{ij,j} = \rho_{Fi}, \tag{1.5.5}$$

$$U_{,k} = \left(\sigma_{ij} u_{i,k}\right)_{,j} + \rho_{\Gamma k} = \sigma_{ij,j} u_{i,k} + \frac{1}{2}\sigma_{ij}\left(u_{i,jk} + u_{j,ik}\right) + \rho_{\Gamma k}. \tag{1.5.6}$$

Here ρ_{Fk} and $\rho_{\Gamma k}$ are some volume densities of internal forces and energy in the solid, when all external loads are zero. These values are of primary importance for some "smart" materials, explosives, common metals after a hard work or treatment, and so on.

We can write Eqs. (1.5.5) and (1.5.6) in terms of the strain tensor components $\varepsilon_{ij} = \frac{1}{2}(u_{i,j} + u_{j,i})$ as follows:

$$\sigma_{ij,j} = \rho_{Fi}, \quad U_{,k} = \sigma_{ij}\varepsilon_{ij,k} + \rho_{\Gamma k} + \rho_{Fi} u_{i,k} \quad (i, j, k = 1, 2, 3). \tag{1.5.7}$$

These are general equations of the theory of elasticity derived from the invariant integrals of Eqs. (1.5.1) and (1.5.2). And so, the invariant integrals provide more general approach to the theory of elasticity.

When $U = U\left(\varepsilon_{ij}\right)$ and $\rho_{Fk} = \rho_{\Gamma k} = 0$, we get the classical theory of elasticity:

$$\sigma_{ij,j} = 0, \quad \sigma_{ij} = \frac{\partial U}{\partial \varepsilon_{ij}}. \tag{1.5.8}$$

In the case when the elastic potential is a quadratic function of strains, these equations are linear. They describe small deformations of most liquid and solid materials within a specific time of relaxation.

Dislocations. All polycrystalline materials like metals have a lot of distortions and defects in the structure. Most important are dislocations, inclusions, and fractures created during the formation and prehistory of the material. The growth, motion, and reproduction of these objects determine life and death of everything made of this material.

Consider first an *elementary screw dislocation*, the simplest linear defect of a lattice. Suppose that it has its front at $x_1 = x_2 = 0, -\infty < x_3 < +\infty$ and its plane at $x_2 = 0, x_1 < 0$ so that its field of stresses and displacements in the infinite homogenous isotropic space is:

$$u_3 = \frac{b_3}{2\pi}\theta, \quad \sigma_{13} = -\frac{\mu b_3}{2\pi r}\sin\theta, \quad \sigma_{23} = \frac{\mu b_3}{2\pi r}\cos\theta \quad (-\pi < \theta < \pi). \tag{1.5.9}$$

Here b_3 is the discontinuity of displacement u_3 of this dislocation; μ is the shear modulus; and $r\theta$ are the polar coordinates in plane Ox_1x_2 (the field is independent of x_3). All other stresses and displacements are zero.

For the general field of the anti-plane strain, the following representation is valid in homogeneous, isotropic, and linearly elastic materials [6]:

$$u_3 = \operatorname{Im} f(z), \sigma_{13} - i\sigma_{23} = -i\mu f'(z), z = x_1 + ix_2 = r\exp(i\theta). \tag{1.5.10}$$

Here $f(z)$ is an analytic function of z.

Let us study the motion of the screw dislocation of Eq. (1.5.9) subject to the action of stress $\sigma_{23} = \sigma_{23}^0$ at infinity which adds the term $\sigma_{23}^0 x_2/\mu$ to the field of displacement u_3 in Eq. (1.5.9).

The dislocation-driving $\Gamma-$ force Γ_1 is determined by the invariant integral of Eq. (1.5.2). In this case, using the invariance, this integral can be written as follows, see Fig. 1.2,

$$\Gamma_1 = \lim \oint \sigma_{23}\frac{\partial u_3}{\partial x_1}n_2 \mathrm{d}S = \frac{b_3}{2\pi}\lim \oint \left(\sigma_{23}^0 + \frac{\mu b_3}{2\pi r}\cos\theta\right)\frac{\partial\theta}{\partial x_1}n_2 \mathrm{d}S = b_3\sigma_{23}^0. \tag{1.5.11}$$

Here the integration path S is chosen along the contour of the long narrow rectangle $x_2 = \pm\delta, x_1 = \pm L$ passed clockwise when $\delta/L \to 0$ and $L \to 0$ (where $n_2 = 1, \mathrm{d}S = \mathrm{d}x_1$ at $x_2 = +\delta$ and $n_2 - -1, \mathrm{d}S = -\mathrm{d}x_2$ at $x_2 = -\delta$). It was taken into account that along this path, we have:

$$\int_{\pi}^{\pi/2} \frac{\cos\theta}{r(\theta)}\mathrm{d}\theta + \int_{-\pi/2}^{-\pi} \frac{\cos\theta}{r(\theta)}\mathrm{d}\theta = 0, \quad \int_{\pi/2}^{0} \frac{\cos\theta}{r(\vartheta)}\mathrm{d}\theta + \int_{0}^{-\pi/2} \frac{\cos\theta}{r(\theta)}\mathrm{d}\theta = 0. \tag{1.5.12}$$

For the same front and plane of the screw dislocation of Eq. (1.5.9), the *elementary edge dislocation* having the displacement discontinuity $[u_1] = b_1$ and the *elementary wedge dislocation* having the displacement discontinuity $[u_2] = b_2$ are possible. At that, the proper field of stresses and displacements for the *elementary edge dislocation* is given by the following equations [6]:

$$\sigma_{11} = -\frac{Cb_1}{2r}(\sin 3\theta + 3\sin\theta), \quad \sigma_{22} = \frac{Cb_1}{2r}(\sin 3\theta - \sin\theta), \tag{1.5.13}$$

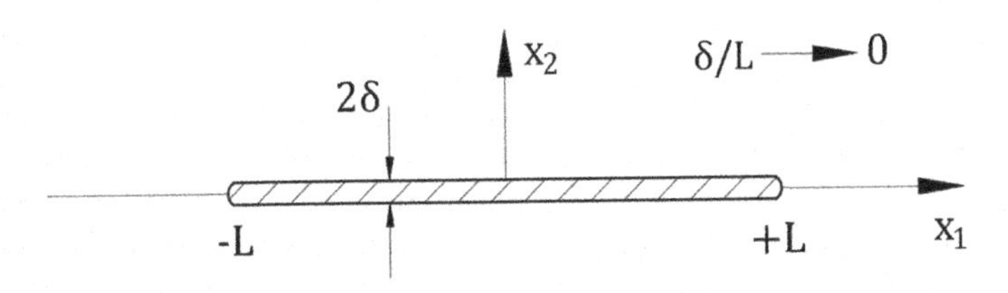

Fig. 1.2 Integration path S passed clockwise

$$\sigma_{12} = \frac{Cb_1}{2r}(\cos 3\theta + \cos\theta), \quad \sigma_{33} = -\frac{2\nu Cb_1}{r}\sin\theta, \tag{1.5.14}$$

$$u_1 = \frac{b_1[\theta(\kappa+1) + \sin 2\theta]}{2\pi(\kappa+1)}, \quad u_2 = -\frac{b_1[(\kappa-1)\ln r + \cos 2\theta]}{2\pi(\kappa+1)}. \tag{1.5.15}$$

Here $C = \dfrac{E}{4\pi(1-\nu^2)}$, $\kappa = 3 - 4\nu$, E is Young's modulus, and ν is Poisson's ratio. All other stresses and strains are zero.

Using the invariant integral of Eq. (1.5.2) similarly to the previous case, we can as well derive the force driving the elementary edge dislocation. In the general case, the elementary dislocation is vector (b_1, b_2, b_3), and the driving force, similar to Eq. (1.5.11), is given by the following Peach–Koehler equation

$$\Gamma_1 = b_1\sigma_{21}^0 + b_2\sigma_{22}^0 + b_3\sigma_{23}^0. \tag{1.5.16}$$

Here $\sigma_{21}^0, \sigma_{22}^0$ *and* σ_{23}^0 are the corresponding stresses at the location, but in absence, of the dislocation.

The values of b_1, b_2 *and* b_3 are specific for each crystal lattice; they have an order of the atomic spacing of this lattice. For example, for cubic lattices, $b_1 = a$ (where a is the atomic spacing). The dislocations of less value are impossible. And so, an arbitrary dislocation represents a sum of elementary dislocations, with the driving force being directly proportional to the sum.

To move a dislocation, the dislocation-driving Γ-force should achieve a certain critical value of Γ_{1C} which depends on the type of lattice and on concentration and sort of foreign atoms of alloys. In pure metals, this critical value is very small for edge and screw dislocations, which explains the high yielding and plasticity of pure metals and, as a result, low strength of pure metals. Much greater critical value is required to move a wedge dislocation in metals and crystals—usually, first comes a rupture so that fracturing is more typical process in this case.

To make metals stronger, alloying and thermal treatment are specifically used to achieve most favorable concentration and distribution of dislocations and alloying atoms in the structure of the metal. The main objective of this treatment is to increase the value of Γ_{1C} which leads to the proportional increase of ultimate strength of the product.

For elementary edge dislocations, we get

$$\Gamma_{1C} = b_1\sigma_S. \tag{1.5.17}$$

Here σ_S is the Schmid stress which characterizes the lattice of the given metal.

Let us provide some simple examples of the interaction of edge dislocations.

Two edge dislocations on one sliding plane. Suppose two elementary edge dislocations are at points $x_1 = 0$ and $x_1 = L$ of infinite space. The field of the first one is given by Eqs. (1.5.13)–(1.5.15), and the field of the other one is provided by same equations where the origin of polar coordinates is at the latter point. According to Eq. (1.5.14), stress σ_{12} produced by the second dislocation at $x_1 = 0$ is equal to $-Cb_1/L$. Hence, based on Eq. (1.5.16), the driving force on the dislocation at $x_1 = 0$ is:

$$\Gamma_1 = -Cb_1^2/L. \tag{1.5.18}$$

The repelling force of same value acts upon the other dislocation, as well. The position of both dislocations does not vary, if the value of the driving force is less than Γ_{1C} so that we have $L > Cb_1^2/\left(\frac{\pi}{2} - \theta\right)_{1C}$. However, if $\Gamma_1 > \Gamma_{1C}$ so that $L < Cb_1^2/\Gamma_{1C}$, the dislocations will move apart until the distance between both achieves $L = Cb_1^2/\Gamma_{1C}$ where $\Gamma_1 = \Gamma_{1C}$.

And so, the interaction $\Gamma-$ force of two arbitrary edge dislocations is equal to

$$\Gamma_1 = -\frac{Eq_1q_2}{4\pi L\left(1 - \nu^2\right)}. \tag{1.5.19}$$

Here q_1 and q_2 are the capacities of the dislocations expressed by some whole numbers of b_1. The interaction force is attractive for dislocations of different signs and repulsive for dislocations of the same sign.

Point inclusions. Most important point defects in the crystal lattice are carbon atoms of alloys in ferrous metals, nitrogen and hydrogen atoms of impurities, alloying ingredients, and others. They can move in the lattice under the action of stresses and self-diffusion.

Let us calculate the Γ-force driving the most typical point defect B.A. Bilby called the interstitial atom and J. D. Eshelby the center of dilatation. This defect placed at the coordinate origin creates the proper spherically symmetrical elastic field:

$$\begin{aligned} u_R^s &= \frac{1+\nu}{2E}\frac{qa^3}{R^2}; \quad (q > 0) \\ \sigma_R^s &= -\frac{qa^3}{R^3}, \quad \sigma_\psi^s = \sigma_\theta^s = \frac{qa^3}{2R^3}. \end{aligned} \tag{1.5.20}$$

Here R, ψ and θ are the spherical coordinates; u_R^s and σ_R^s, σ_ψ^s, σ_θ^s are the corresponding displacement and stresses; a is the radius of the interstitial atom; and q is its pressure on the elastic material. All other stresses and displacements are zero.

Let us apply some constant external stresses at infinity characterizing an arbitrary compression–extension field. Using Eqs. (1.5.2), it can be shown in this case that the force driving this inclusion is zero. The stresses, even, however, strong, cannot move this inclusion.

Now, suppose that the outer elastic field has the following shape:

$$\begin{aligned} \sigma_{33}^0 &= Ax_1; \quad u_1^0 = -\frac{\nu A}{2E}\left(x_1^2 - x_2^2 + \frac{1}{\nu}x_3^2\right); \\ u_2^9 &= -\frac{\nu A}{E}x_1x_2; \quad u_3^0 = \frac{A}{E}x_1x_3. \end{aligned} \tag{1.5.21}$$

Here A is a constant (e.g., directly proportional to the bending moment of a beam).

Let us calculate the Γ-force Γ_1 of this field driving the inclusion of Eq. (1.5.20). Because the integral of Eq. (1.5.2) is invariant with respect to the integration surface S, we can take S as the parallelepiped formed by the faces $x_1 = \pm L$; $x_2 = \pm L$; and $x_3 = \pm\delta$ when $\delta/L \to 0$, $L \to \infty$, and $\delta \to \infty$.

In this case, the integrals of Eq. (1.5.2) are reduced to the following ones

$$\Gamma_1 = -2 \int_{-\infty}^{+\infty}\int_{-\infty}^{+\infty} \left(\sigma_{33}^{s} u_{3,1}^{0} + \sigma_{33}^{0} u_{3,1}^{s} + \sigma_{12}^{s} u_{2,1}^{0} + \sigma_{13}^{s} u_{3,1}^{0}\right) \mathrm{d}x_1 \mathrm{d}x_2,$$
$$\Gamma_2 = \Gamma_3 = 0, \quad (x_3 = \delta). \tag{1.5.22}$$

Here we used the symmetry with respect to the plane $x_3 = 0$ and the following relations: $\sigma_{13}^0 = \sigma_{23}^0 - 0$ when $x_3 = \pm\delta$; $n_3 = 1$ when $x_3 = \delta$; and $n_3 = -1$ when $x_3 = -\delta$.

According to Eq. (1.5.20), we have:

$$\sigma_{33}^{s} = \sigma_{z}^{s} = -\frac{qa^3}{4R^3}(1 + 3\cos 2\psi), \quad \sigma_{rz}^{s} = -\frac{3qa^3}{4R^3}\sin 2\psi,$$
$$\sigma_{13} = \frac{x_1}{r}\sigma_{rz}, \quad \sigma_{23} = \frac{x_2}{r}\sigma_{rz}, \quad u_{3,1}^{s} = -3\frac{1+\nu}{2E}qa^3\frac{x_1 z}{R^5}.$$
$$\left(z = x_3, r^2 = x_1^2 + x_2^2, R^2 = r^2 + z^2\right) \tag{1.5.23}$$

Here r, θ and z are cylindrical coordinates (ψ is measured from the $z-$ axis). Using the equations (for $x_3 = \delta$)

$$R^2 = r^2 + \delta^2, \delta/R = \cos\psi, \quad r/R = \sin\psi, x_1/r = \cos\theta,$$
$$r\mathrm{d}r = R\mathrm{d}R - R^2\tan\psi\,\mathrm{d}\psi, \quad \mathrm{d}x_1\mathrm{d}x_2 = r\mathrm{d}r\mathrm{d}\theta,$$

we calculate integral of Eq. (1.5.22) by means of Eqs. (1.5.21) and (1.5.23)

$$\Gamma_1 = -\frac{2A}{E}\int_0^{\infty}\int_0^{2\pi}\left(\delta\sigma_z^s - \nu r\sigma_{rz}^s + E x_1 u_{3,1}^s\right) r\mathrm{d}r\mathrm{d}\theta$$
$$= \pi\frac{A}{E}qa^3\int_0^{\pi/2}[\sin\psi(1 + 3\cos 2\psi) - 3\nu\sin\psi\tan\psi\sin 2\psi$$
$$+3(1+\nu)\sin^3\psi]\mathrm{d}\psi = 2\pi\frac{1-\nu}{E}Aqa^3. \tag{1.5.24}$$

According to Eq. (1.5.21), we have $A = \partial \sigma_{33}^0 / \partial x_1$ so that

$$\Gamma_1 = 2\pi \frac{1-\nu}{E} q a^3 \frac{\partial \sigma_{33}^0}{\partial x_1}. \tag{1.5.25}$$

Because of spherical symmetry of the point inclusion of Eq. (1.5.21), indices 1 and 33 in Eq. (1.5.25) can be replaced by 2 or 3, and by 11 or 22, correspondingly. And so, the final general result can be written as follows:

$$\Gamma_i = 2\pi \frac{1-\nu}{E} q a^3 \frac{\partial \sigma}{\partial x_i} \quad \left(\sigma = \sigma_{11}^0 + \sigma_{22}^0 + \sigma_{33}^0;\, i = 1, 2, 3\right). \tag{1.5.26}$$

This equation was first derived by J. D. Eshelby using another approach [8]. Thus, the inclusion-driving $\Gamma-$ force is directly proportional to the gradient of the first invariant of the external stress tensor $\sigma_{ij}^0(x_1, x_2, x_3)$.

According to Eshelby's Eq. (1.5.26), point inclusions like atoms of carbon or alloys tend to move into the region stretched by higher tensile stresses and to strengthen a metal. However, atoms of hydrogen penetrating a lattice usually act like wedges and cause embrittlement of the metal (for more detail, see [8]). Like vortices in fluids, the motion of field singularities in solids leads to cardinal changes of the properties of the material.

1.6 Electromagnetic Field

Consider the stationary electromagnetic field in a dielectric medium with zero conductivity, ignoring the medium deformation. The field equations can be specified in the form of the following invariant integrals expressing the energy and momentum conservation laws [3, 4, 6]

$$F_j = \oint_S \left(W n_j - D_i n_i E_j - B_i n_i H_j\right) \mathrm{d}S \quad (i, j = 1, 2, 3) \tag{1.6.1}$$

and the following constitutive equations

$$H_i = \frac{\partial U}{\partial B_i}, \quad E_i = \frac{\partial U}{\partial D_i}, \quad U = U(B_i, D_i). \tag{1.6.2}$$

Here $\boldsymbol{D}, \boldsymbol{E}, \boldsymbol{H}$ and $\boldsymbol{B}$ are the field vectors, U is the potential energy of the field per unit volume (only reversible thermodynamic processes are here taken into account), and W is the following function

$$W(E_i, H_i) = U(B_i, D_i) + E_i D_i + H_i B_i \tag{1.6.3}$$

so that

$$D_i = \frac{\partial W}{\partial E_i}, \quad B_i = \frac{\partial W}{\partial H_i}. \tag{1.6.4}$$

The force F_j in Eq. (1.6.1) is equal to the energy of the field spent to move a field singularity on unit length along the x_j- axis, that is, the j th component of the driving force. Hence, $F_j = 0$ if there are no singularities inside S.

Let us show that Maxwell's equations follow from the invariant integral, Eq. (1.6.1), if $F_j = 0$ inside S. To the end, let us convert Eq. (1.6.1) using the divergence theorem:

$$\oint_S (Wn_j - D_i n_i E_j - B_i n_i H_j)\mathrm{d}S = \int_V \left[W_{,j} - (E_j D_i)_{,i} - (H_j B_i)_{,i}\right]\mathrm{d}V$$
$$= \int_V \left(E_{i,j}\frac{\partial W}{\partial E_i} + H_{i,j}\frac{\partial W}{\partial H_i} - E_j D_{i,i} - E_{j,i} D_i - H_j B_{i,i} - H_{j,i} B_i\right)\mathrm{d}V$$
$$= \int_V \left[D_i(E_{i,j} - E_{j,i}) + B_i(H_{i,j} - H_{j,i}) - E_j D_{i,i} - H_j B_{i,i}\right]\mathrm{d}V = 0. \tag{1.6.5}$$

From here, because V, D_i, B_i, E_j and H_j are arbitrary, it follows that

$$E_{i,j} = E_{j,i}; \quad H_{i,j} = H_{j,i}; \quad D_{i,i} = 0; \quad B_{i,i} = 0. \tag{1.6.6}$$

These are Maxwell's equations in the stationary case.

Point charges in dielectric media. For an electrostatic field in an isotropic linear dielectric, we have:

$$D_i = \varepsilon E_i, \quad H_i = B_i = 0. \quad (i = 1, 2, 3) \tag{1.6.7}$$

Here ε is the dielectric constant supposed to be equal 1 in vacuum.

In this case, the invariant integrals of Eq. (1.6.1) reduce to

$$F_j = \varepsilon \oint_S \left(\frac{1}{2}E_i E_i n_j - E_i n_i E_j\right)\mathrm{d}S \quad (i, j = 1, 2, 3); \tag{1.6.8}$$

$$E_i = -\varphi_{,i}, \quad \varphi_{,ii} = 0. \tag{1.6.9}$$

Here φ is the electrostatic potential.

Consider the following solution of the Laplace's equation singular at the coordinate origin $r = 0$:

$$\varphi = \frac{q}{4\pi\varepsilon r} + E_i^0 x_i. \tag{1.6.10}$$

Here the first term represents the field of the point charge q while the second term the uniform field of intensity $E_i = E_i^0$.

Using Eqs. (1.6.8)–(1.6.10), we calculate the force of this field upon this charge similarly to the calculation of Eqs. (1.3.6)–(1.3.9):

$$F_i = qE_i^0. \tag{1.6.11}$$

From here and Eq. (1.6.10), it follows that the force on the point charge, q_1 at $(0, 0, 0)$ or q_2 at $(L, 0, 0)$, is equal to

$$F_1 = \pm \frac{q_1 q_2}{4\pi \varepsilon L^2} \quad (F_2 = F_3 = 0). \tag{1.6.12}$$

(sign "plus" is for the force on q_2 and "minus" for the force on q_1)

This is Coulomb's Law which is valid only for subrelativistic charges. From Eq. (1.6.12), in particular, it follows that the interaction force is attractive for charges of opposite sign but repelling for charges of the same sign. Relativistic charges can interact differently [6].

Addendum. For a non-stationary electromagnetic field and for irreversible processes in deformable media, invariant integrals were derived in [4]. For ideal and viscous relativistic fluids and for the relativistic heat flow in fluids, invariant integrals of physical fields in the Minkowski space–time were obtained in [7]. Invariant integrals of irreversible thermodynamics were given in [6].

The relativistic theory of electron beams based on the generalized Coulomb's Law of interaction of relativistic electric charges is presented in Chap. 10 of this book.

Literature

1. G.P. Cherepanov, The invariant integral of physical mesomechanics as the foundation of mathematical physics: some applications to cosmology, electrodynamics, mechanics and geophysics. Phys. Mesomech. **18**(3), 203–212 (2015)
2. G.P. Cherepanov, *Mechanics of Brittle Fracture* (McGraw Hill, New York, 1978), pp. 1–940
3. G.P. Cherepanov, The mechanics and physics of fracturing: application to thermal aspects of crack propagation and to fracking. Philos. Trans. A, R. Soc. **A373**, 2014.0119 (2015)
4. G.P. Cherepanov, *Methods of Fracture Mechanics: Solid Matter Physics* (Kluwer Publisher, Dordrecht, 1997), pp. 1–300
5. G.P. Cherepanov, Some new applications of the invariant integrals of mechanics. J. Appl. Math. Mech. (JAMM) **76**(5), 519–536 (2012)
6. G.P. Cherepanov, *Fracture Mechanics* (IKI Publishers, Izhevsk-Moscow, 2012), pp. 1–870
7. G.P. Cherepanov, Invariant integrals in continuum mechanics. Sov. Appl. Mech. **26**(1), 3–16 (1990)
8. G.P. Cherepanov (ed.), *FRACTURE. A Topical Encyclopedia of Current Knowledge* (Krieger, Malabar), pp. 1–870, 1998
9. G.P. Cherepanov, A neoclassic approach to cosmology based on the invariant integral, Chapter 1. In: *Horizons in World Physics* (Nova Publishers, New York, 2017), pp. 1–33
10. R.G. Lerner, G.L. Trigg, (ed.), *Encyclopedia of Physics* (New York, VCH, 1991), pp. 1–1408
11. L.D. Landau, E.M. Lifshitz, *Fluid Mechanics* (Pergamon, New York, 1987), pp. 1–730
12. L.D. Landau, E.M. Lifshitz, *Electrodynamics of Continuous Media* (Addison-Wesley, Reading, 1960), pp. 1–870

Chapter 2
The Floating

Abstract This chapter deals with the floatation of some bodies on the surface of water or other fluids accounting for the surface tension of the fluid. The latter makes an important addition to Archimedes' buoyancy force, especially substantial for small floating objects. Generally, the buoyancy force appears to be equal to the sum of Archimedes' force and the term of summary surface tension forces. Some particular problems illustrating the combined effects of both components are studied. This chapter may be of interest for those who are involved in geophysics or marine science.

Ships and boats navigate seas and oceans of the globe, continents float on the magma of the Earth, and small insects run on the surface of a pond. To understand these multiscale phenomena of floating, we need to investigate the forces acting on rigid hydrophobic bodies located on the surface of a heavy incompressible fluid with some surface tension. For floating continents, the surface layer providing the forces of surface tension is the solid crust of the Earth about 10 km thick, and for icebreakers, it is the polar ice about 1 m thick.

Also, let us neglect inertial forces and assume that the gravitational force is directed downward perpendicularly to the horizontal xy-plane which coincides with the unperturbed liquid surface. The gas (air) over the liquid is assumed to be quiescent and under constant pressure.

The Chapter is based on this author's paper *Some new applications of the invariant integrals in mechanics* published in *J. Appl. Math. Mech. (JAMM)*, 76(5), 2012.

2.1 Invariant Integral of Floating

The nonlinear equation of hydrostatics of the heavy incompressible liquid with surface tension γ is [1–3]:

$$w_{xx}\left(1 + w_y^2\right) - 2w_x w_y w_{xy} + w_{yy}\left(1 + w_x^2\right) = \frac{2w}{\lambda^2}\left(1 + w_x^2 + w_y^2\right)^{3/2}$$

G. P. Cherepanov, *Invariant Integrals in Physics*,
https://doi.org/10.1007/978-3-030-28337-7_2

$$\left(\lambda^2 = \frac{2\gamma}{\rho g}\right) \tag{2.1.1}$$

Here, $w = w(x, y)$ is the equation of the liquid surface, λ is the capillary constant with the dimension of length, and ρg is the specific weight of the liquid (w_x, w_y, w_{xx}, w_{xy} and w_{yy} are the corresponding particular derivatives).

Equation (2.1.1) is obtained from the equilibrium equation of the surface layer of the heavy liquid written by Laplace's Law as follows:

$$\gamma\left(\frac{1}{R_1} + \frac{1}{R_2}\right) = \rho g w + \text{const.} \tag{2.1.2}$$

Here, R_1 and R_2 are the main radii of curvature of the liquid surface, and the constant is zero, if the displacement w is accounted from the unperturbed state.

When $w_x \ll 1$ and $w_y \ll 1$, this equation is reduced to:

$$w_{xx} + w_{yy} = \frac{\rho g}{\gamma} w \tag{2.1.3}$$

Let us introduce the following invariant integral of floating $\left(w_x = w_1, w_y = w_2\right)$:

$$F_i = \frac{1}{2} \oint_L \left(\gamma w_{,j} w_{,j} n_i - 2\gamma w_{,j} n_j w_{,i} + \rho g w^2 n_i\right) \mathrm{d}L, \quad (i, j = 1, 2) \tag{2.1.4}$$

Let us prove that Eq. (2.1.3) follows from this invariant integral, if $F_i = 0$ so that there are no bodies floating inside the closed contour L on the xy-plane. In this case, we apply the divergence theorem to the right-hand part of Eq. (2.1.4) and get:

$$\begin{aligned} &\frac{1}{2} \int_S \left[\gamma\left(w_{,j} w_{,j}\right)_{,i} - 2\gamma\left(w_{,j} w_{,i}\right)_{,j} + \rho g\left(w^2\right)_{,i}\right] \mathrm{d}S \\ &= \int_S \left[\left(\rho g w - \gamma w_{,jj}\right) w_{,i}\right] \mathrm{d}S = 0 \end{aligned} \tag{2.1.5}$$

From here, it follows that Eq. (2.1.3) is valid in S inside L because S is arbitrary. When there is a body inside the contour L, the quantity F_i is equal to the corresponding component of the horizontal force acting on this body.

In what follows, we will restrict the discussion to hydrophobic bodies for which the wetting angle is greater than $\pi/2$. The typical materials of this type are graphite, pure metals, and graphene. (For hydrophilic bodies, the wetting angle is less than $\pi/2$ and all effects will be opposite). We will use the non-wetting angle α which is the contact angle between the unwetted surface of the rigid body and the meniscus of the liquid at their point of intersection, located along the wetting perimeter ($\pi/2 > \alpha > 0$). The angle α is the complementary angle to the wetting angle (Fig. 2.1).

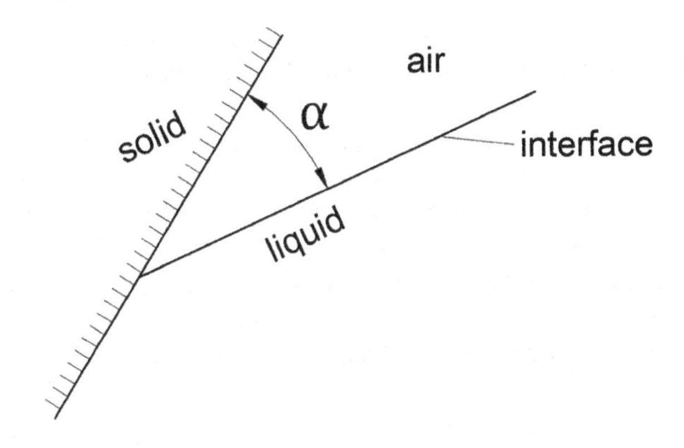

Fig. 2.1 Non-wetting angle α for hydrophobic liquids ($\pi/2 > \alpha > 0$)

Below we show that, when a hydrophobic body comes into contact with the liquid, this liquid resists the penetration of the body like an elastic spring and that some attractive forces act between hydrophobic bodies on the surface of a liquid.

2.2 The Buoyancy Force

Suppose a rigid hydrophobic body, to which a vertical force F_z (the body weight plus a possible external force) is applied, is to be at equilibrium on the surface of a liquid. The surface of the rigid body wetted by the liquid is denoted by S_w and the line separating the wetted and unwetted parts of the body surface is denoted by L_w (the wetting contour).

The buoyancy force of magnitude F_z that acts on the hydrophobic body at rest is directed upward along the z-axis; it is equal to the sum of the Archimedean force F_A and the resistance force of the hydrophobic liquid

$$F_z = F_A + \gamma \oint_{L_w} \sin(\theta - \alpha) \mathrm{d}L \tag{2.2.1}$$

Here, θ called the edge angle is the angle in the normal section of the wetting contour between the horizontal plane and the tangent to the body surface at the point where the solid, liquid, and vapor meet; $\mathrm{d}L$ is an element of the wetting contour. For floating bodies, to calculate the Archimedean and resistance forces, we should know the submerging position of the wetting contour L_w which is determined from the solution of Eq. (2.1.1).

For the Archimedean force to be active, it is important that the vessel, e.g., submarine, to be surrounded by water under its bottom. A submarine can get in trouble if it lies on the bottom so that water and its pressure are diminished underneath—it can never come off its bed. The corresponding equation in paper [3] is valid only in the particular case $\theta = \pi/2$ considered in that paper.

Equation (2.2.1) is the generalized buoyancy law taking account of the surface tension of the liquid. It should be mentioned that this simple law can be applied not only to waters of seas and rivers, but also to many other media that do not seem to be liquids, for example, magma and lava beneath Earth's crust, fluidized beds in chemical reactors, ice at the bottom of glaciers, loose materials, quick sand, extremely viscous materials, etc. To evaluate this property for a material, we need to know the specific time of relaxation of shear stresses as compared to pressure. If the time under consideration is large compared to this specific time, the material can be accounted for as a liquid. And within this time scale, we can use Archimedes's law or the generalized law of Eq. (2.2.1).

2.3 Some Particular Problems

Let us consider some examples.

The penetration of a thin blade (Fig. 2.2). Suppose a semi-infinite rigid blade of zero thickness is embedded along the plane $x = 0$ in a liquid which initially occupies the lower half-space $z < 0$. Two faces and edge of the blade are parallel to the y-axis. This problem is one-dimensional so that $w = w(x)$ and, according to Eq. (2.1.1), the equation of the perturbed surface of the liquid has the form

$$2w\left(1 + w_x^2\right)^{3/2} = \lambda^2 w_{xx} \tag{2.3.1}$$

The solution of this equation should satisfy the following condition at infinity:

$$w = 0, \ \ w_x = 0 \ \ \text{when} \ \ x \to \infty \tag{2.3.2}$$

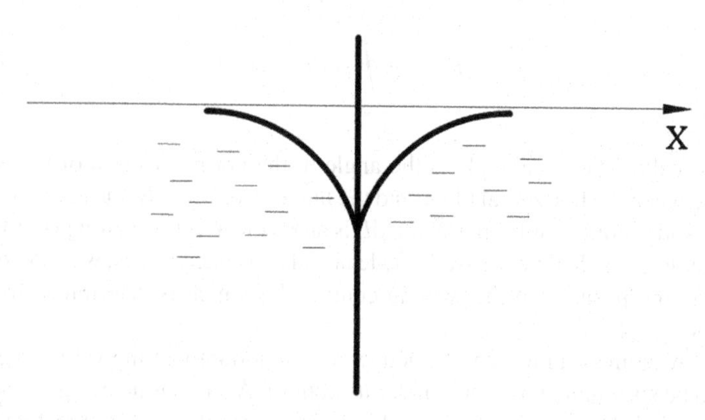

Fig. 2.2 Penetration of a thin blade into a hydrophobic liquid

The integral of Eq. (2.3.1) that satisfies this condition has the form:

$$\left(1-\frac{w^2}{\lambda^2}\right)\sqrt{1+w_x^2}=1 \tag{2.3.3}$$

It is obtained from Eq. (2.3.1) by the simple transformation and solution of the following equation:

$$w_x = u \text{ and } w_{xx} = u\frac{du}{dw}: \quad 2w\left(1+u^2\right)^{3/2} = \lambda^2 u\frac{du}{dw} \tag{2.3.4}$$

The surface of the liquid $w = w(x)$ can be easily found by solving Eq. (2.3.3).

Let us determine the penetration diagram, that is, the function $F_z = F_z(w_0)$ where w_0 is the depth of penetration and F_z is the resistance force which is equal to the sum of the Archimedean force and the force of surface tension of a hydrophobic liquid.

In this case, the Archimedean force is equal to zero and the buoyancy force applied to the faces of the blade is the surface tension which is, according to Eq. (2.2.1), equal to $F_z = 2\gamma l\cos\xi$ where the angle ξ is decreasing from $\pi/2$ to the unwetting angle α in the process of embedding (l is the length of the blade and $\theta = \pi/2$).

Since $\text{cotan}\xi = w_x$ when $x = 0$, using Eq. (2.3.3) we get

$$\sqrt{1-\sin\xi} = \frac{w_0}{\lambda} \quad \text{where } w_0 = w(0) \tag{2.3.5}$$

From here, we find the penetration diagram of the blade (Fig. 2.3):

$$F_z(w_0) = 2\gamma l(w_0/\lambda)\sqrt{2-(w_0/\lambda)^2} \quad \text{as} \quad \lambda\sqrt{1-\sin\alpha} > w_0 > 0 \tag{2.3.6}$$

$$F_z(w_0) = 2\gamma l\cos\alpha \quad \text{as} \quad w_0 > \lambda\sqrt{1-\sin\alpha} \tag{2.3.7}$$

And so, the penetration diagram of a thin blade is similar to the extension diagram of a nonlinearly elastic–ideally plastic bar.

The term $\rho g t l w_0$ should be added to the right-hand part of Eq. (2.3.7), if we take into account the thickness t of the blade. In this case, Eq. (2.3.6) remains the same.

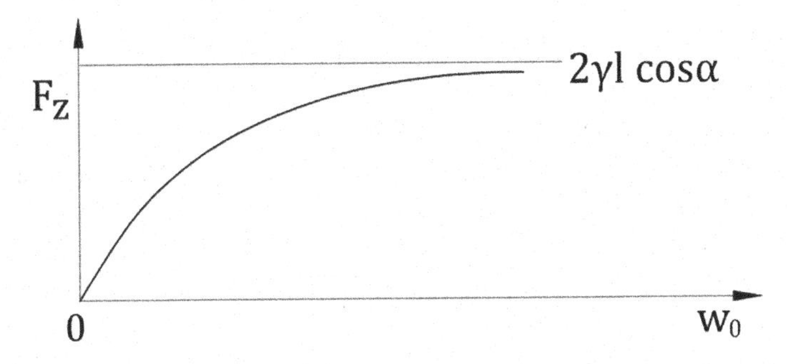

Fig. 2.3 Penetration diagram of a thin blade (F_z is the resistance force)

The penetration of a wedge. Suppose a semi-infinite rigid wedge is embedded along the plane $x = 0$ in a liquid which initially occupies the lower half-space $z < 0$. Two faces of the wedge constitute the angles $\pm\beta$ to the plane $x = 0$ which is the symmetry plane of the problem. The wedge is similar to an icebreaker penetrating the layer of polar ice which can be viewed as a layer of the surface tension in this model. Let us assume that $\alpha + \beta < \pi/2$.

In this problem, the relationship of Eq. (2.3.3) is valid. On the first stage of the penetration, when the contact angle between the face of the wedge and the surface layer is decreasing from $\pi/2 - \beta$ to α, the penetration diagram coincides with that of Eq. (2.3.6). At this stage, the surface layer is not yet broken. On the second stage of the penetration, the surface layer is cut by the wedge and the Archimedean force starts acting.

As a result, we get the following diagram $F_z = F_z(w_0)$ on the second stage:

$$F_z = 2\gamma l\cos\alpha + \rho g l\left(w_0 - \lambda\sqrt{1-\sin\alpha}\right)^2\tan\beta \quad \text{as} \quad w_0 > \lambda\sqrt{1-\sin\alpha} \tag{2.3.8}$$

As to the problem of icebreaker, the value of l should be considered as some function of y depending on the shape of the keel and its position in the work process of the ice cutting, with the nose of the keel being higher than the stern.

The penetration of a paraboloid. Let a paraboloidal cylinder $z = -w_0 + x^2/(2R_0)$ penetrate into a liquid which initially occupies the lower half-space $z < 0$ where R_0 is the radius of curvature of the parabola at $x = 0$ and w_0 is the penetration depth of the top of the cylinder. The plane $x = 0$ is the symmetry plane of the problem.

At the first moment of the contact with the liquid when $w_0 = 0$, the edge angle θ is equal to zero so that to meet the equilibrium the liquid surface instantly takes the slope inclined under the angle $(-\alpha)$ to the horizontal plane at this point. It means that the hydrophobic liquid behaves like a hydrophilic one at that moment, and the surface tension makes the body get immersed in the liquid. And so, at the initial stage, there is no balance and this process of sinking is dynamic until the edge angle θ turns to be equal to α so that the effect of surface tension disappears. At this moment $\theta = \alpha$ we have:

$$x_0 = R_0\tan\alpha, \quad w_0 = \frac{x_0^2}{(2R_0)} \tag{2.3.9}$$

$$F_z = F_A = F_* = \frac{1}{3}\rho g l R_0^2(\tan\alpha)^3 \quad \text{as} \quad \theta = \alpha \tag{2.3.10}$$

These equations provide the initial conditions of the beginning of the stable process of penetration when $\theta > \alpha$, $x_0 > R_0\tan\alpha$ and $F_A > F_*$ characterized by the following penetration diagram:

$$F_z(w_0) = \frac{1}{3}\rho g R_0^2(\tan\theta)^3 + 2\gamma l\sin(\theta - \alpha) \tag{2.3.11}$$

$$x_*^2 = 2R_0(w_0 - z_*), \quad z_*^2 = \lambda^2[1 - \cos(\theta - \alpha)], \quad x_* = R_0 \tan\theta \tag{2.3.12}$$

Here, $(\pm x_*, \ -z_*)$ are two points where liquid meniscus, solid, and air meet together, and θ is the angle between the x-axis and the tangent to the parabola at the point $(+x_*, \ -z_*)$ and $w_0 = -z(0)$ is the depth of penetration. The second equation of Eqs. (2.3.12) follows from the equation of the meniscus, Eq. (2.3.3).

From Eq. (2.3.12), we find that

$$w_0 = \frac{1}{2}R_0(\tan\theta)^2 + \lambda\sqrt{1 - \cos(\theta - \alpha)} \tag{2.3.13}$$

The pair of equations, Eqs. (2.3.11) and (2.3.13), provides parametrically the penetration diagram $F_z - w_0$ with θ being the parameter.

The penetration of a round cylinder into a liquid. For axisymmetric problems of penetration of cylinder, cone or sphere, the equation of liquid surface $w = w(r)$ deformed by surface tension is described by the equation

$$\frac{d^2w}{dr^2} + \frac{1}{r}\frac{dw}{dr}\left[1 + \left(\frac{dw}{dr}\right)^2\right] = \frac{2w}{\lambda^2}\left[1 + \left(\frac{dw}{dr}\right)^2\right]^{3/2} \tag{2.3.14}$$

It can be shown that the solution of this equation tends to zero very fast when $r \gg \lambda$, namely:

$$w = C_0\sqrt{(\pi\lambda)/\left(2r\sqrt{2}\right)}e^{-r\sqrt{2}/\lambda} \quad \text{when} \quad r \to \infty \tag{2.3.15}$$

Here, C_0 is a constant of the dimension of length characterizing the depth of the penetration.

For a round cylinder of radius R_c, using Eq. (2.3.14), it is easy to find the penetration diagram $F_z - w_0$ in the well-developed stage

$$F_z = \pi\rho g w_0 R_c^2 + 2\pi\gamma R_c \cos\alpha \quad \text{when} \quad \frac{2w_0}{\lambda^2} + \frac{1}{R_0} \geq \frac{\cos\alpha}{R_c} \tag{2.3.16}$$

Here, R_0 is the radius of curvature of the radial cross section of the liquid meniscus on the surface of the cylinder. For example, a very thin needle penetrates a liquid without resistance. Using the results of calculations in this section, we can estimate the size and weight of small beings that can run on the surface of a pond.

2.4 The Interaction Between Floating Bodies

Consider first the problem of a heavy sphere floating on the surface of a liquid and find the perturbed liquid surface. The treatment is restricted to the case when $w_x \ll 1$ and $w_y \ll 1$ which is always satisfied at distances that are large compared with the sphere radius. In this case, Eq. (2.1.1) has the form

$$\lambda^2 \frac{1}{r}\frac{\mathrm{d}}{\mathrm{d}r}\left(r\frac{\mathrm{d}w}{\mathrm{d}r}\right) = 2w \tag{2.4.1}$$

Here, r is the distance from the axis of symmetry of the problem.

In the point approximation, that characterizes the field at distances large compared to the radius of the sphere, the solution of Eq. (2.4.1) which vanishes at infinity has the following form

$$w = CK_0\left(\frac{\sqrt{2}}{\lambda}r\right) \tag{2.4.2}$$

Here, C is an arbitrary constant and $K_0(\xi)$ is the zero-order MacDonald function of ξ called also the modified Bessel function. This function is singular at the origin of coordinates so that

$$w = C\ln\left(\frac{\sqrt{2}}{\lambda}r\right) \quad \text{when} \quad r \to 0 \tag{2.4.3}$$

The weight Mg of this point singularity characterizing the heavy sphere is balanced by the surface tension of the liquid, and the equivalent force of which is equal to $2\pi r\gamma C/r$ according to Eq. (2.4.3). From the equilibrium condition, we find that $C = Mg/(2\pi\gamma)$ and

$$w = -\frac{Mg}{2\pi\gamma}K_0\left(\frac{\sqrt{2}}{\lambda}r\right) \tag{2.4.4}$$

Let us study the forces acting on the floating sphere caused by the perturbed liquid surface $w = w_0(x, y)$ made by other sources different from this floating sphere. Evidently, these forces are zero, if $w_0(x, y) = 0$.

The perturbed liquid surface and the floating sphere at the coordinate origin create together the following surface

$$w = w_0(x, y)\frac{Mg}{2\pi\gamma}K_0\left(\frac{\sqrt{2}}{\lambda}r\right) \tag{2.4.5}$$

In what follows we prove that the horizontal force acting on the floating sphere has the following components

$$F_x = -Mg\frac{\partial w_0}{\partial x}, \quad F_y = -Mg\frac{\partial w_0}{\partial y} \tag{2.4.6}$$

Here, $\partial w_0/\partial x$ and $\partial w_0/\partial y$ are taken at $x = y = 0$ where the floating sphere is situated.

Proof The proof is typical and exemplary for the work with invariant integrals and, therefore, will be given further in all detail. Let us calculate the force F_x on the floating sphere at the coordinate origin using Eq. (2.1.4). As the integration contour L, let us use the narrow rectangle $x = \pm\delta$, $y = \pm\varepsilon$ when $\delta \to 0$, $\varepsilon \to 0$ and $\varepsilon/\delta \to 0$. This closed contour passed counterclockwise encompasses the origin of coordinates where the floating sphere is. Because the integral of Eq. (2.1.4) is invariant with respect to any closed contour of integration, the result of this calculation will be valid for any contour L encompassing the coordinate origin.

Since $\varepsilon/\delta \to 0$, we can ignore the integral over the short sides of the rectangle where $x = \pm\delta$. On the horizontal sections of the integration path where $y = \pm\varepsilon$, we have: $n_1 = 0$ and $n_2 = \pm 1$ where sign plus is for the upper path and sign minus for the lower path. With regard of this consideration, Eq. (2.1.4) for F_x is reduced to the following one

$$F_x = -\gamma \oint_L \frac{\partial w}{\partial x}\frac{\partial w}{\partial y} n_2 \mathrm{d}L = -\gamma \oint_L \left(\frac{\partial w_0}{\partial x} + \frac{\partial w_s}{\partial x}\right)\left(\frac{\partial w_0}{\partial y} + \frac{\partial w_s}{\partial y}\right) n_2 \mathrm{d}L \tag{2.4.7}$$

Here, w_s is the singular term of Eq. (2.4.5) given also by Eq. (2.4.4).

According to the rule of Γ-integration of invariant integrals [2], we have:

$$\oint_L \frac{\partial w_0}{\partial x}\frac{\partial w_0}{\partial y} n_2 \mathrm{d}L \to 0 \quad \text{when } \varepsilon \to 0 \text{ and } \delta \to 0 \tag{2.4.8}$$

$$\oint_L \frac{\partial w_s}{\partial x}\frac{\partial w_s}{\partial y} n_2 \mathrm{d}L \to 0 \quad \text{when } \varepsilon \to 0,\ \delta \to 0 \text{ and } \varepsilon/\delta \to 0 \tag{2.4.9}$$

The first equation is evident since $w_0(x, y)$ has no singularity at $x = y = 0$ To prove Eq. (2.4.9), let us calculate using Eq. (2.4.4):

$$\begin{aligned}\frac{\partial w_s}{\partial x} &= -\frac{Mg}{2\pi\gamma}\frac{\partial r}{\partial x}\frac{\partial}{\partial r}K_0\left(\frac{\sqrt{2}}{\lambda}r\right) \\ &= \frac{Mgx\sqrt{2}}{2\pi\gamma\lambda r}K_1\left(\frac{\sqrt{2}}{\lambda}r\right) \to \frac{Mg}{2\pi\gamma}\frac{x}{r^2} \quad \text{as } r^2 = x^2 + y^2 \to 0;\end{aligned}$$

$$\frac{\partial w_s}{\partial y} = \frac{Mg}{2\pi\gamma}\frac{\sqrt{2}}{\lambda}\frac{y}{r}K_1\left(\frac{\sqrt{2}}{\lambda}r\right) \to \frac{Mg}{2\pi\gamma}\frac{y}{r^2} \text{ as } r^2 = x^2 + y^2 \to 0 \tag{2.4.10}$$

Here, $K_1(\xi)$ is the first-order MacDonald function of ξ that satisfies the following relations:

$$K_1(\xi) = -\frac{dK_0(\xi)}{d\xi}, \quad K_1(\xi) \to \frac{1}{\xi} \text{ when } \xi \to 0 \tag{2.4.11}$$

According to Eq. (2.4.10), the integral in Eq. (2.4.9) is reduced to the calculation of the integral of the odd function which is equal to zero:

$$\varepsilon \int_{-\delta}^{+\delta} \frac{x}{r^4} dx = 0 \tag{2.4.12}$$

On the same reason, we get that

$$\gamma \oint_L \frac{\partial w_0}{\partial y}\frac{\partial w_s}{\partial x} n_2 dL \to \frac{Mg}{2\pi}\int_{-\delta}^{+\delta} \frac{\partial w_0}{\partial y}\frac{x}{r^2} dx \to 0 \tag{2.4.13}$$

Since the function $\partial w_0/\partial y$ is limited in the neighborhood of the coordinate origin.

And so, due to Eqs. (2.4.7) to (2.4.13), the component F_x of the force on the floating sphere is equal to

$$\begin{aligned} F_x = -\gamma \oint_L \frac{\partial w_0}{\partial x}\frac{\partial w_s}{\partial y} dL &= -\frac{Mg}{2\pi}\left(\frac{\partial w_0}{\partial x}\right)_{x=y=0} \lim \oint \frac{y}{x^2+y^2} dL \\ &= -Mg\left(\frac{\partial w_9}{\partial x}\right)_{x=y=0} \end{aligned} \tag{2.4.14}$$

The limit of the contour integral over the narrow rectangle in Eq. (2.4.14) is reduced to the following one

$$2\int_{-\infty}^{+\infty} \frac{dt}{1+t^2} = 2\pi \tag{2.4.15}$$

Thus, the first equation in Eqs. (2.4.6) is proven. The second one can be proven similarly using a closed path of integration over the sides of the narrow rectangular stretched along the y-axis near the coordinate origin.

Two floating spheres. Let another small sphere of mass m float on the surface of the liquid at the point $x = R,\ y = 0$ Let us find the interaction force it produces on the sphere of mass M at the coordinate origin.

According to Eqs. (2.4.2) to (2.4.5), this small sphere creates its own field

$$w_0(x, y) = -\frac{mg}{2\pi\gamma} K_0\left(\frac{\sqrt{2}}{\lambda} r_0\right) \quad \text{where} \quad r_0 = \sqrt{(x-R)^2 + y^2} \tag{2.4.16}$$

This sphere makes the following perturbation at the point $x = y = 0$:

$$\left(\frac{\partial w_0}{\partial x}\right)_{x=y=0} = -\frac{mg}{2\pi\gamma}\frac{\sqrt{2}}{\lambda} K_1\left(\frac{\sqrt{2}}{\lambda} R\right), \quad \left(\frac{\partial w_0}{\partial y}\right)_{x=y=0} = 0 \tag{2.4.17}$$

Using Eq. (2.4.6), we find the force on the sphere of mass M produced by the floating mass m

$$F_x = \frac{mMg^2}{\pi\gamma\lambda\sqrt{2}} K_1\left(\frac{\sqrt{2}}{\lambda} R\right), \quad F_y = 0 \tag{2.4.18}$$

By means of Eq. (2.4.5), we can also find a simpler relation in the case of very small floating objects when $R \ll \lambda$

$$F_x = \frac{mMg^2}{2\pi\gamma R} \tag{2.4.19}$$

This equation is characteristic of Ampere's Law of attraction for two parallel linear unidirectional currents.

However, when $R \gg \lambda$, the force of attraction determined by Eq. (2.4.18) decreases even faster than exponentially because

$$K_1(\xi)\sqrt{\pi/(2\xi)}e^{-\xi} \quad \text{when} \quad \xi \to \infty \tag{2.4.20}$$

These results are obviously not only true for spheres but also for any bodies with linear dimensions that are small compared with the distance between the bodies. As the general result, all floating hydrophobic bodies tend to coalesce but the forces of attraction are of an extremely short range.

2.5 Point Masses on a Hard Surface

Let us consider the motion of a heavy point mass on an absolutely smooth hard horizontal surface $w = w(x_1,\ x_2)$ when $\partial w/\partial x_1 \ll 1$ and $\partial w/\partial x_2 \ll 1$.

In this case, the invariant integral is written as

$$F_i = \oint_L (U + K)n_i \mathrm{d}L \quad (i = 1, 2) \tag{2.5.1}$$

Here, $U = -mgw + \text{const}$ is the potential energy; $K = \frac{1}{2}mv_i v_i$ is the kinetic energy of the system; and m and v_i are the mass and velocity components of the point.

Applying the divergence theorem to Eq. (2.5.1) provides

$$F_i = \int_A (U + K)_{,i}\mathrm{d}A = -mgw_{,i} + mv_k v_{k,i} = mgw_{,i} + m\frac{\mathrm{d}v_i}{\mathrm{d}t} \tag{2.5.2}$$

It can be shown that

$$v_k v_{k,i} = \frac{\mathrm{d}v_i}{\mathrm{d}t} \quad \text{where} \quad v_k = \frac{\mathrm{d}x_k}{\mathrm{d}t} \tag{2.5.3}$$

From Eqs. (2.5.2) and (2.5.3), it follows that

$$F_i = -mg\dot{w}_{,i} + m\frac{\mathrm{d}v_i}{\mathrm{d}t} \tag{2.5.4}$$

This is the law of motion of a point heavy mass on a slightly curved hard surface. In the case, if $gw_{,i} \gg dv_i/dt$, this law is reduced to that of Eq. (2.4.6) governing floating bodies on the liquid surface. However, there is no similarity between both problems because there are no forces of interaction between point masses moving on a hard surface.

Literature

1. A.W. Adamson, in *Physical Chemistry of Surfaces*, 4th edn., Chaps. 1 and 2 (Wiley, New York, 1982), pp. 1–730
2. G.P. Cherepanov, in *Fracture Mechanics* (ICS Publ., Moscow-Izhevsk, 2012), pp 1–872
3. G.P. Cherepanov, Some new applications of the invariant integrals in mechanics. J. Appl. Math. Mech. (JAMM) **76**(5), 519–536 (2012)

Chapter 3
The Rolling

Abstract During two hundred years, the Coulomb's coefficient of rolling friction and the force of resistance to rolling were being determined experimentally in expensive natural tests. In this Chapter, using the invariant integral of rolling and new exact solutions of the contact problems of the theory of elasticity, Coulomb's rolling friction coefficient and the resistance force were calculated in all basic cases, including the rolling of: (i) an elastic cylinder on an elastic half-space of another material, or on an elastic plate, or on a sticky membrane; (ii) a ball on an elastic half-space, or an elastic plate, or on a membrane; and (iii) a torus on an elastic half-space, or on an elastic plate, or on a membrane. The theoretical results comply very well with known test probes. This Chapter is a "must-to-know" for automotive engineers.

Man-made transport was born when the wheel was invented. Since then, the study of rolling has started. In 1781, the problem of rolling was formulated by Charles Coulomb (1736–1806) who offered Coulomb's Law of rolling based on the empirical rolling resistance coefficient. In this chapter, using the mathematical theory of elasticity and a new CH rule, the rolling resistance coefficient is calculated in the cases of:

(a) an elastic cylinder rolling on another elastic cylinder of another material in particular on an elastic half-space, and an elastic wheel rolling on the rail of another elastic material;
(b) an elastic ball rolling on another elastic ball of another material, particularly on an elastic half-space;
(c) an elastic torus rolling on an elastic half-space of another material; and
(d) a cylinder, or a ball, or a torus rolling on a tightly stretched membrane or on a thin elastic plate.

Empirical results of the measurement of the rolling resistance coefficient gained earlier by the railroad and automobile engineers appeared to be in excellent agreement with the results of this analytical calculation based on the suggested rule of rolling. The effect of adhesion of rolling bodies was studied using the invariant integral and some exemplary cases.

The chapter is based on this author's paper *The laws of rolling* published in *J. Physical Mesomechanics,* **21**(5), 2018.

G. P. Cherepanov, *Invariant Integrals in Physics*,
https://doi.org/10.1007/978-3-030-28337-7_3

3.1 Introduction

Wheel was one of the earliest and most important inventions of the man because, except for sleds on ice and snow, it was much easier to wheel a weight than to drag it. That is why the man-made transport was born when the wheel was invented. This happened long before the horse domestication. In his fascinating book *Life in Ancient Egypt,* Adolf Erman [1] told us that wheeled carts carried by oxen were common in the ancient Egypt about two thousand years before the conquest of Egypt by proto-Indo-European Hyksos who were the first to invent chariots and breed horses for war.

And so, since the dawn of human history the problem of rolling resistance has existed. Rolling resistance sometimes called also rolling friction or rolling drag is the force resisting the motion when a body rolls on a surface. In mathematical terms, this problem sounds as follows.

What is the value of the rolling resistance coefficient C_{rr} in the following equation?

$$F = C_{rr}N \tag{3.1.1}$$

Here, N is the load on the wheel, for example, the weight of the vehicle, and F is the rolling resistance force.

The greats including *Leonardo de Vinci, Leonhard Euler, Charles de Coulomb, Heinrich Hertz,* and very many others stood before this problem but test seemed to be the only way to find out the rolling resistance coefficient—see, e.g., *Special Report 216 of the US National Academy of Sciences* [2] and sources [3–10], among many thousands of other papers.

For example, in the case of cast iron mine car wheels on steel rails, the following empirical equation is used [2, 9]

$$C_{rr} = 0.0048(9/R)^{1/2}(100/N)^{1/4} \tag{3.1.2}$$

Here, R is the wheel radius in inches, and N is the load on the wheel in lbs.

Coulomb's Law and Dupuit's equation are, probably, the most known theories of rolling that say:

$$C_{rr} = \eta_C/R \quad \text{(Coulomb)} \tag{3.1.3}$$

$$C_{rr} = \eta_D/R^2 \quad \text{(Dupuit)} \tag{3.1.4}$$

Here, η_C and η_D are some empirical constants. These equations work in some cases and do not work in others, as it is common for empirical theories. It is noteworthy that in 1781, the *Theorie des machines simples* won Charles Augustin Coulomb the Grand Prix from the Paris Academie des Sciences for this theory [3].

To get in with what follows, consider at first the elementary problem of equilibrium of an absolutely rigid heavy cylinder of radius R lying in a very small and shallow,

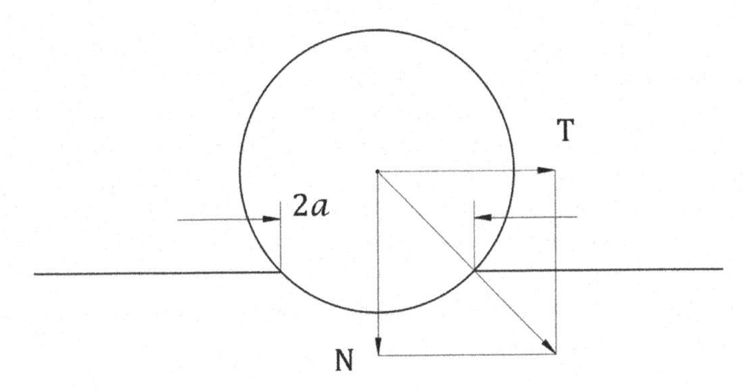

Fig. 3.1 A cylinder rolling out of a rigid socket on the road when $TR \geq aN$ (where $a \ll R$)

cylindrical socket of the same radius in a half-space $y < 0$ of an absolutely rigid material. Let us assume there is no adhesion and apply the thrust force T to the cylinder center directed along the half-space surface $y = 0$, with the gravity force N of the cylinder being directed perpendicularly. Evidently, without a socket, a however small thrust force makes the cylinder roll. It is the socket that provides resistance to rolling so that, if thrust T is less than aN/R, no rolling occurs ($2a$ is the width of the cylindrical socket, $a \ll R$). However, the rolling starts on when $TR > aN$ so that $TR = aN$ is the critical state; see Fig. 3.1.

In reality, a socket is formed by an elastic deformation of the foundation and cylinder caused by the weight and other forces acting upon the cylinder. The socket moves along the surface together with the rolling cylinder. The magnitude of sockets made by rolling bodies should be found from the solution of the corresponding mixed boundary value problems of the elasticity theory. To treat these problems, the Kolosov–Muskhelishvili representations [11] and the Riemann boundary value problem solved by Gakhov still in 1930 [12] are used in this chapter.

And so, the laws of rolling derived below are based on pure reason and mathematics. They accrue from the earlier works of this author [13–28] and well accommodate all previous empirical data.

3.2 An Elastic Cylinder on Another Elastic Cylinder

We study, at first, the mathematical problem of the elasticity theory, and then, based on its solution, we take account of the rolling and adhesion effect.

The Contact Problem of the Theory of Elasticity Let two elastic cylinders of radii R_1 and R_2 generally made of different materials contact each other so that the stresses and strains in both are under plane strain conditions. We confine ourselves by small deformations, which is the most important practical case, so that the width of the contact area $2a$ is always much smaller than R_1 and R_2, that is $R_1 \gg a$. We assume also that $|R_2| > R_1 > 0$.

When $R_2 \to \infty$, the second cylinder turns into a half-space which is the most important particular case. When $R_2 < 0$, it is the infinite elastic space with a cylindrical hole of radius $|R_2|$, inside which an elastic cylinder of radius R_1 is in contact.

Let us use the Cartesian rectangular coordinate system Oxy in a plane normal to the parallel axes of the cylinders so that its origin O is chosen to be at the center of the contact area, and y be the axis of symmetry of the problem directed to the center of the cylinder of radius R_1. This cylinder is pressed to the other cylinder by force N applied to its center.

The boundary value conditions of this plane-strain contact problem of the elasticity theory can be formulated as follows:

$$y = 0,\ |x| > a : \left(\sigma_y - i\tau_{xy}\right)^{\pm} = 0; \quad (i = \sqrt{-1}) \tag{3.2.1}$$

$$y = 0, |x| < a : \left[\sigma_y - i\tau_{xy}\right] = 0, \left[\frac{\partial u}{\partial x} + i\frac{\partial v}{\partial x}\right] = -ix\left(\frac{1}{R_1} + \frac{1}{R_2}\right) + C. \tag{3.2.2}$$

Here, (u, v) is the displacement vector; σ_x, σ_y and τ_{xy} are the stress tensor components; constant C characterizes the difference in displacement u of opposite surfaces of the cylinders, which forms in the process of compression before the adhesion on contact, and $z = x + iy$ is the complex variable.

Besides, the following designation is used everywhere in this chapter:

$$A^{\pm} = \lim A \ when\, z \to x \pm i0; \quad [A] = A^{+} - A^{-}. \tag{3.2.3}$$

The second equation in Eq. (3.2.2) means that neither rupture nor sliding occur on the contact area beyond of some previous local distortion.

Let us use the following Kolosov–Muskhelishvili representations for elastic stresses and strains in the plane elasticity theory [11]:

$$\sigma_x + \sigma_y = 4\mathrm{Re}\Phi_j(z) \quad (j = 1, 2); \tag{3.2.4}$$

$$\sigma_y - i\tau_{xy} = \Phi_j(z) + \overline{\Phi_j(z)} + z\overline{\Phi_j'(z)} + \overline{\Psi_j(z)}; \tag{3.2.5}$$

$$2\mu_j\left(\frac{\partial u}{\partial x} + i\frac{\partial v}{\partial x}\right) = \kappa_j\Phi_j(z) - \overline{\Phi_j(z)} - z\overline{\Phi_j'(z)} - \overline{\Psi_j(z)}. \tag{3.2.6}$$

Here, $j = 1$ and $j = 2$ are subscripts of the upper and lower half-planes, correspondingly; $\Phi(z)$ and $\Psi(z)$ with subscripts 1 or 2 are some functions of the complex variable z which are analytical in the corresponding half-planes; μ_j and ν_j are the shear modulus and Poisson's ratio of the corresponding material, and $\kappa_j = 3 - 4\nu_j$ for plane strain.

The boundary value problem of Eqs. (3.2.1) to (3.2.6) unsolved by Muskhelishvili [11] is being solved here by the approach published by this author in [14, 16].

Let us use the following analytical continuation:

$$\Phi_1(z) = -\overline{\Phi_1}(z) - z\overline{\Phi_1'}(z) - \overline{\Psi_1}(z) \quad (y = \text{Im}z < 0); \tag{3.2.7}$$

$$\Phi_2(z) = -\overline{\Phi_2}(z) - z\overline{\Phi_2'}(z) - \overline{\Psi_2}(z)k4 \quad (y = \text{Im}z > 0). \tag{3.2.8}$$

When $y = 0$, we get from here:

$$\Phi_1^{\pm} = -\overline{\Phi_1^{\mp}} - x\overline{\Phi_1'^{\mp}} - \overline{\Psi_1^{\mp}}, \tag{3.2.9}$$

$$\Phi_2^{\pm} = -\overline{\Phi_2^{\mp}} - x\overline{\Phi_2'^{\mp}} - \overline{\Psi_2^{\mp}}. \tag{3.2.10}$$

Here, either the upper or the lower signs hold.
In view of $\left[\sigma_y - i\tau_{xy}\right] = 0$ everywhere at $y = 0$, we get from Eq. (3.2.5) that

$$\Phi_1^+ + \overline{\Phi_1^+} + x\overline{\Phi_1'^+} + \overline{\Psi_1^+} = \Phi_2^- + \overline{\Phi_2^-} + x\overline{\Phi_2'^-} + \overline{\Psi_2^-}. \tag{3.2.11}$$

Using Eqs. (3.2.9) and (3.2.10), we can rewrite Eq. (3.2.11) as follows:

$$\Phi_1^+ - \Phi_1^- = \Phi_2^- - \Phi_2^+, \tag{3.2.12}$$

or

$$(\Phi_1 + \Phi_2)^+ = (\Phi_1 + \Phi_2)^-. \tag{3.2.13}$$

From here, based on the principle of analytical continuation we conclude that

$$\Phi_1(z) = -\Phi_2(z). \tag{3.2.14}$$

Also, due to Eqs. (3.2.1), (3.2.11), and (3.2.12) we have

$$\Phi_1^+ = \Phi_1^- \quad \text{as} \quad |x| > a,\, y = 0. \tag{3.2.15}$$

Hence, functions $\Phi_1(z)$ and $\Phi_2(z)$ are analytical in the whole plane z cut along $y = 0, -a < x < a$.

Using Eqs. (3.2.9) and (3.2.10), the second boundary condition in Eq. (3.2.2) is written at $y = 0$, $-a < x < a$ as follows:

$$\mu_2\left(\kappa_1\Phi_1^+ + \Phi_1^-\right) = \mu_1\left(\kappa_2\Phi_2^- + \Phi_2^+\right) - 2ix\mu_1\mu_2\left(\frac{1}{R_1} + \frac{1}{R_2}\right) + 2\mu_1\mu_2 C. \tag{3.2.16}$$

From here, using Eq. (3.2.14) we come to the following Riemann problem for one pair of functions [12, 14, 16]:

$$\Phi_2^+ + m\Phi_2^- = 2isx - 2\mu_1\mu_2 C(\mu_1 + \mu_2\kappa_1)^{-1} \quad (y = 0, |x| < a). \tag{3.2.17}$$

Here,

$$m = \frac{\mu_2 + \mu_1\kappa_2}{\mu_1 + \mu_2\kappa_1}, \quad s = \frac{\mu_1\mu_2}{\mu_1 + \mu_2\kappa_1}\left(\frac{1}{R_1} + \frac{1}{R_2}\right). \tag{3.2.18}$$

The general solution of this Riemann problem in the class of functions limited at $z \to \pm a$ and $z \to \infty$ can be written as follows:

$$\Phi_2(z) = \frac{2is}{1+m}z - \frac{2is}{1+m}F(z) - 2\mu_1\mu_2\frac{C}{(1+m)(\mu_1 + \mu_2\kappa_1)}, \tag{3.2.19}$$

where

$$F(z) = (z-a)^{1/2-i\delta}(z+a)^{1/2+i\delta}, \tag{3.2.20}$$

$$\delta = \frac{1}{2\pi}\ln m. \tag{3.2.21}$$

Here, $F(z)$ is the single-valued analytical branch of the function in the z-plane cut along $y = 0, +a > x > -a$, which is equal to

$$F(z) = z + 2ia\delta - \frac{a^2}{2z}(1 + 4\delta^2) + O\left(z^{-2}\right) \quad \text{as} \quad z \to \infty. \tag{3.2.22}$$

Constants a and C are determined by the following behavior of function $\Phi_2(z)$ at infinity [14, 21–23]:

$$\Phi_2(z) = \frac{iN}{2\pi z} \quad \text{as} \quad z \to \infty. \tag{3.2.23}$$

Here, N is the value of the equivalent force of the pressure on the contact area per unit of the cylinder length acting upon the lower half-plane, which is equal to the corresponding weight of the upper cylinder of radius R_1.

According to Eqs. (3.2.19) and (3.2.22), as $z \to \infty$ we have

$$\Phi_2(z) = \frac{2is}{1+m}z - \frac{2is}{1+m}\left(z + ia\delta - \frac{a^2}{2z} - \frac{2a^2\delta^2}{z}\right) - \frac{2\mu_1\mu_2 C}{(1+m)(\mu_1 + \mu_2\kappa_1)}. \tag{3.2.24}$$

By comparing Eqs. (3.2.23) and (3.2.24), we get

$$2sa^2 + 8sa^2\delta^2 = \frac{N}{\pi}(1+m), \quad C = a\delta\left(\frac{1}{R_1} + \frac{1}{R_2}\right). \tag{3.2.25}$$

Using Eqs. (3.2.17) and (3.2.25), we find the width of the contact area

$$a^2 = \frac{N}{2\pi}\frac{\mu_1(1+\kappa_2) + \mu_2(1+\kappa_1)}{\mu_1\mu_2(1+4\delta^2)}\left(\frac{1}{R_1} + \frac{1}{R_2}\right)^{-1}. \tag{3.2.26}$$

It is more convenient to use this equation in terms of common technical elastic constants as follows:

$$a^2 = \frac{4N}{\pi(1+4\delta^2)}\left(\frac{1-\nu_1^2}{E_1} + \frac{1-\nu_2^2}{E_2}\right)\left(\frac{1}{R_1} + \frac{1}{R_2}\right)^{-1}. \tag{3.2.27}$$

Here, E and ν are Young's modulus and Poisson's ratio, with subscripts corresponding to the relative cylinder.

Using Eqs. (3.2.19) to (3.2.21) and (3.2.25), we come to the final solution:

$$\Phi_2(z) = \frac{1}{4}Ai\left\{z - \left(z^2 - a^2\right)^{1/2}\left(\frac{z+a}{z-a}\right)^{i\delta} + ia\delta\right\}. \tag{3.2.28}$$

Here,

$$A = \left(\frac{1}{R_1} + \frac{1}{R_2}\right)\left(\frac{1-\nu_1^2}{E_1} + \frac{1-\nu_2^2}{E_2}\right)^{-1}. \tag{3.2.29}$$

Using Eqs. (3.2.5) and (3.2.28), the stresses on the contact area $y = 0, -a < x < a$ can be written as follows:

$$\sigma_y = -\frac{1}{2}Ach(\pi\delta)\sqrt{a^2 - x^2}\cos\left(\delta\ \ln\left|\frac{a+x}{a-x}\right|\right), \tag{3.2.30}$$

$$\tau_{xy} = -\frac{1}{2}Ash(\pi\delta)\sqrt{a^2 - x^2}\sin\left(\delta\ \ln\left|\frac{a+x}{a-x}\right|\right). \tag{3.2.31}$$

Equations (3.2.27) to (3.2.31) provide the solution of the stated problem in most convenient shape. For the particular case $\delta = 0$, it was given in [25, 26].

The Problem of Rolling Now, let us try to roll the upper cylinder of radius R_1 on the horizontal foundation, that is, on the lower cylinder of radius $R_2 = \infty$. Suppose T is the thrust or driving force per unit of the cylinder length applied to the center of the upper cylinder in the horizontal direction perpendicular to the vertical direction of force N, particularly, the weight of the upper cylinder. If both the cylinder and foundation are rigid, neither socket nor dent can form and no contact area can be made so that the cylinder starts on rolling at a however small thrust because the rolling resistance force F is equal to zero.

But, if either the cylinder or foundation is elastic, a dent forms so that a nonzero contact area provides some resistance to the rolling. If thrust T is small enough so that vector (T, N) of the resultant force points to somewhere inside the contact area, the equilibrium holds and no rolling occurs. This is evident because some change of the contact area size Δa due to elasticity has the order of $a(T/N)^2$ for small values of T/N and, hence, can be ignored.

As soon as vector (T, N) of the resultant force points to somewhere outside the contact area, the equilibrium gets broken and an accelerated rolling begins. This follows from the equilibrium equation of the moments of forces with respect to point $y = 0, x = a$ even for the imaginable situation when both elastic bodies become rigid for small values of T. For real elastic bodies, the loss of balance is even more significant because of an additional elastic reaction of the lower foundation.

Thus, we come to the following rule: *Rolling can start only when the resultant force vector* (T, N) *points to the front of the contact area* so that we have

$$\frac{T}{N} = \frac{a}{R_1}. \tag{3.2.32}$$

This is the necessary rule of rolling called the "chief head" rule, or the CH rule; for short, see Fig. 3.2. It is central for the understanding of the rolling processes and rolling resistance [25, 26].

The state characterized by the CH rule of Eq. (3.2.32) corresponds to the rolling at constant speed when thrust T is equal to the rolling resistance force F. If $T/N < a/R_1$, no rolling occurs, but if $T/N > a/R_1$, the upper cylinder of mass M rolls and moves along the x-axis at the acceleration $\mathrm{d}V/\mathrm{d}t$ so that

$$M\mathrm{d}V/\mathrm{d}t = T - aN/R_1. \tag{3.2.33}$$

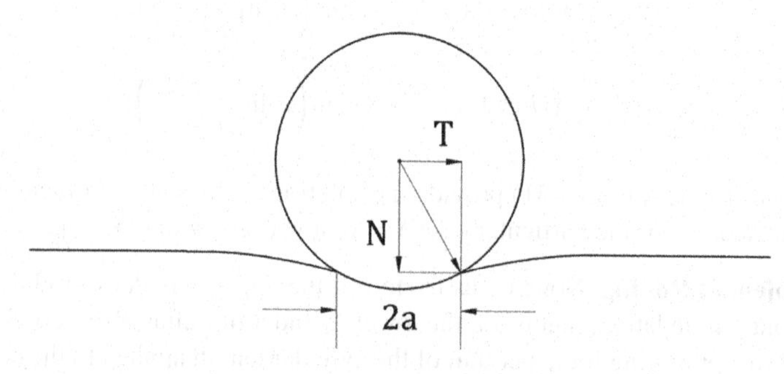

Fig. 3.2 CH rule of rolling $TR = aN$ (where $a \ll R$)

Based on the CH rule, the rolling resistance coefficient is equal to

$$C_{rr} = a/R_1. \tag{3.2.34}$$

Here, the value of a is completely determined by Eq. (3.2.27) written in terms of $N, E_1, E_2, R_1, R_2, \nu_1, \nu_2$ and parameter δ.

And so, Coulomb's problem is solved, and the correct value of C_{rr} proves to be very different from both Coulomb's Law and Dupuit's equation as well as from the empirical law of Eq. 1.1.2.

In the important particular case when $\delta = 0$, $E_1 = E_2 = E$, $\nu_1 = \nu_2 = \nu$, $R_2 = \infty$ and $R_1 = R$, the expressions in Eqs. (3.2.27) to (3.2.32) and Eq. (3.2.34) reduce to the following equations:

$$\Phi(z) = \frac{i(1-\nu^2)}{2ER}\left(z - \sqrt{z^2 - a^2}\right); \tag{3.2.35}$$

$$\sigma_y = -\frac{1-\nu^2}{ER}\sqrt{a^2 - x^2}, \quad \tau_{xy} = 0 \quad (y = 0, |x| < a) \tag{3.2.36}$$

$$C_{rr} = \frac{T}{N} = \frac{a}{R} = 2\sqrt{\frac{2N(1-\nu^2)}{\pi ER}}. \tag{3.2.37}$$

To demonstrate the accuracy of the present analysis based on the CH rule, let us calculate, as an example, the rolling resistance coefficient for railroad steel wheels on steel rails for passenger railcars with eight 36-in. diameter wheels on the 63 kg/m rails used in the New York Central Railroad System.

In this case, the railhead width being contacted with the railcar wheel is equal to 3 in. = 7.62 cm so that using Eq. (3.2.37), we get the following rolling resistance coefficient:

For 10-ton railcar $C_{rr} = 0.00113$ and for 40-ton railcar $C_{rr} = 0.00226$

$$(\nu_1 = \nu_2 = 0.29,\ E_1 = E_2 = 200\,\text{GPa},\ \delta = 0,\ R_2 = \infty, \text{ and } R_1 = 45.72\,\text{cm}).$$

The elastic constants of both the rail and wheel are very close. The plane-strain condition holds well in the process zone of the contact area because its width $2a$ is much smaller than the 3-in. railhead or any other dimension of the structure. Indeed, according to Eq. (3.2.37) we have $a = 0.5$ mm for 10-ton railcar and $a = 1$ mm for 40-ton railcar.

This result of calculation of the rolling resistance coefficient coincides with the corresponding standard data; see, e.g., *Special Report of the US National Academy of Sciences* [2], according to which C_{rr} 0.001–0.0024 for railroad steel wheel on steel rail and $C_{rr} = 0.002$ for passenger railcars.

Similarly, we can calculate that for ordinary car tires on concrete or asphalt the rolling resistance coefficient is equal to $C_{rr} = 0.01$–0.015 in accordance with available data from *Special Report of the US National Academy of Sciences* [2].

Also, let us estimate the practical significance of real values of parameter δ responsible for some anomalous distortion. Since according to Eq. (3.2.18) we have $3 > m > 1/3$, we can derive from Eq. (3.2.2) that

$$-0.175 < \delta = \frac{1}{2\pi} \ln m < 0.175.$$

For common values of Poisson's ratio in the range $(1/4, 1/3)$, the value of δ varies in even more narrow interval $(-0.11, +0.08)$. Small parameter δ causes some specific variations significant only at the very ends of the contact area. It means that when $m \neq 1$ the violation of the condition of non-penetration occurs only in the small neighborhood of the ends so that it can be ignored despite this makes the problem ill-posed [18, 20, 25, 26]. In fact, these variations indicate that when $\delta \neq 0$ some local sliding zones much smaller than a form near the ends of the contact area.

All mathematical calculations of this section are also valid for plane stress by means of the well-known substitution of constants κ_j in Eq. (3.2.6). This is the case of a very thin wheel rolling on a very thin rail. However, this case is of almost no practical value for the rolling because the contact area width a is already very small while the thickness of the plane-stress wheel should be much less than even this width [19, 23].

3.3 Invariant Integral of Rolling

The end of a contact zone is a singular point of the boundary value problems of the theory of elasticity, which requires a special treatment. Because of atomic forces of attraction, all solid surfaces experience the effect of adhesion while getting in contact. This effect can prevail over usual mechanical forces for objects of sufficiently small dimensions. Besides, it can be used in practice for some control of rolling processes by means of special adhesives and lubricants. And so, the effect of adhesion makes it necessary to reconsider the problem of the previous section in a more general case of unlimited stresses at the end of a contact zone.

Let us consider a small neighborhood of the point O where the roller and foundation part, in the scale very small as compared to the radius of curvature of the roller and the contour size of the contact zone. If we look through a microscope at this point, we can see the roller as the upper half-plane $y > 0$ and the foundation as the lower half-plane $y < 0$, with the contact zone being along $y = 0, x > 0$ and with the free traction zone being along $y = 0, x < 0$.

This local plane-strain state is exactly the same as that near the crack-tip, when the open-mode tensile interface crack cutting a body. However, this similarity with this problem of fracture mechanics is valid only for this small neighborhood at the rear end of the contact area under consideration at one moment of rolling. The front of the contact area of the roller that meets the foundation is under the opposite condition of compression.

And so, we come to the following singular problem of the plane elasticity:

$$y = 0, x < 0: \quad \left(\sigma_y - i\tau_{xy}\right)^{\pm} = 0; \tag{3.3.1}$$

$$y = 0, x > 0: \quad \left[\sigma_y - i\tau_{xy}\right] = 0, \quad [u + iv] = 0. \tag{3.3.2}$$

All designations are the same as in the previous section.

The boundary value problem of Eqs. (3.3.1) and (3.3.2) is reduced to the homogenous problem of Eq. (3.2.17):

$$\Phi_2^+ + m\Phi_2^- = 0 \quad \text{as} \quad y = 0, x > 0. \tag{3.3.3}$$

Its solution in the class of analytical functions having a singularity of minimum order at the singular point $z = 0$ has the following shape:

$$\Phi_2(z) = \frac{1}{2\sqrt{2\pi}}(K_I - iK_{II})z^{-1/2+i\delta}, \quad \left(\delta = \frac{1}{2\pi}\ln m\right). \tag{3.3.4}$$

Here, K_I and K_{II} are some constants similar to the stress intensity factors in fracture mechanics which play the same part in the current problem because as a result of adhesion a local zone of tensile stresses forms near the rear point of the contact area of the roller and foundation. In the process of rolling, these constants have some limiting values described in terms of the effective energy loss γ_{fm} per unit of square called the specific energy of adhesion (see [21, 23], for more detail).

To calculate this loss, we study the small neighborhood of the point O where the solution given by Eq. (3.3.4) is valid and calculate the following invariant integral:

$$\Gamma_1 = \oint_S \left(Un_1 - \sigma_{ij}n_j u_{i,1}\right)\mathrm{d}S, \quad (i, j = 1, 2, 3). \tag{3.3.5}$$

Here, the closed contour S encircles the point O. As S, it is convenient to use a narrow symmetrical rectangle like that in Sect. 1.3. Equation (3.3.5) provides the typical Γ force of energy losses spent per unit square to overcome the drag.

In the limit, Eq. (3.3.5) is reduced to the following equation [21, 23]:

$$\Gamma_1 = \frac{\pi}{2}\lim[\sigma_{2i}(+\varepsilon, 0)u_i(-\varepsilon, 0)], \quad (\varepsilon \to 0, i = 1, 2, 3). \tag{3.3.6}$$

This equation allows one to calculate the driving force Γ_1 from the local elastic field of stresses $\sigma_{ij}(x_1, x_2)$ and displacements $u_i(x_1, x_2)$ near the singular point O where $x_1 = x, x_2 = y$ (repeated index means summation).

The calculation using Eqs. (3.3.6), (3.3.4), (3.2.14), and (3.2.4) to (3.2.6) leads to the following result

$$\Gamma_1 = \frac{(\mu_1 + \mu_2\kappa_1)(\mu_2 + \mu_1\kappa_2)}{4\mu_1\mu_2[\mu_1(1+\kappa_2) + \mu_2(1+\kappa_1)]}(K_I^2 + K_{II}^2). \tag{3.3.7}$$

The energy conservation equation $\Gamma_1 = 2\gamma_{fm}$ serves to find the size of the contact area between the roller and foundation under the adhesion conditions.

Particularly, when $\mu_1 = \mu_2 = \mu$ and $\kappa_1 = \kappa_2 = 3 - 4\nu$, we have:

$$\Gamma_1 = \frac{1-\nu}{2\mu}(K_I^2 + K_{II}^2). \tag{3.3.8}$$

This equation is well known also in fracture mechanics [21, 23].

3.4 The Effect of Adhesion on the Rolling

Let us study the effect of adhesion using the results in Sect. 3.3 when $\delta = 0$, i.e., $m=1$, so that $\mu_1(1-\nu_2) = \mu_2(1-\nu_1)$, or in most common case when $\mu_1 = \mu_2 = \mu$ and $\nu_1 = \nu_2 = \nu$. In this case, the stress intensity factor K_{II} at the singular point is equal to zero. By applying the procedure of Sect. 3.2, we come to the following boundary value problem which is similar to Eq. (3.2.17)

$$\Phi_2^+ + \Phi_2^- = 2isx \quad \text{as} \quad y = 0, -b < x < a. \tag{3.4.1}$$

Here, s is given by Eq. (3.2.18).

As a reminder, the vector of the equivalent force of normal tractions on the contact area, which magnitude is equal, for example, to the weight of the upper cylinder per unit length, is directed along the y-axis of this Oxy system. As a result of adhesion, the positive thrust force T applied to the center of the rolling cylinder can create some local extension and concentration of tensile stresses near the trailing point $x = -b$ of the contact area.

The rolling starts when the adhesive bond at point $x = -b$ is broken, that is, when the stress intensity factor at this point achieves a certain limiting value k_{IC} characterizing this bond so that [14, 21–23]

$$\sigma_y = \frac{k_{IC}}{\sqrt{2\pi\varepsilon}} \quad \text{as} \quad \varepsilon \to 0 \quad \text{where} \quad y = 0, x = -b + \varepsilon. \tag{3.4.2}$$

In the case under consideration when $\delta = 0$, the constant k_{IC} called the adhesion toughness or bond toughness is expressed in terms of the specific adhesion bond energy γ_{fm} as follows:

$$16\gamma_{fm} = \frac{\kappa_1\kappa_2 - 1}{\mu_1(\kappa_2 - 1)}k_{IC}^2. \tag{3.4.3}$$

The solution of the Riemann problem (3.4.1), which is limited at point $x = a$ and at infinity but singular at point $x = -b$, has the form [12, 14, 21–23]

$$\Phi_2(z) = si\left(z - \sqrt{(z-a)(z+b)}\right) - iB\sqrt{\frac{z-a}{z+b}}. \tag{3.4.4}$$

Here, a, b, and B are some real constants to be found. As $z \to \infty$, the chosen branches of the root functions behave as follows:

$$\sqrt{(z-a)(z+b)} = z + \frac{1}{2}(b-a) - \frac{1}{8z}(a+b)^2 + O\left(z^{-2}\right), \tag{3.4.5}$$

$$\sqrt{\frac{z-a}{z+b}} = 1 + \frac{a+b}{2z} + O\left(z^{-2}\right). \tag{3.4.6}$$

The tensile bond of adhesion at $z = -b$ is broken, if the critical state described by Eq. (3.4.2) is achieved due to the rolling. In terms of function $\Phi_2(z)$, it means that

$$\Phi_2(z) = \frac{k_{\mathrm{IC}}}{2\sqrt{2\pi(z+b)}} \quad \text{as} \quad z \to -b. \tag{3.4.7}$$

Based on Eqs. (3.4.4) and (3.4.7), we can conclude that

$$2B\sqrt{2\pi(a+b)} = k_{\mathrm{IC}}. \tag{3.4.8}$$

Also, from Eqs. (3.2.23), and (3.4.4) to (3.4.6) we can derive that

$$2B = s(a-b), \quad \pi s(a+b)(3b-a) = 4N. \tag{3.4.9}$$

The equation system, Eqs. (3.4.8) and (3.4.9), serves to determine constants $a, b,$ and B characterizing the effect of adhesion upon the process of rolling.

Due to Eqs. (3.2.2) and (3.4.4) the stresses on the contact area $y = 0$, $-b < x < a$ are equal to

$$\sigma_y = -2s\sqrt{(a-x)(b+x)} + 2B\sqrt{\frac{a-x}{b+x}}, \; \tau_{xy} = 0. \tag{3.4.10}$$

The solution of the equation system, Eqs. (3.4.8) and (3.4.9), is reduced to the following equation:

$$\tau^2 - \Gamma\tau^{1/2} - 1 = 0. \tag{3.4.11}$$

$$\left(\tau = \frac{a}{L} + \frac{b}{L}, \quad L = 2\sqrt{\frac{N}{\pi s}}, \quad \Gamma = \frac{1}{2}k_{\mathrm{IC}}\left(\frac{\pi}{s}\right)^{1/4} N^{-3/4}\right) \tag{3.4.12}$$

The dimensionless number Γ determines the effect of adhesion like the Reynolds number does the effect of viscosity in hydrodynamics.

In the case of a comparatively small effect of adhesion upon the rolling, we get

$$a = (1 + \Gamma)\sqrt{\frac{N}{\pi s}}, \; b = \sqrt{\frac{N}{\pi s}} \quad \text{as} \quad \Gamma \ll 1. \tag{3.4.13}$$

Here are some values of the monotonously growing function $\tau = \tau(\Gamma)$:

Γ	0	1.02	2.12	4.62	7.5
Γ	1	1.5	2	3	4

In the most important scenario, the process proceeds in three stages. Using the CH rule, let us consider it in the case when the second cylinder is a lower half-space.

On the first stage, the elastic cylinder of radius R_1 presses into the half-space of another material by the normal force N so that a symmetrical dent forms along the contact area $(-b, +b)$ with its width $2b$ being determined by Eq. (3.4.13) at $\Gamma = 0$. On the next stage, the thrust T being applied to the center of the cylinder is increasing but having no effect until its value becomes equal to Nb/R_1 so that the vector (T, N) points at the front of the contact area $x = +b$.

On the third stage, the adhesion bond at $x = -b$ is stretching on and the tensile stress singularity is growing at this point while the thrust T increases, and the front of the contact area $x = +a$ moves in order to balance the normal force N. On this stage, the process is stable so that the value of a slowly increases while T grows, with the vector (T, N) always pointing to the front $x = +a$ according to the CH rule.

When the growing stress singularity at $x = -b$ achieves the critical value defined by Eq. (3.4.8), the rolling begins. At this critical state, the equation system, Eqs. (3.4.8) and (3.4.9), becomes valid, with the value of the critical thrust being determined by the same CH rule $T = aN/R_1$. Based on Eq. (3.4.13), for small values of Γ the CH rule provides the following law of rolling:

$$T = (1 + \Gamma)\frac{N^{3/2}}{R_1\sqrt{\pi s}}, \quad C_{rr} = (1 + \Gamma)\frac{1}{R_1}\sqrt{\frac{N}{\pi s}}. \tag{3.4.14}$$

This scenario plays only when, at first, load N increases while $T = 0$ and, then, the thrust T grows, with the normal load being constant. For other loading paths, the results will be different because this process is, evidently, path-dependent.

3.5 An Elastic Ball Rolling on an Elastic Half-Space

Let us consider the problem of an elastic ball of radius R_1 rolling on another elastic ball of another elastic material of greater radius R_2. When $R_2 = \infty$ the latter is an elastic half-space, and when $R_2 < 0$ the latter is an infinite elastic space with a spherical hole of radius $|R_2| > R_1$.

For this purpose, we can use Hertz's solution for the static problem of an axially symmetric contact of two elastic balls of different materials [7]:

$$w = \left(\frac{R_1 + R_2}{R_1 R_2}\right)^{1/3} \left[\frac{3}{4} N \left(\frac{1 - \nu_1^2}{E_1} + \frac{1 - \nu_2^2}{E_2}\right)\right]^{2/3} - \frac{R_1 + R_2}{2R_1 R_2} r^2; \tag{3.5.1}$$

$$\sigma_z = -\frac{3N}{2\pi a^2} \sqrt{1 - \frac{r^2}{a^2}} \quad (R_1 \gg a, |R_2| > R_1 > 0); \tag{3.5.2}$$

$$a^3 = \frac{3}{4} N \left(\frac{1 - \nu_1^2}{E_1} + \frac{1 - \nu_2^2}{E_2}\right) \frac{R_1 R_2}{R_1 + R_2}. \tag{3.5.3}$$

Here, r is the distance from the z-axis which is the axis of symmetry in this problem, a is the radius of the contact area, N is the resultant force of pressure $|\sigma_z|$ on the contact area $z = 0, r < a$ (here, shear stress is equal to zero), and w is the sum of the z-components of the elastic displacements at the opposite points of the contact of the ball surfaces.

Hertz's solution is commonly used to study the impact of two elastic balls. We will use it to study the equilibrium of two heavy elastic balls resting one on the other in the vertical field of gravitation before rolling. The cases when $R_2 \to \infty$ or $R_2 < 0$ are of most interest. They relate to the ball rolling on the surface of a half-space or to the ball rolling on the surface of a spherical cavity in an elastic space.

Let us slowly increase the tangential thrust force T applied to the center of the smaller ball of weight N. This force is perpendicular to the z-axis. If the vector (T, N) points inside the contact area, no motion occurs. In this case, there are no substantial changes in the size of the contact area because T/N is very small.

However, based on the CH rule this ball starts on rolling when vector (T, N) points to the circular front of the contact area so that Eq. (3.2.32) becomes valid. From here, using Eqs. (3.5.3) and (3.1.1) we find the rolling resistance coefficient and the law of rolling

$$C_{rr} = \frac{T}{N} = \frac{a}{R_1} = \frac{1}{R_1} \left[\frac{3}{4} N \left(\frac{1 - \nu_1^2}{E_1} + \frac{1 - \nu_2^2}{E_2}\right) \frac{R_1 R_2}{R_1 + R_2}\right]^{1/3}. \tag{3.5.4}$$

In the case of a ball rolling on a half-space, we get from here

$$C_{rr} = \frac{T}{N} = \frac{a}{R_1} = \left[\frac{3}{4R_1^2} N \left(\frac{1 - \nu_1^2}{E_1} + \frac{1 - \nu_2^2}{E_2}\right)\right]^{1/3}. \tag{3.5.5}$$

Using Eq. (3.5.5), we find, for example, that the rolling resistance coefficient of hardened steel ball bearings on steel is equal to about 0.001–0.0015 within the practical range of parameters, which is in excellent agreement with the known data of the tests [2, 8].

3.6 An Elastic Torus Rolling on an Elastic Half-Space

Let us consider the contact problem of the elasticity theory of the pressure of an elastic solid torus on an elastic half-space of another material. The torus surface is formed by rotation of a circumference of radius r around an axis lying in the plane of this circumference, but not intersecting the latter, so that the circumference center forms another circumference of greater radius $R > r$. The center of the latter is the torus center.

Let this solid torus press on the boundary of an elastic half-space $z < 0$ so that the plane of the major circle of the torus is perpendicular to the half-space boundary $z = 0$. The force N is applied to the torus center and directed perpendicular to the boundary of the half-space (this "load on the wheel" can be transferred to the torus via spokes or a connecting disk).

In the neighborhood of a small contact area, the torus surface coincides with the surface of the following elliptic paraboloid:

$$z = \frac{x^2}{2(r + R)} + \frac{y^2}{2r}. \tag{3.6.1}$$

Here, the directions x and y and the corresponding principal curvature radii $r + R$ and r of this paraboloid coincide with the directions and principal curvature radii of the torus at the initial contact point $x = y = z = 0$.

And so, the problem is reduced to Hertz's problem of the contact of two different elastic paraboloids [7], one of which is the lower half-space $z < 0$. According to Hertz's solution, the small contact area in this problem is the interior of the following ellipse on the plane $z = 0$:

$$\frac{x^2}{L^2} + \frac{y^2}{b^2} = 1 \quad (L > b). \tag{3.6.2}$$

Here, $2L$ and $2b$ are the major and minor axes of the ellipse.

The stresses on the contact area are [7]

$$\sigma_z = -\frac{3N}{2\pi bL}\sqrt{1 - \frac{x^2}{L^2} - \frac{y^2}{b^2}}, \quad \tau_{xz} = \tau_{yz} = 0. \tag{3.6.3}$$

Here, the values of b and L are determined by the following equations:

$$\left(1-e^2\right)\frac{D(e)}{B(e)}=\frac{r}{R+r};\quad 1-e^2=\frac{b^2}{L^2};\tag{3.6.4}$$

$$L^3=\frac{3}{\pi}\left(\frac{1-\nu_1^2}{E_1}+\frac{1-\nu_2^2}{E_2}\right)ND(e)(R+r).\tag{3.6.5}$$

(The subscripts 1 and 2 refer to the torus and base materials, correspondingly.) As a reminder, $B(e)$ and $D(e)$ are the following elliptic integrals:

$$B(e)=\int\limits_0^{\pi/2}\frac{\cos^2\varphi}{\sqrt{1-e^2\sin^2\varphi}}d\varphi,\quad D(e)=\int\limits_0^{\pi/2}\frac{\sin^2\varphi}{\sqrt{1-e^2\sin^2\varphi}}d\varphi.\tag{3.6.6}$$

Now, let us apply the thrust force T to the center of torus in the direction of the major axis x and consider the problem of the rolling of the torus on the surface $z=0$ of the elastic half-space of another material.

As long as the force T is sufficiently small so that vector (T, N) points to somewhere inside the contact area, no rolling occurs. Based on the CH rule, when vector (T, N) points to the edge of the contact area at $x=L, y=0$ the torus starts on rolling.

And so, the law of rolling of an elastic torus on an elastic half-space of another material says that the rolling resistance coefficient in this case is equal to

$$C_{rr}=\frac{T}{N}=\frac{L}{R+r}=\frac{1}{R+r}\left[\frac{3}{\pi}\left(\frac{1-\nu_1^2}{E_1}+\frac{1-\nu_2^2}{E_2}\right)ND(e)(R+r)\right]^{1/3}.\tag{3.6.7}$$

Here, the function $e=e(r/R)$ is determined by Eq. (3.6.4).

This law of rolling is valid as far as Hertz's solution, Eqs. (3.6.3) to (3.6.5), is valid. Its violation can be due to the local stick-and-slip zones on the contact area studied in papers [24, 25]; see also Chap. 5 of the book.

3.7 A Rigid Cylinder Rolling on a Sticky Membrane

Let us study the problem of a rigid cylinder of radius R rolling on the horizontal surface of a flat membrane shell, or film, tightly stretched in all directions by the tension force γ which is equal to the product of the tensile stress in the shell and the shell thickness. At first, we consider the static problem, with the force N per unit length of the cylinder, e.g., its weight, being applied to its center in the vertical direction. Let $2a$ be the width of a small contact area so that $a \ll R$.

The force N is balanced by the film pressure on the contact area and by the tension forces of adhesion applied to the edge of this area. From equilibrium equation of the film $\gamma d^2w/dx^2 = p$, it follows that $\gamma = pR$ because $w = x^2/(2R)$ on the contact area $-a < x < a$ where the shape of the film $w = w(x)$ coincides with the shape of the cylinder (p is the film pressure on the cylinder).

From the energy conservation law, it follows [24] that

$$\gamma(1 - \cos\beta) = \Gamma_C \tag{3.7.1}$$

Here, Γ_C is the specific energy of adhesion of the film and cylinder materials, and β is the angle between the film and the plane tangential to the cylinder at the edge of the contact area. The value of Γ_C can be controlled by special glues or lubricants.

For small β, we get from Eq. (3.7.1)

$$\gamma\beta^2 = 2\Gamma_C. \tag{3.7.2}$$

And so, the balance of forces acting on the cylinder is written as follows:

$$N = 2\gamma(\alpha - \beta). \tag{3.7.3}$$

$$\left(\alpha = \frac{a}{R}, \quad \beta = \sqrt{\frac{2\Gamma_C}{\gamma}}\right). \tag{3.7.4}$$

From Eqs. (3.7.3) and (3.7.4), it follows that

$$a = \frac{R}{2\gamma}\left(N + 2\sqrt{2\gamma\Gamma_C}\right). \tag{3.7.5}$$

Now, let us apply the small thrust force T (per unit length of the cylinder) to the cylinder center in the horizontal direction. It does not cause a rolling, if the vector (T, N) points to somewhere inside the contact area.

By the CH rule, the cylinder starts on rolling, when this vector points to the front of the contact area so that based on Eq. (3.7.5) the law of rolling and the rolling resistance coefficient are given by the following equations:

$$C_{rr} = \frac{T}{N} = \frac{a}{R} = \frac{N}{2\gamma} + \sqrt{\frac{2\Gamma_C}{\gamma}}. \tag{3.7.6}$$

In the case if the adhesion can be ignored so that $N^2 \gg 8\gamma\Gamma_C$, we can get $C_{rr} = N/2\gamma$ where $\gamma \gg N$. In the opposite case, when the weight is supported by the adhesion only so that $N^2 \ll 8\gamma\Gamma_C$, we have $C_{rr} = \sqrt{2\Gamma_C/\gamma}$.

3.8 A Rigid Ball Rolling on a Sticky Film

Let us study the problem of a rigid ball of radius R rolling on a flat, sticky membrane, or a film, tightly stretched by tension γ in all directions. Again, we need, at first, to solve the corresponding static problem, with the force N, being applied to the ball center (e.g., its weight). Let Orz be the cylindrical coordinate system, the z-axis of which is the vertical coinciding with the axis of symmetry of the problem. The contact area is inside a circle of small radius $a \ll R$.

The displacement $w(r)$ of the film is described by the following equation:

$$\gamma\left(\frac{\mathrm{d}^2 w}{\mathrm{d}r^2}+\frac{1}{r}\frac{\mathrm{d}w}{\mathrm{d}r}\right)=p\theta(r). \tag{3.8.1}$$

Here, $\theta(r) = 0$ when $r > a$ and $\theta(r) = 1$ when $r < a$, and p is the pressure in the contact area to be found.

From (3.8.1), we can find

$$w = C\ln\frac{r}{a} \text{ when } r > a, \quad \text{and} \quad w = \frac{p}{4\gamma}\left(r^2 - a^2\right) \text{ when } r < a. \tag{3.8.2}$$

Here, C is a constant to be found.

For $r < a$, the function $w(r)$ coincides with the surface of the ball so that using Eq. (3.8.2), we get from here

$$pR = 2\gamma. \tag{3.8.3}$$

Similar to the problem in Sect. 3.7, because of the adhesion with the ball surface the film makes up the angle β with the plane tangential to the ball at any point of the front of the contact area at $z = 0,\ r = a$. This angle is determined by Eq. (3.8.1), or by Eq. (3.8.2) for small β.

The force N, for example, the weight of the ball, is balanced by the film tension so that:

$$N = 2\pi a\gamma(\alpha - \beta) = 2\pi a\gamma\left(\frac{a}{R} - \sqrt{\frac{2\Gamma_C}{\gamma}}\right); \tag{3.8.4}$$

$$C = a(\alpha - \beta). \tag{3.8.5}$$

Using Eqs. (3.8.4), (3.8.5) and the CH rule, we find the radius of the contact area, the law of rolling, and the rolling resistance coefficient:

$$C_{rr} = \frac{T}{N} = \frac{a}{R} = \sqrt{\frac{N}{2\pi\gamma R} + \frac{\Gamma_C}{2\gamma}} + \sqrt{\frac{\Gamma_C}{2\gamma}}. \tag{3.8.6}$$

In the case of a negligibly small adhesion, when $\pi R\Gamma_C \ll N$, we get a simpler relation $C_{rr} = \sqrt{N/(2\pi\gamma R)}$ from Eq. (3.8.6). In the opposite case of a very strong adhesion, when $\pi R\Gamma_C \gg N$, the formula $C_{rr} = \sqrt{2\Gamma_C/\gamma}$ is valid.

It should be kept in mind that the notion of membrane shell, or film, is rather about the stress state in the shell than about its mechanical property. If tensile stresses in a shell are much greater than bending stresses, the shell can be treated as a membrane, even if it is made, for example, of the hardest steel.

3.9 A Rigid Torus Rolling on a Membrane Shell

Let us study the problem of a rigid torus rolling on a flat membrane shell, tightly stretched by tension γ in all directions so that the size of the contact area is very small as compared to the main radii r and R of the torus. In this case, the contact problem for the torus is reduced to the contact problem for the elliptic paraboloid given by Eq. (3.6.1), the main radii of curvature at the apex of this paraboloid being equal to r and $r + R$.

And so, let us solve the contact problem for this rigid paraboloid pressing by its apex onto the flat membrane shell tightly stretched by a very strong tension so that the contact area is small compared to the minor radius of curvature at the apex [18]. We use the $Oxyz$ coordinate system, with its origin being at the apex and the z-axis being the symmetry line inside this paraboloid; see Eq. (3.6.1). Each cross section $z = \text{const}$ of the paraboloid is an ellipse, with its major axis lying on the x-axis. The resistance force N is directed along the z-axis since the paraboloid presses down in the opposite direction (no rolling so far!).

The purpose of our next calculation is to find the shape and size of the unknown contact area D on the xy plane, which is evidently symmetrical with respect to the x-axis and y-axis. The displacement w of the membrane shell satisfies the following equation:

$$\frac{\partial^2 w}{\partial x^2} + \frac{\partial^2 w}{\partial y^2} = \frac{p}{\gamma}\theta. \tag{3.9.1}$$

Here, p is the pressure on the paraboloid inside the domain D where $\theta = 1$, while $\theta = 0$ outside the domain D.

Inside the contact area D, the displacement w coincides with z in Eq. (3.6.1). Using Eqs. (3.6.1) and (3.9.1), we find

$$p = \gamma\frac{2r + R}{r(r + R)} \tag{3.9.2}$$

Let us introduce the complex variable $\xi = x + iy$. The general solution of Eq. (3.9.1) inside the domain D is

$$w = \mathrm{Ref}(\xi). \quad (3.9.3)$$

Here, $f(\xi)$ is the analytical function to be found.

The derivative of this function is

$$f'(\xi) = \frac{\partial w}{\partial x} - i\frac{\partial w}{\partial y}. \quad (3.9.4)$$

Functions $\partial w/\partial x$ and $\partial w/\partial y$ are continuous on the sought contour C_D of the contact area because the normal and tangential derivatives of the membrane displacement on this contour should coincide with the corresponding values of the paraboloid. From here, using Eqs. (3.6.1) and (3.9.4) it is easy to derive the following equation of this boundary value problem on the unknown contour C_D

$$f'(\xi) = \frac{x}{r+R} - i\frac{y}{r} \cdot (\xi \varepsilon C_D) \quad (3.9.5)$$

Let us apply the conformal mapping $\xi = \omega(\zeta)$ which converts the unknown domain D into the exterior of the unit circle $|\zeta| > 1$ on the parametric plane of the complex variable ζ, with the x-axis being converted to the real axis on the ζ-plane. According to the Riemann theorem, the analytical function $\omega(\zeta)$ is uniquely defined by this way.

Since $\omega(\zeta) = x + iy$, we can transform Eq. (3.9.5) into the following boundary value problem for the exterior of the unit circle $|\zeta| > 1$ on the parametric ζ-plane:

$$2r(r+R)F(\zeta) = (R+2r)\overline{\omega(\zeta)}; \quad (|\zeta| = 1) \quad (3.9.6)$$

$$F(\zeta) = f'(\omega(\zeta)) + \frac{R}{2r(R+r)}\omega(\zeta). \quad (3.9.7)$$

And so, we arrived at the boundary value problem (3.9.6) for two unknown analytical functions $\omega(\zeta)$ and $F(\zeta)$. Both have the first-order pole at infinity.

To solve this problem, let us apply the method of functional equations [13, 15–17]. First, we continue Eq. (3.9.6) analytically from the unit circle onto the whole ζ-plane. As a result, we get the following functional equation in the ζ-plane

$$2r(r+R)F(\zeta) = (R+2r)\bar{\omega}(\zeta^{-1}). \quad (3.9.8)$$

Let us show that this functional equation has the following exact solution:

$$\omega(\zeta) = c_0\zeta + c_1\zeta^{-1}; \quad (3.9.9)$$

$$2r(r+R)F(\zeta) = (R+2r)\left(\overline{c_1}\zeta + \overline{c_0}\zeta^{-1}\right). \quad (3.9.10)$$

This solution satisfies Eq. (3.9.8) for any values of coefficients c_0 and c_1.

To determine these coefficients, let us use the behavior of the solution at infinity where the function $f'(\xi) \to 0$. Hence, according to Eqs. (3.9.7), (3.9.9), and (3.9.10), we get

$$Rc_0 = (R + 2r)\overline{c_1}. \tag{3.9.11}$$

From the equilibrium equation of the shell at infinity, it follows that

$$f'(\xi) = \frac{N}{2\pi\gamma\xi} \quad \text{as} \quad \xi \to \infty. \tag{3.9.12}$$

Now, study the pole at $\zeta \to 0$ and eliminate the singularity in Eq. (3.9.7) using Eqs. (3.9.9), (3.9.10), and (3.9.12). As a result, we get one more relation

$$(R + 2r)\overline{c_0} = c_1 R + \frac{r(r+R)}{\pi\gamma c_0} N. \tag{3.9.13}$$

From Eqs. (3.9.11) and (3.9.13), we can find that

$$c_0^2 = \frac{R+2r}{4\pi\gamma} N, \quad c_1 = \frac{R}{R+2r} c_0. \tag{3.9.14}$$

And so, the conformal mapping is done by the following function

$$\omega(\zeta) = \sqrt{\frac{R+2r}{4\pi\gamma} N} \left(\zeta + \frac{R}{R+2r} \frac{1}{\zeta} \right). \tag{3.9.15}$$

The displacement field of the membrane shell has the following shape:

$$\frac{\partial w}{\partial x} - i \frac{\partial w}{\partial y} = \frac{1}{\zeta} \sqrt{\frac{N}{\pi\gamma(R+2r)}}. \tag{3.9.16}$$

Equations (3.9.15) and (3.9.16) provide the parametric solution to the contact problem of a heavy body lying on the membrane shell. The body can be a torus, a paraboloid or any smooth symmetrical body with a small contact area.

As a matter of fact, this solution fits also for the hydrodynamic problem of a heavy vessel on the surface of a liquid in the case of no wetting for small values of the dimensionless number $\delta R^2/\gamma$, where R is a specific dimension of the body, δ is the specific weight of the liquid, and γ is the surface tension of the liquid.

From Eq. (3.9.15), it is easy to derive the following equation of the contour C_D of the contact area

$$\frac{x^2}{(R+r)^2} + \frac{y^2}{r^2} = \frac{N}{\pi\gamma(R+2r)}. \tag{3.9.17}$$

And so, the contact area represents the ellipse which major radius is equal to

$$a = (R + r)\sqrt{\frac{N}{\pi \gamma (R + 2r)}}. \tag{3.9.18}$$

Now, let us consider the original problem of the torus rolling on the flat, tightly stretched membrane shell under the action of the thrust force applied to the torus center along the major axis of the ellipse on the contact area. Based on the CH rule, the steady-state rolling takes place when the resultant force vector (T, N) points at the frontal edge $x = a,\ y = z = 0$ of the contact area. From here, using Eq. (3.9.18) we get the law of rolling of the rigid torus on the flat membrane and the rolling resistance coefficient

$$C_{rr} = \frac{T}{N} = \frac{a}{R + r} = \sqrt{\frac{N}{\pi \gamma (R + 2r)}}. \tag{3.9.19}$$

The problem is solved.

3.10 A Cylinder Rolling on an Elastic Plate

Plates considered in this and next sections are assumed to be isotropic, elastic, thin, of constant thickness, and to have a small lateral deflection w which should be less than a few tenths of the plate thickness t. As distinct from membranes, plates carry loads essentially by bending stresses.

Contact problems for plates have some peculiarities that often do not allow both practitioners and mathematicians to solve or use them. For example, in the case of the problem of a flat stamp, plates can carry a heavy stamp only by concentrated forces applied at the edges of the stamp, with no load being carried inside flat sides.

Even worse, it can be proven that the contact problem of indentation of any heavy parabolic cylinder onto an elastic plate has no solutions with a contact area, except for two trivial ones. In the first solution, all weight is carried by concentrated forces at the ends of the contact area, with the distance between the ends being undetermined. In the second solution, all weight is carried by one contact point at the cylinder vertex so that the zero thrust force can make the parabolic cylinder of any weight roll on the plate, which is absurd. Figuratively speaking, plates "do not like parabolic cylinders and flat stamps."

Therefore, the simple solution of the contact problem given below is of principal interest. The point is that the cross-sectional shape of the rolling cylinder on the contact area should be replaced not by a parabola but by a more complicated figure described by the following equation:

$$y = \frac{x^2}{2R} + \frac{x^4}{8R^3} \cdot (x \ll R) \tag{3.10.1}$$

This follows from the exact equation of the cylinder surface

$$y = R\left(1 - \sqrt{1 - \frac{x^2}{R^2}}\right). \tag{3.10.2}$$

Equation (3.10.1) provides two first terms of the expansion of function $y = y(x)$ in Eq. (3.10.2) for small x/R so that we get a more precise, second-order approximation of the cylinder of radius R in the contact area. In the contact problems of cylinders on plates, it is insufficient to use only the first term of this equation like it was done for solids and membranes.

Let us write down the equations of one-dimensional bending of flat plates:

$$q = D\frac{\mathrm{d}w^4}{\mathrm{d}x^4}, \quad M = -D\frac{\mathrm{d}^2 w}{\mathrm{d}x^2}, \quad Q = -D\frac{\mathrm{d}^3 w}{\mathrm{d}x^3}. \tag{3.10.3}$$

Here, $w(x)$ is the plate deflection; q, M and Q are the load, bending moment, and transverse shear force; and D is equal to

$$D = \frac{1}{12}Et^3\left(1 - \nu^2\right)^{-1} \tag{3.10.4}$$

Suppose the rigid cylinder of Eq. (3.10.1) presses onto the plate by the force N per unit length of the cylinder so that a contact area of width $2a$ forms. We assume that the y-axis is the symmetry line of the problem, and that $R \gg a \gg t$. The plate deflection $w = y(x)$ in the contact area $-a < x < +a$ is determined by Eq. (3.10.1) so that using the first equation in Eq. (3.10.3) we get the load on the plate

$$q = 3D/R^3 \quad (-a < x < a). \tag{3.10.5}$$

From the equilibrium equation and Eq. (3.10.5), we find the width of the contact area

$$a = \frac{NR^3}{6D}. \tag{3.10.6}$$

Then, using the CH rule and Eq. (3.10.6) we derive the law of rolling and the rolling resistance coefficient for a rigid cylinder of radius R rolling on an elastic plate

$$C_{rr} = \frac{T}{N} = \frac{a}{R} = \frac{NR^2}{6D} = \frac{2NR^2\left(1 - \nu^2\right)}{Et^3}. \tag{3.10.7}$$

The elastic line of the plate outside the contact area can be found from the first equation in Eq. (3.10.3):

$$w = -\frac{N}{12D}x^3 + \frac{x^2}{2R}\left(1 + \frac{NaR}{2D}\right) - \frac{Na^2}{3D}x + w_m \cdot (x > a) \tag{3.10.8}$$

Here, w_m is defined by the boundary condition at the end support. At $x = \pm a$, the deflection $w(x)$ and its first two derivatives are equal to the function $y(x)$ and its corresponding derivatives.

3.11 A Ball Rolling on an Elastic Plate

To solve the problem of a rigid ball of radius R rolling on an elastic plate, we apply the approach of the previous section. To this purpose, we replace the ball by the axially symmetric body which for small r/R approximates the shape of the ball more accurately than any paraboloid

$$z = \frac{r^2}{2R} + \frac{r^4}{8R^3}. \tag{3.11.1}$$

The axially symmetric problems of plate bending are described by the following equation in terms of their lateral deflection $w(r)$

$$\frac{1}{r}\frac{\mathrm{d}}{\mathrm{d}r}\left\{r\frac{\mathrm{d}}{\mathrm{d}r}\left[\frac{1}{r}\frac{\mathrm{d}}{\mathrm{d}r}\left(r\frac{\mathrm{d}w}{\mathrm{d}r}\right)\right]\right\} = \frac{q}{D}. \tag{3.11.2}$$

Since $w(r) = z(r)$ in the contact area, using Eqs. (3.11.1) and (3.11.2) we find

$$q = \frac{8D}{R^3}. \tag{3.11.3}$$

In this case, the contact area is circular. Using Eq. (3.11.3) and the equilibrium equation, we get the radius a of the contact area

$$a^2 = \frac{NR^3}{8\pi D} = \frac{3(1-\nu^2)}{2\pi}\frac{N}{E}\left(\frac{R}{t}\right)^3. \tag{3.11.4}$$

By means of Eq. (3.11.4) and the CH rule, the law of the ball rolling on an elastic plate and the rolling resistance coefficient can be written as follows:

$$C_{rr} = \frac{T}{N} = \frac{a}{R} = \sqrt{\frac{3(1-\nu^2)NR}{2\pi Et^3}}. \tag{3.11.5}$$

The lateral deflection of the plate outside the contact area can be found with the help of Eq. (3.11.2) similar to the previous section. The homogeneous equation of Eq. (3.11.2) has four particular solutions, namely, $\ln r$, r^2, $r^2 \ln r$ and a constant. Hence, the general solution of Eq. (3.11.2) can be written as their superposition

$$w = c_1 r^2 \ln r + c_2 r^2 + c_3 \ln r + c_4. \tag{3.11.6}$$

Here, c_1, c_2, c_3 and c_4 are some constants. The first three of them are defined by two boundary conditions at the edge of the contact area and by the equilibrium equation. Constant c_4 depends only on the support outside the contact area.

If the support is very far from the contact area, then based on Saint-Venant's principle, at the distance much greater than a, there exists the following intermediate field corresponding to the concentrated force N at the center [21–23]:

$$w = -\frac{N}{16\pi D}\left(2r^2 \ln\frac{r}{c} + c^2 - r^2\right); \tag{3.11.7}$$

$$M_r = D\left(\frac{\mathrm{d}^2 w}{\mathrm{d}r^2} + \frac{\nu}{r}\frac{\mathrm{d}w}{\mathrm{d}r}\right) = \frac{N}{4\pi}\left[1 + (1+\nu)\ln\frac{r}{c}\right]; \tag{3.11.8}$$

$$Q_r = \Delta\frac{\mathrm{d}w}{\mathrm{d}r} = \frac{N}{2\pi r}, \quad M_r - M_\theta = \frac{1-\nu}{4\pi}N. \tag{3.11.9}$$

Here, M_r, M_θ, and Q_r are the corresponding bending moments and transverse shear force (Δ is Laplace's operator). The lateral deflection in Eq. (3.11.7) counts from the support at $r = c$ where it is assumed that $\mathrm{d}w/\mathrm{d}r = 0$.

3.12 A Rigid Torus Rolling on an Elastic Plate

In the case of a torus rolling on an elastic plate, it can be shown that, similar to more simple problems in Sects. 3.10 and 3.11, it is insufficient to approximate the surface of the torus by the elliptic paraboloid of Eq. (3.6.1) which represents only the first-order expansion of the torus surface equation at and near the contact area. Although we can construct two mathematical solutions for this first-order approximation, they do not satisfy some evident physical requirements.

In the first solution, with the one-point contact at the base of the rolling torus, the torus of any weight can be rolled on any elastic plate by zero thrust force, which is absurd. In the second solution, the load-free contact area represents an ellipse of some constant eccentricity but of any undetermined dimension, with the weight being carried by concentrated transverse shear forces distributed along the contour of the ellipse [18, 20]. The latter can be proven by a simple dimensional consideration.

And so, to get the physically reasonable solution of this problem of rolling we should use the following second-order approximation of the torus surface at and near the contact area

$$z = \frac{x^2}{2(R+r)} + \frac{y^2}{2r} + \frac{x^4}{8(R+r)^3} + \frac{y^4}{8r^3}. \tag{3.12.1}$$

Here, evidently, the fourth-order terms are much less than the second-order terms so that this change of the surface of the ellipsoidal paraboloid is very small.

The lateral deflection of the plate should satisfy the biharmonic equation

$$\Delta^2 w = q/D. \tag{3.12.2}$$

Therefore, inside the contact area where $z(x) = w(x)$ the load is equal to

$$q = 3D\left[\frac{1}{r^3} + \frac{1}{(r+R)^3}\right]. \tag{3.12.3}$$

And so, a very small change in the shape of the pressing body caused a very significant effect, namely a redistribution of the bearing load from the concentrated transverse shear force at the edge of the contact area to the constant load throughout all contact area. This is, again, a bad signal for the solution with concentrated transverse shear forces—it suffers a big change by a small variation of boundary conditions and is easily "washed away" by the continuous solution [18, 20–23]. Such unstable solutions are called incorrect and usually ignored. It is the way we follow as well.

Unfortunately, the exact solution of the contact problem for a body described by Eq. (3.12.1) is very complicated and cumbersome for practice. Instead, we apply an approximate solution which is much simpler and easier to use.

For this purpose, we approximate the sought contour of the contact area by an ellipse which is coaxial and similar to the basic ellipse defined by the first two terms of Eq. (3.12.1) and which supports weight N of the torus by the distributed load of Eq. (3.12.2). As a result, we get the following two equations:

$$\pi abq = N; \tag{3.12.4}$$

$$\frac{a}{b} = \sqrt{\frac{R+r}{r}}. \tag{3.12.5}$$

Here, a and b are, respectively, the major and minor radii of the sought ellipse (πab is its area).

Using Eqs. (3.12.4) and (3.12.5), we find the major radius of the contact area that determines the rolling law of the torus

$$a = \left(\frac{N}{\pi q}\right)^{1/2}\left(1 + \frac{R}{r}\right)^{1/4}. \tag{3.12.6}$$

Substituting q here by Eq. (3.12.2), we find the major radius of the contact area in terms of original data

$$a = \left(\frac{N}{3\pi D}\right)^{1/2}\left(1+\frac{R}{r}\right)^{1/4}\left[\frac{1}{r^3}+\frac{1}{(r+R)^3}\right]^{-1/2}. \qquad (3.12.7)$$

From here, it follows that the law of rolling of a torus on a plate and the rolling resistance coefficient can be written as follows:

$$C_{rr} = \frac{T}{N} = \frac{a}{R+r} = \left(\frac{N}{3\pi D}\right)^{1/2}\frac{r^{5/4}(r+R)^{3/4}}{\sqrt{r^3+(r+R)^3}}. \qquad (3.12.8)$$

Here, D is given by Eq. (3.10.4) in terms of Young's modulus, Poisson's ratio, and the thickness of the plate. As a reminder, r and R are the minor and major radii of the torus.

Conclusion. The exact laws of rolling and the accurate rolling resistance coefficients so much needed in industries and for so long gained only in very expensive tests have gotten now known for all basic round bodies. This knowledge should become common to all professionals and graduate students of mechanical and automotive engineering. The present chapter can serve as a semester course for this purpose.

Literature

1. A. Erman, in *Life in Ancient Egypt* (Macmillan Co., London, 1894; Dover, New York, 1971) 570 pp. ISBN 0-486-22632-8
2. Special Report 216 (National Academy of Sciences, Transportation Research Board, Washington, 2006)
3. A.A. Coulomb, in *Theorie des machines simples* (Academie des Sciences, Paris, 1781)
4. R. Cross, Coulomb's Law for rolling friction. Am. J. Phys. **84**(3), 221–230 (2016)
5. Wikipedia: *Rolling Resistance*
6. B.M. Javorsky, A.A. Detlaf, *The Guide on Physics*, 7th edn. (Nauka, Moscow, 1979) 942 pp (in Russian)
7. H.R. Hertz, Uber die Beruhrung Fester Elastischer Korper. Zeitschrift fur Reine Angew. Math. **92**, 156 (1882)
8. J.A. Williams, *Engineering Tribology* (Oxford University Press, London, 1994)
9. R.C. Hibbeler, *Engineering Mechanics: Statics and Dynamics* (Prentice Hall, Pearson, 2007)
10. V.L. Popov, *Contact Mechanics and Friction: Physical Principles and Applications* (Springer-Verlag, Berlin, 2010)
11. N.I. Muskhelishvili, *Some Basic Problems of the Mathematical Theory of the Elasticity* (Noordhoff, Groningen, 1963), p. 718
12. F.D. Gakhov, *Boundary Value Problems* (Pergamon Press, London, 1966), p. 584
13. G.P. Cherepanov, A non-linear problem in the theory of analytical functions. Doklady USSR Acad. Sci. (Math.) **147**(3), 566–568 (1962)
14. G.P. Cherepanov, Stresses in an inhomogeneous plate with cracks. Not. USSR Acad. Sci. (Mech.) **1**, 131–138 (1962). (in Russian)
15. G.P. Cherepanov, On a method of the solution of elastic-plastic problems. J. Appl. Math. Mech. (JAMM) **27**(3), 428–435 (1963)
16. G.P. Cherepanov, The Riemann-Hilbert problem for cuts along a straight line or circumference. Doklady USSR Acad. Sci. (Math.) **156**(2), 275–277 (1964)

17. G.P. Cherepanov, Boundary value problems with analytical coefficients. Doklady of the USSR Academy of Sciences (Mathematics) **161**(2), 312–314 (1965)
18. G.P. Cherepanov, Some problems concerning the unknown body boundaries in the theory of elasticity and plasticity, in *Applications of the Theory of Functions in Continuum Mechanics*, vol. 1. (Nauka, Moscow, 1965), pp. 135–150 (in Russian)
19. G.P. Cherepanov, On modeling in linear reology, in *Problems of Hydrodynamics and Continuum Mechanics*, L. I. Sedov Anniversary Volume, (Nauka, Moscow, 1969), pp. 553–560
20. L.A. Galin, G.P. Cherepanov, Contact elastic-plastic problems for plates. Doklady USSR Acad. Sci. (Mech.) **177**(1), 56–58 (1967)
21. G.P. Cherepanov, On the non-uniqueness problem in the theory of plasticity. Doklady USSR Acad. Sci. (Mech.) **218**(4), 1124–1126 (1974)
22. G.P. Cherepanov, *Mechanics of Brittle Fracture* (McGraw Hill, New York, 1978), p. 950
23. G.P. Cherepanov, *Methods of Fracture Mechanics: Solid Matter Physics* (Kluwer, Dordrecht, 1997), p. 300
24. G.P. Cherepanov, *Fracture Mechanics* (ICR, IzhevsK-Moscow, 2012), p. 872
25. G.P. Cherepanov, Some new applications of the invariant integrals in mechanics. J. Appl. Math. Mech. (JAMM) **76**(5), 519–536 (2012)
26. G.P. Cherepanov, Theory of rolling: solution of the Coulomb problem. J. Appl. Mech. Tech. Phy. (JAMT) **55**(1), 182–189 (2014)
27. G.P. Cherepanov, The contact problem of the mathematical theory of elasticity with stick-and-slip areas: the theory of rolling and tribology. J. Appl. Math. Mech. (JAMM) **79**(1), 81–101 (2015)
28. G.P. Cherepanov, The laws of rolling. J. Phys. Mesomech. **23**(5), 25–48 (2018)

Chapter 4
The Theory of Flight

Abstract The basic equations of gas dynamics are written in the form of invariant integrals describing the laws of conservation. The Kutta–Joukowski equation and the lift force of wings were derived from Joukowski' profiles using the invariant integrals and complex variables. The optimal shape of airfoils is suggested and calculated. Method of discrete vortices applied to turbulent flows with large Reynolds number appeared to be useful for the characterization of hurricanes. This chapter may be of special interest for aerodynamics and meteorology.

Both the fluid dynamics and the theory of functions of a complex variable originated in the works of Leonhard Euler almost three centuries ago. However, he proved that the drag and lift of a body moving in a perfect (inviscid) fluid equals zero, and so, this theory seemed to be useless for a long time. It is only in the beginning of the twentieth century that became clear that vortices and separated flows play the major part in the creation of the forces on the body. Because the fluid viscosity transferred these forces on the body and triggered shooting vortices, most of the twentieth century efforts were concentrated on the attempts to solve the Navier–Stokes equations for viscous fluids. Today, it is clear that the Navier–Stokes equations do not have a reasonable solution for large Reynolds numbers, most important in applications, and for any practical purpose of calculation of drag/lift forces, the turbulent flow can be effectively described by some system of vortices or vortex sheets in a perfect fluid.

The simplest version of the method of discrete vortices is given below as well as the application of the invariant integrals of gas dynamics to the theory of flight. The chapter is based on this author's publications [4–13].

4.1 Introduction

In this section, we provide the basic information about singular integral equations and the Riemann problem used in this book.

Cauchy integral is the complex function $\phi(z)$ of a complex variable z of the following shape

G. P. Cherepanov, *Invariant Integrals in Physics*,
https://doi.org/10.1007/978-3-030-28337-7_4

$$\phi(z) = \frac{1}{2\pi i} \int\limits_{L} \frac{\varphi(\tau) \mathrm{d}\tau}{\tau - z}, \quad (\tau \epsilon L). \tag{4.1.1}$$

Here, L is a contour in the z-plane and $\varphi(\tau)$ is a complex function.

The function of a complex variable is said to be analytic at a point if it is differentiable any number of times and expandable into a convergent power series at the point. The Cauchy integral in Eq. (4.1.1) provides a function $\phi(z)$ that is analytic at any point z outside L and discontinuous on L, that is, $\phi(z)$ tends to different values $\phi^{+}(t)$ and $\phi^{-}(t)$ when z tends to the same point t on L from the left-hand and right-hand side of L, respectively, if one goes along L.

A *singular integral* is the complex function $\phi(t)$ of a complex variable t defined by the divergent improper integral

$$\frac{1}{2\pi i} \int\limits_{L} \frac{\varphi(\tau)}{\tau - t} \mathrm{d}\tau, \quad (t \epsilon L), \tag{4.1.2}$$

understood in the meaning of the *Cauchy Principal Value* as

$$\phi(t) = \frac{1}{2\pi i} \text{P.V.} \int\limits_{L} \frac{\varphi(\tau) \mathrm{d}\tau}{\tau - t} = \lim \frac{1}{2\pi i} \int\limits_{L_{\epsilon}} \frac{\varphi(\tau) \mathrm{d}\tau}{\tau - t}, \quad (\epsilon \to 0, t \epsilon L, \tau \epsilon L). \tag{4.1.3}$$

Here, L_{ϵ} is L without its vicinity of $z = t$ cut off by a circle of a small radius $\in$ centered at $z = t$. For example, calculating the improper integral

$$\int\limits_{a}^{b} \frac{\mathrm{d}x}{x - c} = \lim \left(- \int\limits_{a}^{c - \in_1} \frac{\mathrm{d}x}{c - x} + \int\limits_{c + \in_2}^{b} \frac{\mathrm{d}x}{x - c} \right) = \ln \frac{b - c}{c - a} + \lim \frac{\in_1}{\in_2} \tag{4.1.4}$$

where $\in_1 \to 0$, $\in_2 \to 0$ and $a < c < b$, we get ($\in_1 = \in_2 = \in$)

$$\text{P.V.} \int\limits_{a}^{b} \frac{\mathrm{d}x}{x - c} = \ln \frac{b - c}{c - a}. \tag{4.1.5}$$

Hence, the improper integral does not exist (diverges) because the limit depends on the way of vanishing $\in_1$ and $\in_2$. However, the Cauchy principal value of this singular integral exists and is determined by Eq. (4.1.5).

Still in 1873, Sokhotsky derived the following basic equations rederived by Plemelj in 1908:

$$\phi^{+}(t) - \phi^{-}(t) = \varphi(t), \quad t \in L; \tag{4.1.6}$$

$$\phi^+(t) + \phi^-(t) = \text{P.V.}\frac{1}{\pi i}\int_L \frac{\varphi(\tau)\mathrm{d}\tau}{\tau - t}. \quad (4.1.7)$$

Sokhotsky equations allow mutual connection of the following famous boundary value problems:
Riemann Problem:

$$\phi^+(t) = G(t)\phi^-(t) + g(t), \quad t \in L. \quad (4.1.8)$$

Here, $G(t)$ and $g(t)$ are some given functions, and $\phi^+(z)$ and $\phi^-(z)$ are analytic functions required to be found (in the case of unclosed curve L, it is one function that suffers a discontinuity on the line L).
Hilbert Problem:

$$a(s)u(s) + b(s)v(s) = c(s). \quad (4.1.9)$$

Here, s is the length of arc on L, $a(s)$ and $b(s)$ are some given real functions, and $u(s)$ and $b(s)$ are the real and imaginary parts of an analytic function required to be found.
Singular Integral Equation Problem (with a Cauchy kernel):

$$\mathrm{d}(t)\varphi(t) + \frac{1}{\pi i}\text{P.V.}\int_L \frac{M(t,\tau)}{\tau - t}\varphi(\tau)\mathrm{d}\tau = g(t), \quad t \in L. \quad (4.1.10)$$

Here, $d(t)$, $g(t)$ and $M(t, \tau)$ are some given complex functions, and $\varphi(t)$ is a sought complex function.
Singular Integral Equation Problem (with a Hilbert kernel):

$$a(s)u(s) - \frac{1}{2\pi}b(s)\int_0^{2\pi} u(\alpha)\cot\frac{\alpha - s}{2}\mathrm{d}\alpha = c(s). \quad (4.1.11)$$

All these equations can be reduced one to another—e.g., Eq. (4.1.11) to Eq. (4.1.9) in the same designations.

Poincare, Hilbert, and Noether reduced these equations to Fredholm integral equation while Hilbert, Plemelj, and Carleman found explicit solutions to some particular cases of these problems. In 1941, Gakhov gave the full explicit solution of the problems as formulated in Eqs. (4.1.8)–(4.1.11). Similar problems for a system of singular integral equations reduced to the matrix Riemann problem are not solved, as yet, in explicit form, except for some particular cases found by Gakhov, Khrapkov, and this author.

Carleman's singular integral equation. As an illustration of the solution method, let us consider Carleman's singular integral equation

$$\mu\varphi(x) + \frac{\lambda}{\pi}\int_0^1 \frac{\varphi(\tau)d\tau}{\tau - x} = g(x), \quad (0 < x < 1). \tag{4.1.12}$$

Here, μ and λ are some constants, $g(x)$ is a known function, and $\varphi(x)$ is the function required to be found (the sign P.V. before the integral is omitted as usual).

Let us introduce the Cauchy function $\phi(z)$ as follows

$$\phi(z) = \frac{1}{2\pi i}\int_0^1 \frac{\varphi(\tau)d\tau}{\tau - z}. \tag{4.1.13}$$

It is analytic outside the cut along the segment (0, 1) of the real axis on the complex z-plane. On this segment, $\phi(z)$ suffers a discontinuity.

Using Sokhotsky relations, Eqs. (4.1.6) and (4.1.7), we reduce Carleman's equation, Eq. (4.1.12), to the following Riemann problem

$$\phi^+(x) + m\phi^-(x) = g(x)/(\mu + i\lambda), \quad (0 < x < 1). \tag{4.1.14}$$

$$m = -\frac{\mu - i\lambda}{\mu + i\lambda} = \mathrm{e}^{i\Delta}, \quad \Delta = \pi - 2\arctan\frac{\lambda}{\mu}, \quad (0 < \Delta < \pi). \tag{4.1.15}$$

First, let us find an auxiliary (canonical) solution $X(z)$ to the problem for the case where $g(x) = 0$ and m is an arbitrary complex number:

$$X^+(x) + mX^-(x) = 0, \quad (0 < x < 1). \tag{4.1.16}$$

By knowing the properties of the power-law functions of a complex variable, it is easy to obtain a solution to Eq. (4.1.16) in the form

$$X(z) = z^\delta(z-1)^{1-\delta}, \tag{4.1.17}$$

$$\delta = \frac{1}{2} - \frac{1}{2\pi i}\ln m, \quad \left(\lim \frac{X(z)}{z} = 1 \text{ as } z \to \infty\right). \tag{4.1.18}$$

Function $X(z)$ suffers a discontinuity on the cut (0, 1) of the complex z-plane because the argument of $X(z)$ equals $\pi i(1-\delta)$ on the upper bank of the cut and $-\pi i(1-\delta)$ on the lower bank of the cut.

In the case of Carleman's integral equation, due to Eqs. (4.1.15) and (4.1.18), the constant δ is real and equal to

$$\delta = \frac{1}{2} - \frac{\Delta}{2\pi} = \frac{1}{\pi}\arctan\frac{\lambda}{\mu}. \tag{4.1.19}$$

However, in more complicated problems of the theory of elasticity, this constant is usually a complex number, see, e.g., Sect. 3.1.

Using Eqs. (4.1.16) and (4.1.17), the Riemann problem of Eq. (4.1.14) can be written as follows

$$\left(\frac{\phi}{X}\right)^{+} - \left(\frac{\phi}{X}\right)^{-} = \frac{g(x)}{(\mu + i\lambda)X^{+}(x)}, \quad 0 < x < 1. \tag{4.1.20}$$

From physical consideration, it is important to specify the behavior of the solution near the ends $x = 0$ and $x = 1$. Function $F(z)$

$$F(z) = \frac{\phi(z)}{X(z)} - \frac{1}{2\pi i(\mu + i\lambda)} \int_{0}^{1} \frac{g(\tau)\mathrm{d}\tau}{X^{+}(\tau)(\tau - z)} \tag{4.1.21}$$

in view of Eq. (4.1.6) will satisfy the condition equation

$$F^{+}(x) = F^{-}(x), \quad 0 < x < 1. \tag{4.1.22}$$

Hence, $F(z)$ is analytic everywhere in the z-plane with the possible exception of singular points $z = 0$ and $z = 1$ where $F(z)$ can have poles, dependent on the specified solution behavior at these points. For example, if the only integrable singularities of the solution are physically permitted, these poles are simple; then, according to the Liouville theorem, we obtain

$$F(z) = \frac{C_1}{z} + \frac{C_2}{z - 1}, \tag{4.1.23}$$

where C_1 and C_2 are some constants.

Considering Eqs. (4.1.21) and (4.1.23), we obtain the solution in the form

$$\phi(z) = \frac{X(z)}{2\pi i(\mu + i\lambda)} \int_{0}^{1} \frac{g(\tau)\mathrm{d}\tau}{X^{+}(\tau)(\tau - z)} + X(z)\left(\frac{C_1}{z} + \frac{C_2}{z - 1}\right), \tag{4.1.24}$$

where

$$C_1 + C_2 = \frac{1}{2\pi i(\mu + i\lambda)} \int_{0}^{1} \frac{g(\tau)\mathrm{d}\tau}{X^{+}(\tau)}. \tag{4.1.25}$$

Equation (4.1.25) is stipulated by the fact that $\phi(z) \to 0$ as $z \to \infty$ according to the definition of $\phi(z)$ by Eq. (4.1.13).

The sought solution $\varphi(x)$ of Carleman's equation is given by Eqs. (4.1.24), (4.1.25), and (4.1.17), and by the Sohotsky relation, Eq. (4.1.7).

From this analysis, it follows also that, when $g(x) = 0$, Carleman's integral equation, Eq. (4.1.12), can have many solutions of the class

$$\varphi(x) = \phi^{+} - \phi^{-} \quad \text{where } \phi(z) = X(z)P\left(z, \frac{1}{z}, \frac{1}{z-1}\right). \tag{4.1.26}$$

Here, $P\left(z, \frac{1}{z}, \frac{1}{z-1}\right)$ is a polynomial function of its arguments which can be determined only by specific physical requirements.

4.2 Invariant Integrals of Gas Dynamics

Let us study the irrotational steady polytropic flow of an inviscid compressible gas. In this case, the governing equations of gas dynamics can be written in the form of the following invariant integrals [5]:

$$\oint_S \rho v_i n_i \mathrm{d}S = 0, \quad (i = 1, 2, 3), \tag{4.2.1}$$

$$F_k = \oint_S (\rho v_k v_i n_i + p n_k)\mathrm{d}S, \quad (i, k = 1, 2, 3). \tag{4.2.2}$$

where

$$p/p_\infty = (\rho/\rho_\infty)^\kappa. \tag{4.2.3}$$

Here, ρ and p are the density and pressure of gas, v_i are the Cartesian components of gas velocity, ρ_∞ and p_∞ are the density and pressure of unperturbed gas flow, κ is the polytrope coefficient equal to the ratio of specific heat capacities (c_p/c_V) in the case of adiabatic processes, S is an arbitrary closed surface in gas, n_i are outer unit normal components to S, and F_k are the Cartesian components of the equivalent outer force applied to bodies or singularities inside S. The forces F_k are equal to zero if there are no field singularities (e.g., vortices) or shock waves inside S.

Equation (4.2.1) is the mass conservation law. Equation (4.2.2) can be considered as both the energy or momentum conservation law. Equation (4.2.3) is valid for a locally polytropic (e.g., adiabatic) process. When there is only gas and no bodies inside S, by applying the divergence theorem to Eqs. (4.2.1) and (4.2.2), one can easily derive the differential equation system of gas dynamics. When there is a body inside S, using Eq. (4.2.2) allows computation of the drag and lift forces acting upon the body. Because of invariance property, the S can be chosen in any way convenient for calculation.

As an example, let us find the forces of lift and drag acting upon a body of finite dimensions moving with a subsonic speed in a gas that is at rest at infinity. Suppose there are no vortices and sources anywhere in the flow. In this case, the perturbed flow at infinity has an order

$$\Delta\rho = O\left(r^{-3}\right), \quad \Delta p = O\left(r^{-3}\right) \text{ and } \Delta v_i = O\left(r^{-2}\right). \tag{4.2.4}$$

Here, r is the distance from the body.

Using the invariance of the integral of Eq. (4.2.2), let us move the integration surface S to infinity. Then, from Eqs. (4.2.2) and (4.2.4), it follows that $F_k = 0$. This is the famous Euler paradox. It embarrassed scientists for more than a century before they understood the role of vortices as carriers of lift and drag. The use of sources and their thrust forces led to the advent of rocketry and jets.

4.3 Lift Force of a Thin Aerofoil

Let us consider the two-dimensional problem of a thin aerofoil moving in a gas. This is an important stage of studies of smooth profiles of aircraft real wings, see Sect. 1.4 of this book for more detail. Engineering applications require the minimum possible perturbations of gas by a moving aerofoil, or maximum lift and minimum drag. The latter is provided by a thin-shaped profile close to a flat, infinitely thin plate moving with a small angle of attack.

In this case, we designate:

$$\begin{aligned} &v_1 = V(1+\bar{u}_1), \quad v_2 = V\,\bar{u}_2, \quad v_3 = 0, \\ &p = p_\infty(1+\bar{p}), \quad \rho = \rho_\infty(1+\bar{\rho}), \end{aligned} \tag{4.3.1}$$

where the dimensionless perturbed quantities are small:

$$\bar{u}_1 \ll 1, \quad \bar{u}_2 \ll 1, \quad \bar{p} \ll 1, \quad \bar{\rho} \ll 1. \tag{4.3.2}$$

Here, p_∞, ρ_∞ and V are the pressure, density, and velocity of unperturbed gas flow at infinity with respect to the aerofoil (the gas velocity direction at infinity coincides with the direction of the x_1-axis).

Substituting the corresponding quantities in Eqs. (4.2.1)–(4.2.3) by Eq. (4.3.1) and omitting higher-order terms, we get

$$\oint\limits_L \bar{u}_i n_i \mathrm{d}L = M_\infty^2 \oint\limits_L \bar{u}_1 n_1 \mathrm{d}L, \quad (i = 1, 2), \tag{4.3.3}$$

$$F_2 = \rho_\infty V^2 \oint_L (\bar{u}_2 n_1 - \bar{u}_1 n_2) \mathrm{d}L, \tag{4.3.4}$$

$$\begin{gathered} \bar{p} = \kappa\bar{\rho} = -\kappa M_\infty^2 \bar{u}_1, \quad F_1 = 0, \quad \bar{\rho} = -M_\infty^2 \bar{u}_1. \\ \left(M_\infty = U/c_\infty, c_\infty^2 = \kappa p_\infty/\rho_\infty\right) \end{gathered} \tag{4.3.5}$$

Here, M_∞ and c_∞ are the Mach number and sound speed of unperturbed gas flow at infinity, L is an arbitrary closed contour in gas, and F_1 and F_2 are the drag and lift forces acting upon a thin aerofoil. The equation $\overline{\rho} = -M_\infty^2 \bar{u}_1$ is the Cauchy–Bernoulli equation in this case.

Equations (4.3.3)–(4.3.5) provide the gas dynamics equation system that holds for an arbitrary body at a sufficient large distance from it or for very thin aerofoils in the entire flow domain. Using the divergence theorem in the integrals of Eqs. (4.3.3) and (4.3.4) along a contour encompassing only fluid domain, we arrive at the following differential equation system:

$$\left(1 - M_\infty^2\right) \frac{\partial \bar{u}_1}{\partial x_1} + \frac{\partial \bar{u}_2}{\partial x_2} = 0, \quad \frac{\partial \bar{u}_1}{\partial x_2} - \frac{\partial \bar{u}_2}{\partial x_1} = 0. \tag{4.3.6}$$

The general solution of this system can be written as

$$u_1 = Re W'(z), \quad u_2 = -\sqrt{1 - M_\infty^2} \mathrm{Im} W'(z), \tag{4.3.7}$$

where

$$z = x_1 + i x_2 \sqrt{1 - M_\infty^2}. \tag{4.3.8}$$

Here, $W'(z)$ is an analytic function in the flow domain. It behaves at infinity as

$$W'(z) = \frac{\Gamma i}{2\pi z}, \quad (z \to \infty). \tag{4.3.9}$$

Here, Γ is the circulation of the flow that should be found.

Let us write the aerofoil contour L equation as follows

$$L: x_2 = -x_1 \tan\alpha + f_\pm(x_1), \quad 0 < x_1 < c. \tag{4.3.10}$$

Here, the functions $f_+(x)$ and $f_-(x)$ describe the upper and lower sides of the aerofoil which contour is close to the straight linear segment of length c (the chord of the aerofoil), inclined to the x_1-axis with angle α (the angle of attack), and issuing the coordinate origin. The angle of attack is positive when it counts clockwise from the direction of the unperturbed flow velocity which coincides with the direction of the x_1-axis.

The condition of non-penetration on the aerofoil contour means that the normal component of the fluid velocity equals zero on L

$$v_2 = v_1 \tan\varphi \quad \text{where} \quad \tan\varphi = x_2'(x_1) = -\tan\alpha + f_\pm'(x_1). \tag{4.3.11}$$

Using Eqs. (4.3.6)–(4.3.8) and (4.3.1), the boundary condition, Eq. (4.3.11), takes the following form on L

$$-\sqrt{1-M_\infty^2}\,\mathrm{Im}W'(z) = V\big[1 + ReW'(z)\big]\big[f_\pm' - \tan\alpha\big]. \tag{4.3.12}$$

In view of $\alpha \ll 1$ and Eq. (4.3.2) for thin profiles, the boundary condition of Eq. (4.3.12) can be simplified and held on the cut $(0, c)$ along the x_1-axis. Hence, we get

$$\mathrm{Im}W'(z) = V\big[\alpha - f_\pm'(x_1)\big]/\sqrt{1-M_\infty^2} \quad \text{as } x_2 = \pm 0, \ \ 0 < x_1 < c. \tag{4.3.13}$$

Using this boundary condition, it is required to find $W'(z)$ that is analytic outside the cut $(0, c)$ of the real axis and vanishes at infinity according to Eq. (4.3.9).

Let us introduce new functions $\Psi(z)$ and $\Omega(z)$:

$$\Psi(z) = \frac{1}{2}\big[W'(z) + \overline{W'}(z)\big],$$
$$\Omega(z) = \frac{1}{2}\big[W'(z) - \overline{W'}(z)\big]. \tag{4.3.14}$$

Here,

$$\overline{W'}(z) = \overline{W'(\bar{z})}. \tag{4.3.15}$$

It is evident that $\bar{z} \to x_1 - i0$ when $z \to x_1 + i0$. The functions $\Psi(z)$ and $\Omega(z)$ are analytic outside the same cut on the z-plane. In addition, at $x_2 = 0$ and $x_1 < 0$ or $x_1 > c$, we get:

$$\Psi(x_1) = ReW'(x_1), \quad \text{that is } \mathrm{Im}\Psi(x_1) = 0; \tag{4.3.16}$$

$$\Omega(x_1) = i\mathrm{Im}W'(x_1), \quad \text{that is } Re\Omega(x_1) = 0. \tag{4.3.17}$$

According to Eqs. (4.3.13) and (4.3.14), at $x_2 = 0$ and $0 < x_1 < c$, we have:

$$\mathrm{Im}\Psi^\pm(x_1) = \mp\frac{1}{2}V\big[f_+'(x_1) - f_-'(x_1)\big]/\sqrt{1-M_\infty^2}, \tag{4.3.18}$$

$$\mathrm{Im}\Omega^\pm(x_1) = V\left[\alpha - \frac{1}{2}f_+'(x_1) - \frac{1}{2}f_-'(x_1)\right]/\sqrt{1-M_\infty^2}. \tag{4.3.19}$$

Equations (4.3.16) and (4.3.18) represent the classic Dirichlet problem for the upper or lower half-plane of the complex variable $z = x_1 + i x_2 \sqrt{1 - M_\infty^2}$. Its solution vanishing at infinity is:

$$\Psi(z) = -\frac{V}{2\pi\sqrt{1 - M_\infty^2}} \int_0^c \frac{f_+'(t) - f_-'(t)}{t - z} dt. \tag{4.3.20}$$

It is easy to verify this solution using the Sokhotsky relations, Eqs. (4.1.6) and (4.1.7).

The mixed problem of Eqs. (4.3.17) and (4.3.19) is reduced to another Dirichlet problem for the function $\Omega(z)X(z)$ where

$$X(z) = \sqrt{\frac{z - c}{z}}, \quad X(z) = 1 - \frac{c}{2z} \quad \text{as } z \to \infty. \tag{4.3.21}$$

And so, we can find

$$\Omega(z) = \frac{V X(z)}{2\pi\sqrt{1 - M_\infty^2}} \int_0^c \frac{2\alpha - f_+'(t) - f_-'(t)}{X^+(t)(t - z)} dt. \tag{4.3.22}$$

According to Eq. (4.3.14), we have

$$W'(z) = \Psi(z) + \Omega(z), \tag{4.3.23}$$

so that using Eqs. (4.3.20) and (4.3.22), we get the sought solution

$$W'(z) = -\frac{V}{2\pi\sqrt{1 - M_\infty^2}} \int_0^c \frac{f_+'(t) - f_-'(t)}{t - z} dt + \frac{V X(z)}{2\pi\sqrt{1 - M_\infty^2}} \int_0^c \frac{2\alpha - f_+'(t) - f_-'(t)}{X^+(t)(t - z)} dt. \tag{4.3.24}$$

It vanishes at infinity and is limited at the rear end of the profile owing to the trailing vortex, since it is required by the Joukowski theory.

This solution has a peculiarity caused by the linearization of the initial problem, namely it has a sink singularity at the frontal point $z = 0$ of the profile and a source of the same intensity at infinity; the first one produces some thrust while the other the drag of the same value so that the Euler–Bernoulli paradox of zero drag holds.

From Eq. (4.3.24) and the condition at infinity, Eq. (4.3.9), it follows that the circulation Γ of thin profiles required by the Joukowski rule is equal to

$$\Gamma = \frac{V}{\sqrt{1 - M_\infty^2}} \int_0^c \frac{2\alpha - f_+'(t) - f_-'(t)}{|X(t)|} dt. \tag{4.3.25}$$

In the limiting particular case of a flat aerofoil of a very small thickness, when $f'_+(t) = f'_-(t) = 0$, we get from Eq. (4.3.25)

$$\Gamma = \frac{2\alpha c V}{\sqrt{1 - M_\infty^2}} \int_0^1 \sqrt{\frac{t}{1-t}} \mathrm{d}t = \frac{\pi \alpha c V}{\sqrt{1 - M_\infty^2}}. \tag{4.3.26}$$

Paticularly, we have $\Gamma = \pi \alpha c V$ when $M_\infty = 0$ (compare with the corresponding exact solution $\Gamma = -\pi c U \sin\alpha$ for incompressible fluid, Eq. (1.4.21), with $U = -V$ in this case).

Thus, we proved *the Prandtl–Glauert law of similarity*, according to which the values of Γ, $\overline{u_1}$, $\bar{p}$ and $\bar{\rho}$ of the subsonic flight are directly proportional to $\left(1 - M_\infty^2\right)^{-1/2}$. This law allows one to obtain important information using experimental data for low-speed wings when the air behaves like an incompressible fluid. *The Ackeret–Buseman law* extends this rule for supersonic speeds, as the direct proportionality to $\left(M_\infty^2 - 1\right)^{-1/2}$. However, both laws do not work for transonic velocities.

By calculating the lift of a wing, we should take into account the span s of the wing and its shape in the plan view. The x_3-coordinate will contract, like the x_2-coordinate in the 2*D* consideration, proportionally to $\sqrt{1 - M_\infty^2}$ because instead of the wave equation equivalent to Eq. (4.3.6), we have for the finite span:

$$\left(1 - M_\infty^2\right)\frac{\partial^2 \varphi}{\partial x_1^2} + \frac{\partial^2 \varphi}{\partial x_2^2} + \frac{\partial^2 \varphi}{\partial x_3^2} = 0 \quad \text{and} \quad \bar{u}_i = \frac{\partial \varphi}{\partial x_i}, \quad (i = 1, 2, 3). \tag{4.3.27}$$

Therefore, the lift of the wing is directly proportional to $\left(1 - M_\infty^2\right)^{-1}$. In particular, for the thin flat wing of the rectangular shape in the plan view with the chord c and span s, the lift is equal to (Fig. 4.1)

$$F_2 = \frac{\pi \alpha s c \rho_\infty V^2}{1 - M_\infty^2}. \tag{4.3.28}$$

For the thin flat wing of the triangular shape in plan, with the maximum chord c and the maximum span s, the lift is twice less (Fig. 4.1).

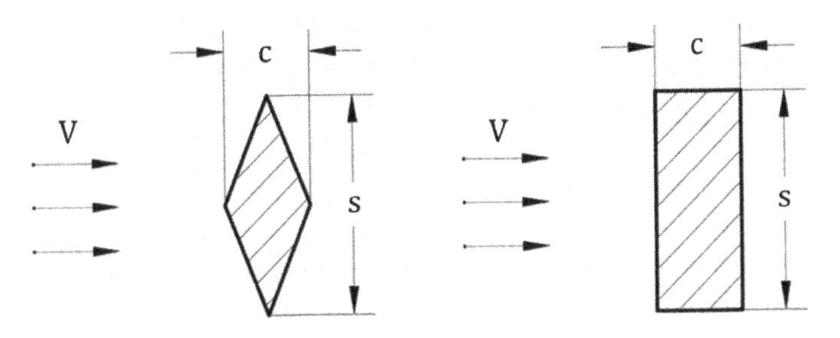

Fig. 4.1 Triangular and rectangular wings in the plan view

This theory of N. E. Joukowsky is well confirmed by practice. It is important to emphasize that it singles out one of many possible mathematical solutions based just on some common sense.

4.4 Optimal Aerofoil Problem

The lift of a thin aerofoil increases when the flight speed grows until the local Mach number achieves 1 at a critical point on the upper side of the aerofoil. This occurs at a critical Mach number $M_\infty = M_* < 1$. If $M_\infty > M_*$, a local supersonic zone and local shock waves develop near the critical point. At this regime, the drag is sharply increased.

If a moving body is "well-streamlined," like a thin plate moving with a very small angle of attack α, so that there are no separation of a boundary layer and no formation of drag vortices in the fluid, the drag is defined solely by viscous friction in the boundary layer. For example, in the case of a thin plate, the dimensionless coefficients of drag D and lift F_2 are equal to:

$$c_L = \frac{F_2}{\frac{1}{2}A\rho_\infty V^2} = \frac{2\pi\alpha}{1 - M_\infty^2}, \quad c_D = \frac{D}{\frac{1}{2}A\rho_\infty V^2} = 2.66\sqrt{\frac{\nu}{aV}}, \quad \left(M_\infty = \frac{V}{c_\infty}\right). \tag{4.4.1}$$

Here, ν is the kinematic viscosity of the fluid, a and A are the width and area of the plate, and V is the speed of the plate.

Thus, when the flight speed grows, the coefficient of lift is increased and the coefficient of drag is decreased. Hence, for every thin aerofoil, there exists a certain optimal flight speed before the drag vortices separate from the aerofoil or local shock waves form. From here, it follows that the optimal aerofoil shape should provide for a maximum flight speed without local shock waves and drag vortices separation.

The greater is the critical Mach number, the closer is the aerofoil to the perfect design. In the limit, the greatest critical Mach number is achieved, if the local sonic speed in the air flow is simultaneously achieved at all points of the upper side of the profile so that it becomes critical as a whole. It is clear that we could not achieve more in this direction. And so, we can call this profile optimal [4, 5].

There are three main requirements to an optimal aerofoil:

1. The tail of the aerofoil should be a trailing edge, i.e., separation locus of a vortex sheet. To achieve this aim at different regimes of flight, the tail part should be a controlled hinged flap with a sharp cusped end.
2. The nose of the aerofoil should coincide with the stagnation point and be sharply cusped, which helps localize some problems connected with large velocity gradients and separation of drag vortices. There should be no other stagnation points

different from the nose. To achieve this aim at different regimes of flight, the nose part should be a controlled hinged flap with a sharp cusped spike.

3. The upper side of the middle part of the aerofoil should provide for one and the same pressure at all the points at a virtual velocity of flight, in order to maximize the critical Mach number at this flight regime.

A sketch of the optimal aerofoil is shown in Fig. 4.2. One and same streamline doubles at the nose and emerges at the tail of the optimal aerofoil.

Let us formulate the problem of an optimal aerofoil in this approach for thin profiles. This is a mathematical problem with an unknown boundary requested to be found in the process of a solution. In this case, the unknown boundary is the upper side of the profile where the additional boundary value condition $p = p_0 = \text{const}$ according to Eqs. (4.3.2) and (4.3.5) can be written as

$$\bar{u}_1 = u_* \quad \text{where } u_* = \frac{p_\infty - p_0}{\kappa p_\infty M_\infty^2}. \tag{4.4.2}$$

Constant u_* or p_0 should be chosen from additional physical and commercial considerations (so that $1 + u_*$ is close to c/V). For passenger jets, they should correspond to the height of the flight at the cruising speed where most of the fuel is spent. Besides, the boundary condition of non-penetration holds along the entire surface of the aerofoil. It means that on the upper arc of uniform pressure being sought, two boundary condition equations have to be satisfied.

Using Eq. (4.4.2) and a consideration analogous to that used for deriving Eq. (4.3.13), we can get the corresponding boundary value problem of an optimal aerofoil for the analytical function $W'(z)$:

Upper side of nose flap:

$$z = x_1 + i0, 0 < x_1 < \varepsilon_n a: \quad \text{Im}W'(z) = \frac{\alpha_n - f'_{n+}(x_1)}{\sqrt{1 - M_\infty^2}}; \tag{4.4.3}$$

Upper side of middle part of aerofoil:

$$z = x_1 + i0, \quad \varepsilon_n a < x_1 < a(1 - \varepsilon_t): \quad ReW'(z) = u_*; \tag{4.4.4}$$

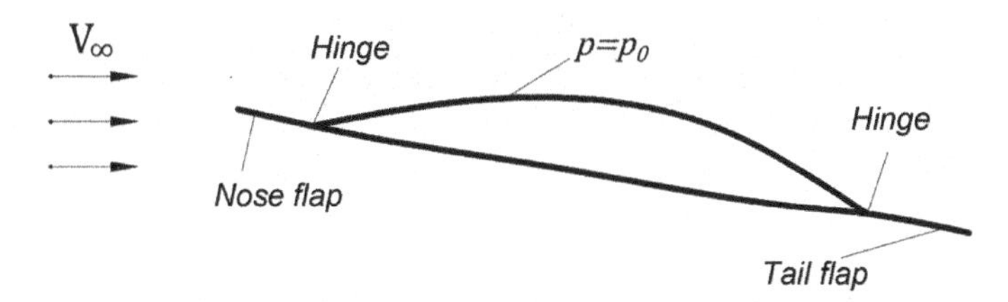

Fig. 4.2 Sketch of an optimal airfoil having a nose flap and a tail flap. Pressure is uniform along the upper middle part of the airfoil. One and same streamline doubles at the nose and trails from the tail of the airfoil

Upper side of tail flap:

$$z = x_1 + i0, \quad a(1-\varepsilon_t) < x_1 < a: \quad \operatorname{Im} W'(z) = \frac{\alpha_t - f'_{t+}(x_1)}{\sqrt{1-M_\infty^2}}; \tag{4.4.5}$$

Lower side of nose flap:

$$z = x_1 - i0, \quad 0 < x_1 < \varepsilon_n a: \quad \operatorname{Im} W'(z) = \frac{\alpha_n - f'_{n-}(x_1)}{\sqrt{1-M_\infty^2}}; \tag{4.4.6}$$

Lower side of middle part of aerofoil:

$$z = x_1 - i0, \quad \varepsilon_n a < x_1 < a(1-\varepsilon_t): \quad \operatorname{Im} W'(z) = \frac{\alpha_m - f'_m(x_1)}{\sqrt{1-M_\infty^2}}; \tag{4.4.7}$$

Lower side of tail flap:

$$z = x_1 - i0, \quad a(1-\varepsilon_t) < x_1 < a: \quad \operatorname{Im} W'(z) = \frac{\alpha_t - f'_{t-}(x_1)}{\sqrt{1-M_\infty^2}}. \tag{4.4.8}$$

Here, the functions $f'_{n+}(x_1), f'_{n-}(x_1), f'_m(x_1)$ and $f'_{t-}(x_1)$ and the constants $\varepsilon_n, \varepsilon_t, \alpha_n, \alpha_m$ and α_t are assumed to be given. They can be used as fitting parameters in the search for a better design.

This mixed problem, Eqs. (4.4.3)–(4.4.8), is a particular case of the general Riemann–Hilbert problem for any number of cuts along one and same straight line in the complex plane, which was solved in an explicit form [6, 9, 10]. But in the case of one cut, this problem can also be solved by the traditional method using the conformal mapping onto the half-plane or the unit circle exterior.

As an example, using the direct method [6, 9, 10], we provide the solution result to the problem, Eqs. (4.4.3)–(4.4.8), only in the particular case when both the nose and tail flaps are not available, that is when $\varepsilon_n = \varepsilon_t = 0$:

$$W'(z) = u_* - \frac{u_*}{\sqrt{2}}\left[e^{\pi i/4} X(z) + e^{-\pi i/4} X^{-1}(z)\right]$$
$$+ \frac{1}{2\pi i\sqrt{1-M_\infty^2}} \int_0^a \left[\frac{X(z)}{X^+(x)} - \frac{X^+(x)}{X(z)}\right] \frac{\alpha_m - f'_m(x)}{x-z} \mathrm{d}x. \tag{4.4.9}$$

Here,

$$X(z) = \left(\frac{z-a}{z}\right)^{1/4} \quad \text{where} \quad \lim X(z) = 1 \text{ as } z \to \infty. \tag{4.4.10}$$

The function $W'(z)$ is finite at the nose $z = 0$ and the tail $z = a$, if the following equations are met:

$$\int_0^a \frac{\alpha_m - f_m'(x)}{x^{3/4}(a-x)^{1/4}} \mathrm{d}x = \pi u_* \sqrt{2(1 - M_\infty^2)}.$$

$$\int_0^a \frac{\alpha_m - f_m'(x)}{x^{1/4}(a-x)^{3/4}} \mathrm{d}x = \pi u_* \sqrt{2(1 - M_\infty^2)}. \quad (4.4.11)$$

These condition equations should be satisfied during the optimal regime of the virtual flight.

The optimal shape of the upper middle part of the aerofoil is computed from the non-penetration condition which according to Eqs. (4.3.1) and (4.3.7) can be written as follows

$$\frac{\mathrm{d}y}{\mathrm{d}x_1} = \frac{v_2}{v_1} = \frac{\bar{u}_2}{1+\bar{u}_1} = -\frac{\sqrt{1-M_\infty^2}\,\mathrm{Im}W'(z)}{1+ReW'(z)} = -\frac{\sqrt{1-M_\infty^2}\,\mathrm{Im}W'(z)}{1+u_*}. \quad (4.4.12)$$

In the particular case of no tail and nose flaps, we can obtain the upper boundary of the optimal aerofoil in the optimal flight regime from Eqs. (4.4.9) and (4.4.12):

$$y = \frac{1}{1+u_*} \int_0^{x_1} \left\{ F(x) + \frac{u_*}{\sqrt{2}} \sqrt{1-M_\infty^2} \left[\left(\frac{a-x}{x} \right)^{1/4} - \left(\frac{x}{a-x} \right)^{1/4} \right] \right\} \mathrm{d}x,$$

$$F(x) = \frac{1}{2\pi} \int_0^a \frac{\sqrt{x(a-t)} - \sqrt{t(a-x)}}{[t(a-t)]^{1/4}[x(a-x)]^{1/4}} \cdot \frac{f_m'(t) - \alpha_m}{t-x} \mathrm{d}t. \quad (4.4.13)$$

Evidently, the optimal properties of each optimal design can be displayed only under some terms projected beforehand.

4.5 Method of Discrete Vortices

According to the method of discrete vortices, the surface of a moving body or the boundary layer and vortex sheets are replaced by a finite system of many vortices, so that the original problem is replaced by the problem of a large number of vortices that are produced, accumulated, and shed in the infinite boundless domain. This method is justified by the following philosophy:

1. Separation past bluff bodies arise roughly at the Reynolds number $Re > 10$ when the interaction of vortices with one another and with the surface of a body is much more substantial than the separation of boundary layer and formation of

vortices. As a reminder, $Re = VL/\nu$ where ν is the kinematic viscosity, and V and L are the characteristic velocity and linear size of the flow.

2. Hence, for large Reynolds numbers, when a great number of vortices are generated in a fluid, a many-vortices problem of inviscid fluid is practically the problem of turbulence. For any practical purpose of calculation of drag and lift, turbulent flows can be effectively described by some system of vortices or vortex sheets in a perfect fluid, the circulation of vortices being determined in the process of solution.

This philosophy allows one to generate practical solutions for separated and turbulent flows without to use the Navier–Stokes equations. Meanwhile, namely such flows mostly take place in nature and industry. Below, we provide the simple mathematics of this approach for the 2D or plane flows.

The fluid velocity $\boldsymbol{v}$ of a plane flow can be described by a vector function of time t and space coordinates x_1 and x_2

$$v = v_1(x_1, x_2, t) + i v_2(x_1, x_2, t). \tag{4.5.1}$$

The vorticity of the flow is described by the curl vector $\boldsymbol{\omega} = \nabla \times \mathbf{v}$. This vector is perpendicular to the x_1, x_2-plane and its magnitude is equal to

$$\omega(x_1, x_2, t) = \frac{\partial v_2}{\partial x_1} - \frac{\partial v_1}{\partial x_2}. \tag{4.5.2}$$

The relation between velocity and vorticity is analogous to that between magnetic field and electric current density. Vorticity can be interpreted physically as an angular momentum density; a round fluid particle, instantaneously frozen without loss of angular momentum, rotates with angular velocity $\omega/2$. The term "vortex motion" refers to flows in which the vorticity is confined to finite regions, called vortices, inside which the motion is said to be rotational. A fluid particle without vorticity can acquire it by viscous diffusion or the action of non-conservative outer forces. Hence, the study of vortices is of interest, first of all, for incompressible gases and fluids because vorticity can be associated with fluid particles.

Most famous examples of two-dimensional vortex motion are hurricanes, the Karman vortex street in the wake of a cylinder, the row of vortices formed by the Kelvin–Helmholtz instability at the interface between two streams of different velocity, dust devils and tornadoes, and the quantized vortices of He II. In incompressible gases and fluids, vortices are created in the process of the separation of a boundary layer. Therefore, the terms "separated flows" and "vortex flows" are practically equivalent. However, vortices can also be due to quantum mechanical effects (for He II), to non-conservative external forces, and to barocline density variations (in compressible gases).

There are vortex filaments, vortex patches, and vortex sheets. A vortex filament is a vortex in the shape of a tube of small cross section. Below, we consider only the limit of straight vortex filaments of finite strength and zero cross section called rectilinear line vortices or, shortly, vortices. The Helmholtz laws imply that a rectilinear line vortex moves with the fluid unless external forces act on the core. The vortex is said to be bound if external forces keep it at rest or fixed to the boundary of a body. The vortex is said to be free if it moves with the fluid.

Vortex patches are finite areas of vorticity in two-dimensional flows. A vortex sheet is the limit in which the vorticity domain is compressed into a line of zero thickness in a 2D flow. The velocity component normal to the sheet is continuous; the tangential component is discontinuous, the magnitude of the discontinuity being the strength of the sheet.

According to the Joukowsky–Chaplygin theory, flow past a wing can be described by bound vortices in the wing together with a trailing vortex sheet downstream, which rolls up into an opposite rotating vortex. As a reminder, a rectilinear line vortex having a core at the coordinate origin in the unbounded fluid has the velocity field

$$v_1 = -\frac{\Gamma x_2}{2\pi r^2}, \quad v_2 = \frac{\Gamma x_1}{2\pi r^2}, \quad v_3 = 0 \tag{4.5.3}$$

or, in polar coordinates r and θ

$$v_r = 0, \quad v_\theta = \frac{\Gamma}{2\pi r}, \quad v_3 = 0. \tag{4.5.4}$$

Here, Γ is the strength of the vortex called circulation. The streamlines of the field are concentric circles, centered at the coordinate origin.

According to Eq. (4.5.4), a vortex with circulation Γ_k having a core at point A with coordinates $x_1 = x_{1k}$ and $x_2 = x_{2k}$ induces the fluid velocity, the value of which at point $B(x_{1B}, x_{2B})$ is equal to

$$|v_{kB}| = \frac{\Gamma_k}{2\pi R_{kB}}, \quad R_{kB} = \sqrt{(x_{1B} - x_{1k})^2 + (x_{2B} - x_{2k})^2} \tag{4.5.5}$$

and the velocity vector is directed perpendicularly to the radius vector $\boldsymbol{R}_{kB}$ coinciding in sign with the vortex rotation (see Fig. 4.3).

The x_1 and x_2-components of the induced velocity $\mathbf{v}_{\mathrm{kB}}$ at any point B designated as $(v_{kB})_1$ and $(v_{kB})_2$ can be written as follows:

$$(v_{kB})_1 = -\frac{\Gamma_k \sin\theta_{kB}}{2\pi R_{kB}}, \quad \sin\theta_{kB} = \frac{x_{2B} - x_{2k}}{R_{kB}}; \tag{4.5.6}$$

$$(v_{kB})_2 = \frac{\Gamma_k \cos\theta_{kB}}{2\pi R_{kB}}, \quad \cos\theta_{kB} = \frac{x_{1B} - x_{1k}}{R_{kB}}. \tag{4.5.7}$$

Hence, for any number N of vortices in the unbounded fluid, the fluid velocity components $(v_B)_1$ and $(v_B)_2$ at point B are equal to:

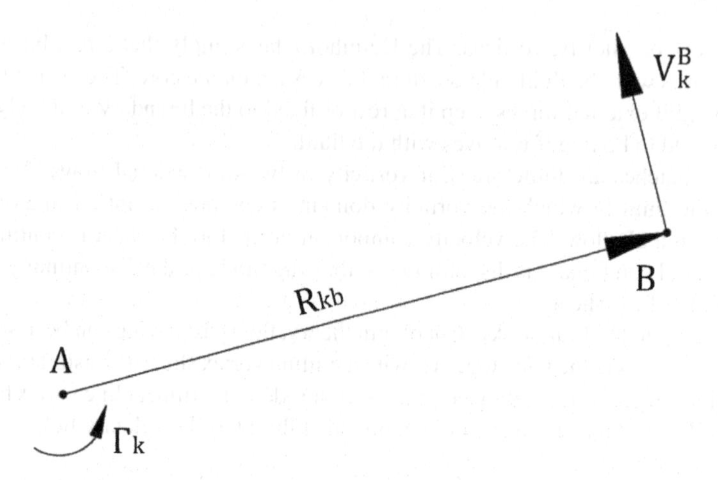

Fig. 4.3 Velocity v_K^B induced by the kth vortex of circulation Γ_K

$$(v_B)_1 = -\frac{1}{2\pi}\sum_{k=1}^{N}\frac{\Gamma_k \sin\theta_{kB}}{R_{kB}}; \tag{4.5.8}$$

$$(v_B)_2 = \frac{1}{2\pi}\sum_{k=1}^{N}\frac{\Gamma_k \cos\theta_{kB}}{R_{kB}}. \tag{4.5.9}$$

These equations provide the velocity field of any number of vortices in unbounded fluid, if the vortex circulations Γ_k and the vortex motion laws $x_{1k}(t)$ and $x_{2k}(t)$ are known for all values of k.

When the body boundary and vortex sheets are replaced by a number of discrete vortices of unknown circulations, the problem is reduced to that of unbounded fluid with N vortices, so that the values of Γ_k, as well as the values of $x_{1k}(t)$ and $x_{2k}(t)$ of free vortices, should be determined in the process of solution.

The motion equations of vortices can be written as follows:

$$\Gamma_k \frac{\mathrm{d}x_{1k}}{\mathrm{d}t} = \Gamma_k U_{1k} - G_{1k}; \tag{4.5.10}$$

$$\Gamma_k \frac{\mathrm{d}x_{2k}}{\mathrm{d}t} = \Gamma_k U_{2k} - G_{2k}. \quad (k = 1, 2, \ldots N) \tag{4.5.11}$$

Here, U_{1k} and U_{2k} are the fluid velocity components at the core of the kth vortex produced by all other vortices, and G_{1k} and G_{2k} are the components of the external force applied to the core of the kth vortex. The values of U_{1k} and U_{2k} are defined by Eqs. (4.5.8) and (4.5.9) in which the kth addend should be excluded.

The equation system, Eqs. (4.5.10) and (4.5.11), includes N_F free vortices and N_B bound vortices so that

$$3N_F + N_B = 2N. \tag{4.5.12}$$

And so, the number of equations is equal to the number of unknowns which are the circulations of all vortices and the coordinates of free vortices. The problem is reduced to the computation of this equation system for large N. It seems that this simple numerical approach can practically solve the problem of turbulence by avoiding the Navier–Stokes equations and the fancy nature of fluids. However, the instability ignored in [4, 5] may become the main problem in this approach, as well.

When the external forces are conservative or absent, Eqs. (4.5.10) and (4.5.11) are Hamiltonian, and the methods of statistical mechanics can be applied for very large N. A temperature can be defined that when positive characterizes a well-mixed state and when negative leads to states where vortices of like rotation clump together.

The problem of instability. Let us consider the original 2D problem of a uniform fluid flow with the characteristic velocity V around a body of the characteristic size L. Using the method of discrete vortices, this problem is reduced to the solution of the differential equation system, Eqs. (4.5.10) and (4.5.11), in the unbounded space with only one new parameter, which is the characteristic circulation Γ of discrete vortices. Since it has the dimension of the kinematic viscosity, the only dimensionless number is $Ch = VL/\Gamma$ in this problem.

In vortex flows of inviscid fluid, it plays the part the Reynolds number $Re = VL/\nu$ plays in the flow of viscous fluid. Hence, at a high Ch, the instability of any vortex flow is unavoidable.

Let us consider the Karman vortex street as an example of vortex systems. This consists of two parallel rows of the same spacing, but of opposite vorticities Γ and $-\Gamma$, so arranged that each vortex of the upper row at $z = an + ib/2$ when $t = 0$ is directly above the midpoint of the line joining two vortices of the lower row at $z = \left(n + \frac{1}{2}\right)a - \frac{1}{2}ib$ when $t = 0$ $(n = 0, \pm 1, \pm 2, \ldots)$.

The complex potential of this vortex flow at the instant $t = 0$ is equal to

$$W = \frac{i}{2\pi}\Gamma \log \sin \frac{\pi}{a}\left(z - \frac{1}{2}ib\right) - \frac{i}{2\pi}\Gamma \log \sin \frac{\pi}{a}\left(z - \frac{1}{2}a + \frac{1}{2}ib\right). \tag{4.5.13}$$

The upper and lower rows advance with one and same velocity V

$$V = \frac{1}{2a}\Gamma \tanh \frac{\pi b}{a}. \tag{4.5.14}$$

The rows will advance the distance a in time $\tau = a/V$ and the configuration will be the same after this interval as at the initial instant. This row system is generally unstable, except for some special values of b/a and $\Gamma/(aV)$. In the limit $a/b \to 0$, these vortex rows turn into two vortex sheets along $z = \pm\frac{ib}{2}$ where the tangential velocity of fluid suffers a discontinuity.

Suppose a cylindrical body is placed in a stream, so that vortices leave the opposite edges alternatively, with a Karman vortex street being created behind the body. In this case, Karman derived the following approximate equation for the drag D (in our designation)

$$\frac{D}{\rho b U^2} = \frac{1}{Ch}\left(1 + \frac{a}{2\pi b Ch}\right) \text{ where } U = U_0 - 2V, \quad Ch = \frac{aU}{\Gamma}. \tag{4.5.15}$$

Here, U_0 is the velocity of the cylinder with respect to liquid at rest, Ch is the characteristic dimensionless number similar to the Reynolds number Re, and V is the velocity given by Eq. (4.5.14).

The function $D(Ch)$ of Eq. (4.5.15) reminds the similar function $D(Re)$ for the drag of a body moving in a viscous fluid, until the flow crisis occurs at the Reynolds number of the order of 2000–5000. Evidently, for vortex flows of perfect fluid similarly to the flow of viscous fluid, the following theorem is valid:

Any vortex system becomes unstable at some critical value of the number $Ch = VL/\Gamma$, *where* V *is the characteristic speed of the uniform fluid flow,* L *is the characteristic size of the body, and* Γ *is the characteristic circulation of vortices.*

The study of large Reynolds number flows, especially separated flows, using the method of discrete vortices suggests some new, attractive avenues of research.

Hurricanes and tornados. Most majestic vortices in nature are tornados and hurricanes called typhoons if they happen in the Pacific or Indian oceans. They are characterized by the speed V of maximum sustained winds, by the specific radius R and height H of their penetration in atmosphere, by the circulation Γ measuring their strength, and by their energy $E \approx \rho V^2 H R^2 \sim \rho H \Gamma^2$. The dimensionless number $Ch = VR/\Gamma$ provides the opportunity to study hurricanes and tornadoes using the laws of similarity.

A great base of data collected for all hurricanes of the last hundred years made it possible to create a dozen of well-working empirical models that predict the trajectory and strength of any hurricane very accurately. Each hurricane has a male name and a detailed biography carefully printed in the computer memory. The height of hurricanes is about the thickness of the Earth layer of atmosphere. Therefore, based on the similarity law, the main characteristic of the strength of a hurricane is the maximum speed of sustained winds which is directly proportional to the circulation of the hurricane.

There are five categories of strength, with the maximum speed of sustained winds being in the range of: 33–42 m/s (1st category), 42–49 m/s (2nd category), 49–58 m/s (3rd category), 58–70 m/s (4th category), and 70+ m/s (5th category). Some gusts can be much stronger; during the hurricane Andrew in Florida in 1991, the recorded speed of some gusts exceeded 200 m/s. The wind then levelled cities with the ground on the path of the hurricane; it tossed up railway wagons and cars. The energy of a major hurricane can exceed the energy of some thousands of Hiroshima atomic bombs.

Literature

1. G.K. Batchelor, *An Introduction to Fluid Dynamics* (Cambridge University Press, London, 1977), p. 750
2. G. Birkhoff, E.H. Zarantonello, *Jets, Wakes and Cavities* (Academic Press, New York, 1957), 280 pp
3. S.A. Chaplygin, *On Gas Jets* (Moscow University Press, 1902), 180 pp
4. G.P. Cherepanov, An introduction to singular integral equations in aerodynamics, in *Method of Discrete Vortices*, ed. G.P. Cherepanov, S.M. Belotserkovsky, I.K. Lifanov (CRC Press, Boca Raton, 1993), 450 pp
5. G.P. Cherepanov, An introduction to two-dimensional separated flows, in *Two-Dimensional Separated Flows*, ed. G.P. Cherepanov, S.M. Belotserkovsky, et al. (CRC Press, Boca Raton, 1993), 320 pp
6. G.P. Cherepanov, The solution to one linear Riemann problem. J. Appl. Math. Mech. (JAMM) **26**(5), 623–632 (1962)
7. G.P. Cherepanov, The flow of an ideal fluid having a free surface in multiple connected domains. J. Appl. Math. Mech. (JAMM) **27**(4), 508–514 (1963)
8. G.P. Cherepanov, On stagnant zones in front of a body moving in a fluid. J. Appl. Mech. Tech. Phys. (JAMTP) (3), 374–378 (1963)
9. G.P. Cherepanov, The Riemann-Hilbert problems of a plane with cuts. Dokl. USSR Acad. Sci. (Math.) **156**(2) (1964)
10. G.P. Cherepanov, On one case of the Riemann problem for several functions. Dokl. USSR Acad. Sci. (Math.) **161**(6) (1965)
11. G.P. Cherepanov, Invariant Γ-integrals and some of their applications to mechanics. J. Appl. Math. Mech. (JAMM) **41**(3) (1977)
12. G.P. Cherepanov, Invariant Γ-integrals. Eng. Fract. Mech. **14**(1) (1981)
13. G.P. Cherepanov, Invariant integrals in continuum mechanics. Soviet Appl. Mech. **26**(7) (1990)
14. F.D. Gakhov, *Boundary Value Problems* (Pergamon Press, London, 1980)
15. M.I. Gurevich, *The Theory of Jets in an Ideal Fluid* (Academic Press, New York, 1965)
16. J. Hapel, H. Brenner, *Low Reynolds Number Hydrodynamics* (Prentice Hall, New York, 1965)
17. A.A. Khrapkov, Problems of elastic equilibrium of infinite wedge with asymmetrical cut at the vertex, solved in explicit form. J. Appl. Math. Mech. (JAMM) **35**(6) (1971)
18. L.M. Milne-Thomson, *Theoretical Hydrodynamics* (Dover, New York, 1996)
19. N.I. Muskhelishvili, *Singular Integral Equations* (Noordhoff, Groningen, 1946)
20. F.S. Sherman, *Viscous Flow* (McGraw Hill, New York, 1990)

Chapter 5
The Strength of Adhesion

Abstract A series of most significant problems concerning the motion of a sharp punch on an adhesive, elastic foundation, plate, or membrane are studied using the invariant integral of adhesion of different materials. A film covered most of the bodies is also taken into account. This is, in fact, a new contact mechanics accounting for the adhesion and predicting the resistance forces. Also, in this chapter, the basic mathematical problem of stick and slip on the contact area was, at last, solved after many unsuccessful attempts of such great individuals as Hertz, Prager, Muskhelishvili, and others; the solution of this nonlinear problem appeared to depend on six dimensionless parameters! This chapter is a must for mechanical engineers.

Adhesion is a word of the Latin origin meaning the clinging of dissimilar materials one to another (while "cohesion" means the clinging of the like materials). The strength of adhesion is characterized by the specific adhesion energy which is the work spent to liberate a unit of the common surface of two materials. The "liberation" or the interface fracturing is a complicated process which essentially depends on the mechanical and geometrical properties of the structure.

In this chapter, we consider some basic problems of adhesion for static elasticity and small deformations represented by the corresponding invariant integrals. Especially complicated are the stick and slip problems accounting for various scenarios of contact interactions.

The chapter is based on this author's publications *The contact problem of the mathematical theory of elasticity with stick-and-slip areas. The theory of rolling and tribology* and *Some new applications of the invariant integrals in mechanics* published in *J. Appl. Math. Mech. (JAMM),* **76**(5),2012 and **79**(1), 2015.

5.1 The Resistance Forces Upon a Punch—The Γ−Force

Let us consider the problem of a smooth rigid punch with an angular leading edge moving along the surface of a half-space and confine ourselves to the case of plane strain. According to the classical theory, the smooth punch does not experience

G. P. Cherepanov, *Invariant Integrals in Physics*,
https://doi.org/10.1007/978-3-030-28337-7_5

resistance forces opposing its motion, since the shear stress in the contact area is equal to zero. However, this conclusion does not take into account the stress singularity on the leading edge of the punch which causes the energy dissipation on this edge. The latter gives rise to a head resistance force at the leading edge of the punch we will study in the next Sects. 5.2–5.4.

Single smooth punch. Let the boundary conditions of the problem have the following shape (Fig. 5.1):

$$\sigma_y = 0, \tau_{xy} = 0 \quad \text{when} \quad y = 0, \text{and} \quad x < -b \text{ or} x > 0; \tag{5.1.1}$$

$$u_y = f(x), \quad \tau_{xy} = 0 \quad \text{when} \quad y = 0, -b < x < 0. \tag{5.1.2}$$

Here, Oxy are Cartesian coordinates with the origin on the leading edge of a punch moving in the direction of the $x-$ axis along the boundary $y = 0$ of the elastic half-plane $y < 0$, and $f(x)$ is the shape of the punch surface, and $(-b, 0)$ is the contact area.

The displacement components u_x and u_y and the stress components $\sigma_x, \sigma_y,$ *and* τ_{xy} are expressed by the Kolosov–Muskhelishvili formulae:

$$\sigma_x + \sigma_y = 4Re\Phi(z), \sigma_y - \sigma_x + 2i\tau_{xy} = 2\left[\bar{z}\Phi'(z) + \Psi(z)\right]; \tag{5.1.3}$$

$$2\mu\left(u_x + iu_y\right) = \kappa\varphi(z) - z\overline{\varphi'(z)} - \overline{\psi(z)}, z = x + iy, \kappa = 3 - 4\nu. \tag{5.1.4}$$

Here, $\varphi(z)$ *and* $\psi(z)$ are analytic functions in the elastic domain, $\Phi(z) = \varphi'(z)$ and $\Psi(z) = \psi'(z)$, and μ is the shear modulus, and ν is Poisson's ratio.

The solution of the boundary value problem, Eqs. (5.1.1) and (5.1.2), which is unbounded in the stresses on the leading edge but bounded on the trailing edge, and which disappears at infinity, has the following form

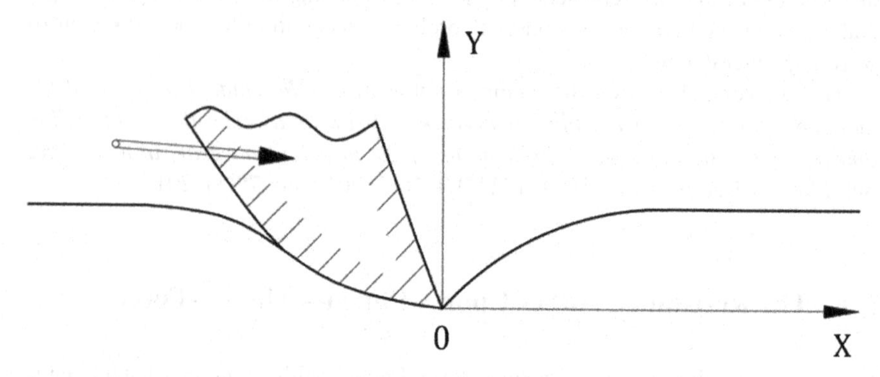

Fig. 5.1 A sharp punch moving on the surface of an elastic body

$$\Phi(z) = -\frac{2\mu X(z)}{\pi(\kappa+1)} \int_{-b}^{0} \frac{f'(x)}{X^{-}(x)} \frac{dx}{x-z}, \quad \Psi(z) = -z\Phi'(z). \tag{5.1.5}$$

$$X(z) = \sqrt{1+(b/z)}, \quad (X(z) \to 1 \text{ when } z \to \infty). \tag{5.16}$$

From the equilibrium condition, it follows that

$$\Phi(z) = \frac{iP}{2\pi z} \text{ when } z \to \infty. \tag{5.1.7}$$

Here, P is the magnitude of the force indenting the punch into the half-plane. From Eqs. (5.1.5) and (5.1.7), we can find that

$$-\mu \int_{-b}^{0} f'(x) \left| \sqrt{\frac{x}{x+b}} \right| dx = (1-\nu)P. \tag{5.1.8}$$

This equation serves to determine the length b of the contact area.

In a small neighborhood of the leading edge when $z \to 0$, from Eq. (5.1.5), we get:

$$\Phi(z) = 2\Psi(z) = \frac{iK}{2\sqrt{2\pi z}}, \quad (z \to 0) \tag{5.1.9}$$

$$K = \frac{\mu}{1-\nu} \sqrt{\frac{2b}{\pi}} \int_{-b}^{0} \frac{f'(x)dx}{\left|\sqrt{x(x+b)}\right|}. \tag{5.1.10}$$

These equations follow from Gakhov's equations describing behavior of the Cauchy-type integral at the end of the contour of integration in the case of a power-law singularity. Functions of Eq. (5.1.9) also correspond to the solution of the problem of a semi-infinite tensile mode crack along $y = 0, /; x > 0$ where K is the stress intensity factor. Since the drag is applied to the leading edge of a moving punch, the value of the head resistance for punches of arbitrary shape is determined by Eqs. (5.1.9) and (5.1.10).

In the problem under study, the field singularity of the leading edge moves along the $x-$ axis. The Γ–Force of drag acting on this singularity is given by the invariant integral

$$\Gamma_x = \int_S (U n_x - \sigma_{ij} n_j u_{i,x}) dS \ , \quad (i, j = 1, 2) \tag{5.1.11}$$

Here, as distinct from Eq. (3.3.5), S is any contour traversed counterclockwise from an initial point on the $x-$ axis to the left of the coordinate origin to a final point on the $x-$ axis to the right of the origin. The value of Γ_x does not depend on the shape of the contour and the choice of its initial and final points because Eq. (5.1.11) expresses the energy conservation law.

To calculate the Γ–Forceof drag by Eq. (5.1.11), we can use Eq. (3.3.6) in the form:

$$\Gamma_x = \frac{\pi}{2} \lim\left[\sigma_y(-\varepsilon, 0)u_y(+\varepsilon, 0) + \tau_{xy}(-\varepsilon, 0)u_x(+\varepsilon, 0)\right], \quad (\varepsilon \to 0). \quad (5.1.12)$$

For the potential functions of Eq. (5.1.9), the stresses and displacements can be found using Eqs. (5.1.3) and (5.1.4) so that when $y = 0$, we get:

$$\begin{aligned} &u_y = 0 \quad \text{when } x < 0;\ u_y = 2K\left(1 - \nu^2\right)E^{-1}(2\pi)^{-1/2}x^{1/2} \quad \text{when } x > 0; \\ &\sigma_y = 0 \text{ when } x > 0;\ \sigma_y = -2K(2\pi|x|)^{-1/2} \quad \text{when} \quad x < 0. \end{aligned} \quad (5.1.13)$$

Since $\tau_{xy} = 0$ when $y = 0$, from Eqs. (5.1.13) and (5.1.12) it follows that

$$\Gamma_x = -\frac{1 - \nu^2}{E} K^2. \quad (5.1.14)$$

This is the drag force applied to the leading edge of an arbitrarily shaped punch. The values of K and b are given by Eqs. (5.1.8) and (5.1.10) so that the problem is solved. This force is concentrated; it plays, evidently, the main part in wear and tear of any cutting tools, but it is, amazingly, ignored in the scientific literature.

We can also derive Eq. (5.1.14) using another method. An advance of the punch to the right by an amount Δ can be represented as the result of applying a normal loading of the pressure from zero to $2K(2\pi|x - \Delta|)^{-1/2}$ in the interval $(0, \Delta)$, under the action of which the boundary in this part undergoes a displacement from $2K\left(1 - \nu^2\right)E^{-1}(2\pi)^{-1/2}x^{1/2}$ to zero. Using Clapeyron's theorem, we can calculate the work done in this process. By the energy conservation law, this work is equal to the loss $\Gamma_x\Delta$. By this way, we again come to Eq. (5.1.14).

The third way of calculating the value of Γ_x is to use its invariance with respect to contour S in Eq. (5.1.11) and to deform it into another contour where the integration is easier (like it is common in the theory of analytical functions of a complex variable). The value of Γ_x in Eq. (5.1.11) does not change, if contour S is deformed into the sum of intervals $(-\infty, -b)$, $(-b, -\varepsilon)$ and $(+\varepsilon, +\infty)$ of the real axis and the half of the circle $|z| = R$ in the lower half-plane when $\varepsilon \to 0$, $R \to \infty$. From here, by using the boundary conditions in Eqs. (5.1.1) and (5.1.2), we can derive:

$$\Gamma_x = -\int_{-b}^{0} \sigma_y \frac{\partial u_y}{\partial x} dx \quad \text{when} \quad y = 0. \quad (5.1.15)$$

In particular, for the flat punch, when $u_y = -\beta x$ we get:

$$\Phi(z) = -\frac{2\beta\mu i}{\kappa+1}\left[1-\sqrt{1-\frac{b}{z}}\right],$$
$$\Gamma_x = -\beta P,\quad b = \frac{2(1-\nu)}{\pi\mu\beta}P,\quad K = -\frac{\mu\beta}{1-\nu}\sqrt{\frac{1}{2}\pi b}. \tag{5.1.16}$$

The concentrated head force of resistance applied at the leading edge of the punch is balanced by the distributed force of the normal pressure under the punch.

In the case, when the shape of the punch is the parabola $f(x) = ax^2$, the calculations using Eqs. (5.1.5), (5.1.7), (5.1.9), and (5.1.13) lead to the following results:

$$\Phi(z) = \frac{4ia\mu}{\kappa+1}\left[z-\left(z-\frac{b}{2}\right)\sqrt{1+\frac{b}{z}}\right], \tag{5.1.17}$$

$$b^2 = \frac{\kappa+1}{3\pi a\mu}P,\ K = -\frac{4\mu ab}{\kappa+1}\sqrt{2\pi b},\quad \Gamma_x = -\frac{\pi a^2b^2E}{2(1-\nu^2)}. \tag{5.1.18}$$

In this case, the relation between the drag and the indenting force is:

$$\Gamma_x = \sqrt{\frac{a(1-\nu^2)}{\pi E}}\left(\frac{2}{3}P\right)^{3/2}. \tag{5.1.19}$$

It should be kept in mind that the concentrated force Γ_x applied to the leading edge is balanced by the total horizontal component of the distributed force from the pressure under the punch, which is equal to the integral of $\sigma_y\partial u_y/\partial x$ with respect to dx. Hence, the total resistance force acting on a punch moving over an elastic body is equal to zero.

However, all real materials possess plastic and other inelastic properties that manifest themselves under high stresses. On account of this, the stress singularity on the leading edge leads to irreversible deformations in the body close to this edge so that a large part of the work of the force on the edge is irreversible and it is converted into heat (as well as into the energy of the residual deformations in the trail behind the moving punch). In the case of a fairly large indenting force, the reversible part of this work that is balanced by the work of the distributed forces of the pressure under the punch is negligibly small compared with the irreversible work of this force.

Adhesion effect. Now, let us evaluate the effect of adhesion in the limiting case, when the resistance to rupture is much greater than the resistance to shear. In this case, we can ignore the shear stress in the contact area and let the tensile stresses be near the trailing edge of the punch. The latter necessitate the growth of a stress singularity at this edge, the strength of which is determined by the adhesion energy on the punch-material contact.

In this case, the solution of the boundary problem of Eqs. (5.1.1) and (5.1.2), which is unbounded both on the leading and trailing edge and behaves at infinity according to Eq. (5.1.7), has the form:

$$\Phi(z) = \frac{2\mu}{\pi(\kappa+1)X(z)} \int_{-b}^{0} f'(x)X^{-}(x)\frac{dx}{x-z} + \frac{iP}{2\pi X(z)}, \tag{5.1.20}$$
$$X(z) = \sqrt{z(z+b)},\ \Psi(z) = -z\Phi'(z),\ X(z) \to z \quad \text{when } z \to \infty.$$

In a small neighborhood of the trailing edge when $z \to -b$, the field should correspond to that at the tensile crack end, that is,

$$\Phi(z) = 2\Psi(z) = \frac{K_A}{2\sqrt{2\pi(z+b)}}, \quad z \to -b. \tag{5.1.21}$$

Here, K_A is the adhesion toughness.

Using Eqs. (5.1.20) and (5.1.21) when $z \to -b$, we find the equation to determine the value of b

$$K_A\sqrt{2\pi b} = 2P + \frac{2\mu}{1-\nu}\int_{-b}^{0} f'(x)\left|\sqrt{\frac{x}{x+b}}\right|dx. \tag{5.1.22}$$

Similarly, an analysis of a small neighborhood of the leading edge when $z \to 0$, provides the elastic field and head resistance Γ_x

$$\Phi(z) = 2\Psi(z) = \frac{iK}{2\sqrt{2\pi z}}, \quad \Gamma_x = \frac{1-\nu^2}{E}K^2, \tag{5.1.23}$$

$$K\sqrt{2\pi b} = 2P + \frac{2\mu}{1-\nu}\int_{-b}^{0} f'(x)\left|\sqrt{\frac{x+b}{x}}\right|dx. \tag{5.1.24}$$

In the simplest case of a flat punch when $f'(x) = -\beta$, we obtain:

$$\Phi(z) = -\frac{2\mu\beta i}{\kappa+1}\left(1 - \frac{z+C}{\sqrt{z(z+b)}}\right) \quad \text{where } C = \frac{1}{2}b - \frac{\kappa+1}{4\pi\mu\beta}P; \tag{5.1.25}$$

$$\Gamma_x = -\beta P - \frac{1-\nu^2}{E}K_A^2, \quad P - 2\pi\frac{\mu\beta b}{\kappa+1} = K_A\sqrt{\frac{\pi}{2}b}. \tag{5.1.26}$$

The adhesion at the trailing edge increases the force Γ_x of the head resistance.

5.2 A Punch on a Body Covered with an Inextensible Film

Young's modulus for a monatomic film of graphene is roughly ten times greater than Young's modulus of steel, and the tensile strength of this film is of the order of

Young's modulus of steel. These films, for the experimental development of which K. S. Novoselov and A. K. Geim were awarded the Nobel Prize in Physics, can serve as some promising coatings on the critical elements of structures. A study of stresses and strains in elastic bodies covered with an inextensible membrane of zero thickness, by which a film of graphene is well simulated, is therefore of great interest. We will consider the problem of the pressure of a punch on an elastic half-plane covered with an inextensible film of zero thickness bonded to the material of the half-plane by adhesion. We will assume the plane-strain conditions.

According to Eq. (5.1.4), if the boundary of the half-plane is inextensible, that is $\partial u_x/\partial x = 0$ when $y = 0$, then the following relation is valid at $\mathrm{Im} z = 0$ and everywhere based on the principle of analytic continuation:

$$z\Phi'(z) + \Psi(z) = (\kappa - 1)\Phi(z). \tag{5.2.1}$$

From this and Eqs. (5.1.3) and (5.1.4), it follows that:

$$\sigma_y = (\kappa + 1)Re\Phi(z), \quad \tau_{xy} = (\kappa - 1)\mathrm{Im}\Phi(z) \quad \text{when} \quad y = 0; \tag{5.2.2}$$

$$\mu \partial u_y/\partial x = \kappa Im\Phi(z), \quad \sigma_x = (3 - \kappa)Re\Phi(z) \quad \text{when} \quad y = 0. \tag{5.2.3}$$

Using Eqs. (5.2.2) and (5.2.3), the solution of the contact problem

$$y = 0 : u_y = f(x) \quad \text{when} \quad |x| < b; \quad \sigma_y = 0 \quad \text{when} \quad |x| > b \tag{5.2.4}$$

can be written as:

$$\Phi(z) = \frac{i\mu}{\pi\kappa\sqrt{z^2 - b^2}} \int_{-b}^{+b} f'(x)\frac{\sqrt{b^2 - x^2}}{x - z}dx + \frac{iP}{\pi(\kappa + 1)\sqrt{z^2 - b^2}}. \tag{5.2.5}$$

Here, P is the value of the indenting force and $\sqrt{z^2 - b^2} \to z$ when $z \to \infty$.

The tension $T(x)$ in the film on the free surface is found by integrating the equilibrium equation of the film

$$\frac{\mathrm{d}T}{\mathrm{d}x} = \tau_{xy}. \tag{5.2.6}$$

Here, τ_{xy} is the shear stress under the film, as it is defined by Eqs. (5.2.2) and (5.2.5).

We shall first study the auxiliary problem of the motion of a semi-infinite rectangular punch on the boundary of the half-plane $y = 0$ along the $x-$ axis:

$$u_y = \text{const} \quad \text{when} \quad x < 0; \quad \sigma_y = 0 \quad \text{when} \quad x > 0. \tag{5.2.7}$$

This problem characterizes the elastic field close to the leading edge of any punch moving along the surface of an elastic body which, in this case, is covered with an inextensible film of almost zero thickness. According to Eqs. (5.2.1)–(5.2.3), the elastic field of this auxiliary problem will be as follows:

$$\Phi(z) = \frac{iK}{2\sqrt{2\pi z}}, \quad \Psi(z) = \left(\kappa - \frac{1}{2}\right)\Phi(z), \quad z = x + iy. \tag{5.2.8}$$

Here, K is a constant, and the corresponding stresses and displacement at $y = 0$ are equal to:

$$x > 0: \quad u_y = \frac{K}{\mu}\sqrt{\frac{x}{2\pi}}, \quad \tau_{xy} = \frac{1-2\nu}{\sqrt{2\pi x}}K, \quad \sigma_x = \sigma_y = 0; \tag{5.2.9}$$

$$x < 0: \quad \sigma_x = -\frac{2\nu}{\sqrt{2\pi|x|}}K, \quad \sigma_y = -\frac{2(1-\nu)}{\sqrt{2\pi|x|}}K, \quad \tau_{xy} = 0. \tag{5.2.10}$$

Based on the boundary conditions of Eq. (5.2.7) and the condition of the inextensibility of the boundary, the integral Γ_x defined by Eq. (5.1.11) will be invariant with respect to any open contour in the lower half-plane with an origin at any point lying on the $x-$ axis to the left of the coordinate origin and with an end at any point on the $x-$ axis to the right of the origin. Evaluation of this invariant integral again leads to Eq. (5.1.14) in which, however, the coefficient K is defined by Eqs. (5.2.8)–(5.2.10). In this auxiliary problem, the magnitude of Γ_x in Eq. (5.1.14) provides the drag opposing the motion of the punch along the surface of an elastic body covered by an inextensible film of almost zero thickness. In the case of an arbitrary punch, the coefficient K is determined by the punch shape, and Eq. (5.1.14) provides the drag force acting on its leading edge.

In the case of a problem that is symmetrical about the $y-$ axis, according to Eq. (5.2.5) the coefficient K at the edges $x = \pm b$ of the punch of arbitrary shape is equal to

$$K = -\frac{2P}{(\kappa+1)\sqrt{\pi b}} + \frac{\mu}{\kappa\sqrt{\pi b}}\int_{-b}^{+b} f'(x)\left|\sqrt{\frac{x+b}{x-b}}\right| dx. \tag{5.2.11}$$

If the function $f'(x)$ is continuous, and the adhesion/interaction between the materials of the punch and the film is negligibly small, then there is no singularity at the edges of the punch, that is, $K = 0$. In this case, Eq. (5.2.11) serves to define the size of the contact area in terms of the indenting force.

If, however, the function $f'(x)$ is continuous, but considerable attractive forces act between the punch and film materials, characterized by an adhesion energy Γ_A, then, according to Eq. (5.1.14), the magnitude of K will be equal to $K_A = (2E\Gamma_A)^{1/2}(1-\nu^2)^{-1/2}$. In this case, the size of the contact area is found from Eq. (5.1.14), a solution of which exists even in the case of tensile stresses and negative values of P. In the latter case, the process will obviously be unstable so that Eq. (5.1.14) provides the relation between the limiting delamination force and the size of the initial contact area.

Let us discuss two special cases of the plane and parabolic punches.

Plane punch. In this case, $f'(x) = 0$, and using Eqs. (5.2.2)–(5.2.5), we obtain on the boundary of the half = plane:

$$|x| > b: \quad \tau_{xy} = \mu\left(1 - \frac{1}{\kappa}\right)\frac{\partial u_y}{\partial x} = \frac{(\kappa - 1)P}{\pi(\kappa + 1)\sqrt{x^2 - b^2}}, \quad \sigma_x = \sigma_y = 0; \tag{5.2.12}$$

$$|x| < b: \quad \sigma_y = \frac{1 + \kappa}{3 - \kappa}\sigma_x = -\frac{P}{\pi\sqrt{b^2 - x^2}}, \quad \tau_{xy} = 0, \partial u_y/\partial x = 0. \tag{5.2.13}$$

Using Eqs. (5.2.12) and (5.2.6), we find the tension in the film

$$T = \frac{\kappa - 1}{\pi(\kappa + 1)}P\ln\left(\frac{x}{b} + \sqrt{\frac{x^2}{b^2} - 1}\right) \quad \text{when} \quad |x| > b. \tag{5.2.14}$$

Under the punch, the tension in the film is obviously equal to zero. It can be seen that, as one moves away from the punch, the tension in the film increases without limit so that for a fairly large size the film will be torn. This imposes some limitations on the maximum size of structural elements covered with a graphene film. Equation (5.2.14) enables one to establish a safe size of structural elements. The limiting tension in graphene films has an order of $6n$ kG/m where n is the number of monatomic layers in the film.

Parabolic punch. According to Eq. (5.2.5), in the case of a parabolic punch $f(x) = ax^2$ we find

$$\Phi(z) = \frac{ia\mu(b^2 - 2z^2)}{\kappa\sqrt{z^2 - b^2}} + 2ia\frac{\mu}{\kappa}z + \frac{iP}{\pi(\kappa + 1)\sqrt{z^2 - b^2}}. \tag{5.2.15}$$

In the case of the punch-film adhesion, using Eq. (5.2.15) at the edge of the contact area, we have

$$K_A = \sqrt{\frac{\pi}{b}}\left[-\frac{1}{\pi}P + a\mu b^2\left(1 + \frac{1}{\kappa}\right)\right]. \tag{5.2.16}$$

If the adhesion can be ignored, then the adhesion toughness K_A is equal to zero, and we get

$$P = \pi a\mu b^2\left(1 + \kappa^{-1}\right). \tag{5.2.17}$$

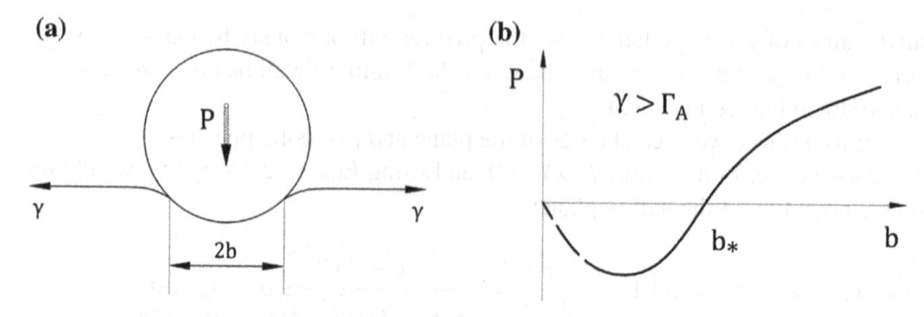

Fig. 5.2 **a, b** A ball rolling on a membrane

The corresponding result for the material without a film coincides with that of Eq. (5.2.17) if its right-hand part is multiplied by $4\kappa(1+\kappa)^{-2}$.

If $P < 0$, Eq. (5.2.16) also provides the solution of the problem of the limiting force tearing the punch off the foundation. According to Eq. (5.2.16), the relation $P = P(b)$ has the form shown in Fig. 5.2.a, b. If $P > 0$, the contact area, starting from a certain value $b = b_*$ for which $P = 0$, increases as the indenting force increases. This value is equal to

$$b_* = \frac{\lambda^{1/3}(1+\nu)^{1/3}(3-4\nu)^{2/3}}{a(1-\nu)} \quad \text{where} \quad \lambda = \frac{a(1-\nu^2)}{4\pi E}K_A^2. \tag{5.2.18}$$

The dimensionless parameter λ characterizes the role of adhesion in the process, which is obviously unstable when $0 < b < b_m = b_*/8$ because in this range the value of tearing force increases, but the size of the contact area also increases. The limiting tearing force $P = -P_m$ is attained at $b = b_*/8$ while the value of the force grows along the stable path when $b_* > b > b_*/8$:

$$64P_m = 16K_A\sqrt{2\pi b_*} - \pi a\mu b_*^2(1+\kappa^{-1}). \tag{5.2.19}$$

This analysis also holds qualitatively in the case of a material without a film if the adhesion of the material to the punch is significant.

5.3 A Half-Space Covered with an Inextensible Film

Let us consider an elastic half-space $z < 0$ with a boundary covered with an inextensible film subjected to the action of a normal concentrated force F. We have

$$z = 0: \quad u_x = u_y = 0, \quad \sigma_z = F\delta(x, y). \tag{5.3.1}$$

In this and next sections, u_x, u_y *and* u_z are the components of the displacement vector along the axes of the Cartesian coordinates x, y *and* z; σ_z is the normal stress, and $\delta(x, y)$ is the Dirac delta function. This problem is an analog of the classical Boussinesq problem for a half-space without a film.

The result of solving the boundary value problem of Eq. (5.3.1) can be represented as follows:

$$u_x = F\frac{xz}{\mu_* r^3}, \quad u_y = F\frac{yz}{\mu_* r^3}, \quad \text{and} \quad u_z = F\frac{r^2(3-4\nu)+z^2}{\mu_* r^3}$$
$$r^2 = x^2 + y^2 + z^2, \quad \mu_* = 8\pi\mu(1-\nu). \tag{5.3.2}$$

The strains and stresses are determined from here using the common equations of the theory of elasticity. The solution of Eq. (5.3.2) can serve as Green's function for an arbitrary distributed load.

We will now consider the problem of the interaction of two heavy masses m and M lying on the horizontal surface of the half-space $z < 0$ covered with an inextensible film. The size of the masses is assumed to be small compared with the distance between them, and the force of gravity is assumed to be directed vertically downwards. In this case, using Eq. (5.3.2) and the invariant integral of Eq. (5.1.11), we can find that an attractive force acts between these masses, which is equal to

$$F = \frac{(3-4\nu)mMg^2}{\mu_* R^2}. \tag{5.3.3}$$

Here, R is the distance between the masses, and g is the gravitational acceleration. This force is $(3-4\nu)g^2/[8\pi\mu G(1-\nu)]$ times greater than the Newtonian gravity force (G is the gravitational constant).

As a comparison, the attractive force between two heavy masses lying on the horizontal surface of an elastic half-space without a film is equal to

$$F = \frac{(1-\nu^2)mMg^2}{\pi E R^2}. \tag{5.3.4}$$

As seen, the film slightly, $4(1-\nu)^2/(3-4\nu)$ times, decreases the force.

5.4 A Ball Lying on a Membrane

We will now consider the contact problem of a heavy rigid sphere lying on a horizontal, flexible film (membrane) that is stretched by a tension γ uniformly in all directions; see Fig. 5.2a, b. Some bonding forces, characterized by an adhesion energy Γ_A, act between the sphere and film materials. The gravity force is directed downward. We will confine ourselves to the case when the contact area can be assumed to be a paraboloid of revolution: $w = r^2/(2R)$ where R is the sphere radius, w is the vertical displacement of the membrane, and r is the horizontal distance from the center of the circular contact area (the displacement is measured relative to the center).

The displacement of the membrane outside the contact area is obviously equal to:

$$w = C\ln\frac{r}{b} + \frac{b^2}{2R}, \quad r > b,\ R \gg b. \tag{5.4.1}$$

Here, b is the radius of the circular contact area, and C is a constant. The sphere experiences a pressure $p = 2\gamma/R$ exerted by the membrane that is constant everywhere in the contact area.

The driving force Γ of the active contact front per unit length of the front is equal to $\frac{1}{2}\gamma\left[(\partial w/\partial n)^2\right]$ at this contact front of single moment less shells, where the square brackets mean the discontinuity of the quantity in brackets, and see the theory of delamination of multilayered shells and plates in [2]. In this case, the contact front is at $r = b$, and $\partial w/\partial n$ is equal to dw/dr at $r = b+$ and b/R at $r = b-$, so that we have

$$\Gamma = \frac{1}{2}\gamma\left(\frac{dw}{dr} - \frac{b}{R}\right)^2 = \frac{1}{2}\gamma\left(\frac{C}{b} - \frac{b}{R}\right)^2. \tag{5.4.2}$$

Since $\Gamma = \Gamma_A$ (the adhesion energy of the ball-film materials), the equilibrium equation has the form

$$P = 2\pi\gamma b^2/R - 2\pi C\Gamma_A \quad \text{where} \quad C = b^2/R + b\sqrt{2\Gamma_A/\gamma}. \tag{5.4.3}$$

Here, P is the force of the ball pressure on the membrane.

When $\gamma > \Gamma_A$, the diagram $P = P(b)$ has the form shown qualitatively depicted in Fig. 5.2a, b. When $P = 0$, we have

$$b = b_* = R\sqrt{2}\frac{(\Gamma_A/\gamma)^3}{1 - (\Gamma_A/\gamma)} \quad \text{where} \quad \gamma > \Gamma_A. \tag{5.4.4}$$

In this position, the weight of the ball is balanced by the delamination force directed vertically upwards. The maximum delamination force that has to be applied to the sphere in order to separate it completely from the membrane is equal to the force

$$P = 3\pi\gamma R\frac{(\Gamma_A/\gamma)^3}{1 - (\Gamma_A/\gamma)} \quad \text{where} \quad \gamma > \Gamma_A, \tag{5.4.5}$$

which corresponds to the point of the minimum $b = b_m = b_*/2$ in the diagram.

For any $b > b_m$, the diagram $P = P(b)$ made using Eq. (5.4.3) uniquely determines the size of the contact area as a function of the applied force and, in particular, of the sphere weight. When $0 < b < b_m$, the equilibrium position is unstable.

In the case when $\gamma < \Gamma_A$, the force $P = P(b)$ decreases monotonically when b grows. Hence, any state in the diagram is unstable, and always $P < 0$.

In the special case of no adhesion when $\Gamma_A = 0$, $b = b_* = 0$, we have

$$b = \sqrt{(PR)/(2\pi\gamma)}, \quad C = b^2/R. \tag{5.4.6}$$

In this case, the well-known condition of smooth joining on the edge of the contact area follows from here.

The present analysis of this problem is also important for the design of optimal armored waistcoat as the first approximation of the impact of a bullet upon the waistcoat.

In the case of two heavy masses m and M lying on an horizontal membrane stretched from both sides by the tension, the following force of attraction F acts between the masses (which can be calculated using the invariant integral):

$$F = \frac{mMg^2}{2\pi\gamma R}. \tag{5.4.7}$$

The linear size of the masses is assumed to be small compared with the distance R between them. This law is similar to Ampere's Law of interaction of parallel linear currents of one and same direction.

5.5 Stick-and-Slip Problems

By any contact of two elastic bodies, there are always some contact areas where opposite parts glide one on the other (slip areas) and where they are fixed one to the other (stick areas). This is the case in all real contact problems. However, the stick-and-slip problems are exceptionally difficult and, therefore, their solutions are almost absent in the rich library of many thousand papers and books on contact problems. Needless to say that any contact is accompanied also by an adhesion due to the atomistic nature of all solids; however, adhesion was usually ignored, as well. Luckily, invariant integrals help us to determine the size and development of the contact areas in these complicated mixed problems.

Below, in Sects. 5.5–5.9 some basic stick-and-slip problems of contact mechanics are studied with account of the adhesion effect. Bearing this in mind, we will consider the following contact boundary value problem of the mathematical elasticity theory under plane strain, when one elastic material occupies the upper half-space, while the other occupies the lower half-space so that:

$$y = 0, \quad x \epsilon L_1: \quad \left(\sigma_y - i\tau_{xy}\right)_{z=x\pm 0} = p^{\pm}(x), \tag{5.5.1}$$

$$y = 0, \quad x \epsilon L_2: \quad [v] = 0, \tau_{xy} + f\sigma_y = \tau_s, \left[\sigma_y - i\tau_{xy}\right] = 0, \tag{5.5.2}$$

$$y = 0, \quad x \epsilon L_3: \quad [u + iv] = e(x) + iw(x), \left[\sigma_y - it_{xy}\right] = 0. \tag{5.5.3}$$

Here and below, f and τ_s are dry Coulomb friction constants on the slip areas L_2; $p^{\pm}(x)$ are complex vectors of the normal and shear loads on the opposite banks L_1 of the slit; and $e(x)$ and $w(x)$ are the distribution of the edge and wedge dislocations on the stick areas L_3 (they are equal to zero if the opposite banks are glued without any dislocations). The latter can also take into account the wear and tear of the opposite contact surfaces. The value of $[A] = A^+ - A^-$ denotes the difference of function $A(x, y)$ on the upper and lower banks of the slit.

In Eq. (5.5.2), it is assumed that in the slip area the normal stress σ_y is always negative, while the shear stress τ_{xy} is positive; if τ_{xy} is negative, we must put the minus sign in front of the constants f and τ_s . The position of points of discontinuity of boundary conditions is usually unknown beforehand and must be found in the process of solution.

The infinitely distant point is treated differently depending on whether the principal force and moment of the stresses σ_y and τ_{xy} at $y = 0$ are limited or unlimited. If they are limited, the common St. Venant principle holds. If they are unlimited, the St. Venant principle is not valid, and some special conditions must be used at infinity [1–11].

To solve the boundary value problem of Eqs. (5.5.1)–(5.5.3), we will use, instead of four functions of Kolosov–Muskhelishvili, only two functions $\Phi(z)$ and $\Omega(z)$ introduced by this author in 1962, see also Sect. 4.2. These functions are analytic over the whole $z-$ plane, with the exception of the $x-$ axis, where they undergo a discontinuity. The basic representations of the elasticity theory in terms of these functions have the following form:

in the upper half-plane where $\mathrm{Im} z > 0$:

$$\begin{aligned}
&\sigma_x + \sigma_y = 4Re\Phi(z), \quad (z = x + iy) \\
&\sigma_y - i\tau_{xy} = \Phi(z) + \Omega(\bar{z}) + (z - \bar{z})\overline{\Phi'(z)}, \\
&2\mu_1\left(\frac{\partial u}{\partial x} + i\frac{\partial v}{\partial x}\right) = \kappa_1\Phi(z) - \Omega(\bar{z}) - (z - \bar{z})\overline{\Phi'(z)};
\end{aligned} \tag{5.5.4}$$

in the lower half-plane where $\mathrm{Im} z < 0$:

$$\begin{aligned}
\sigma_x + \sigma_y &= 4\frac{\mu_1 + \kappa_1\mu_2}{\mu_1(1+\kappa_2)} Re\left\{\Phi(z) + \frac{\mu_1 - \mu_2}{\mu_1 + \kappa_1\mu_2}\Omega(z)\right\}, \\
\sigma_y - i\tau_{xy} &= \frac{\mu_1 + \kappa_1\mu_2}{\mu_1(1+\kappa_2)}\Phi(z) + \frac{\mu_1 - \mu_2}{\mu_1(1+\kappa_2)}\Omega(z) + \frac{\kappa_2\mu_1 + \mu_2}{\mu_1(1+\kappa_2)}\Omega(\bar{z}) \\
&+ \frac{\kappa_2\mu_1 - \kappa_1\mu_2}{\mu_1(1+\kappa_2)}\Phi(\bar{z}) + (z - \bar{z})\left\{\frac{\mu_1 + \kappa_1\mu_2}{\mu_1(1+\kappa_2)}\overline{\Phi'(z)} + \frac{\mu_1 - \mu_2}{\mu_1(1+\kappa_2)}\overline{\Omega'(z)}\right\}, \\
2\mu_2\left(\frac{\partial u}{\partial x} + i\frac{\partial v}{\partial x}\right) &= \kappa_2\frac{\mu_1 + \kappa_1\mu_2}{\mu_1(1+\kappa_2)}\Phi(z) + \kappa_2\frac{\mu_1 - \mu_2}{\mu_1(1+\kappa_2)}\Omega(z) - \frac{\kappa_2\mu_1 + \mu_2}{\mu_1(1+\kappa_2)}\Omega(\bar{z}) \\
&- \frac{\kappa_2\mu_1 - \kappa_1\mu_2}{\mu_1(1+\kappa_2)}\Phi(\bar{z}) - (z - \bar{z})\left\{\frac{\mu_1 + \kappa_1\mu_2}{\mu_1(1+\kappa_2)}\overline{\Phi'(z)} + \frac{\mu_1 - \mu_2}{\mu_1(1+\kappa_2)}\overline{\Omega'(z)}\right\}.
\end{aligned} \tag{5.5.5}$$

Here, μ and ν are the shear modulus and Poisson's ratio, and $\kappa = 3 - 4\nu$ for the plain strain. The subscripts 1 and 2 denote the upper and lower half-planes, respectively.

The functions $\Phi(z)$ and $\Omega(z)$ are expressed in terms of analytic functions $\varphi_1(z), \psi_1(z), \varphi_2(z)$ and $\psi_2(z)$ introduced by Kolosov and Muskhelishvili:

$$\Phi(z) = \varphi_1'(z) \quad \text{for} \quad \mathrm{Im} z > 0,$$
$$\Phi(z) = \frac{\mu_1(1+\kappa_2)}{\mu_1+\kappa_1\mu_2}\varphi_2'(z) + \frac{\mu_2-\mu_1}{\mu_1+\kappa_1\mu_2}\{\bar{\varphi}_1'(z) + z\bar{\varphi}_1''(z) + \bar{\psi}_1'(z)\} \text{ for } \mathrm{Im} z < 0,$$
$$\Omega(z) = \bar{\varphi}_1'(z) + z\bar{\varphi}_1''(z) + \bar{\psi}_1'(z) \quad \text{for} \quad \mathrm{Im} z < 0,$$
$$\Omega(z) = \frac{\mu_1(1+\kappa_2)}{\mu_2+\kappa_2\mu_1}\{\bar{\varphi}_2'(z) + z\bar{\varphi}_2''(z) + \bar{\psi}_2'(z)\} - \frac{\kappa_2\mu_1-\kappa_1\mu_2}{\mu_2+\kappa_2\mu_1}\varphi_1'(z) \quad \text{for} \quad \mathrm{Im} z > 0. \tag{5.5.6}$$

Conditions at infinity when Eq. (5.5.1) *hold at infinity.* Suppose all stresses vanish at infinity. In this case, the Kolosov–Muskhelishvili functions behave as follows:

in the upper half-plane

$$\varphi_1'(z) = -\frac{X-iY}{2\pi z} + \frac{Mi}{2\pi z^2}, \quad \psi_1'(z) = \frac{X+iY}{2\pi z} + \frac{Mi}{\pi z^2}, \quad (z \to \infty); \tag{5.5.7}$$

in the lower half-plane

$$\varphi_2'(z) = -\frac{X+iY}{2\pi z} - \frac{Mi}{2\pi z^2}, \quad \psi_2'(z) = \frac{X-iY}{2\pi z} - \frac{Mi}{\pi z^2}, \quad (z \to \infty). \tag{5.5.8}$$

Here, (X, Y) is the resultant force vector of stresses applied to the real axis of the lower half-plane, and M is the resultant moment of these stresses with respect to the coordinate origin (assumed negative for clockwise rotation).

According to Eqs. (5.5.6)–(5.5.8), the functions $\Phi(z)$ and $\Omega(z)$ undergo a discontinuity and behave at infinity as follows:

in the upper half-plane

$$\Phi(z) = -\frac{X-iY}{2\pi z} + \frac{Mi}{2\pi z^2} \quad \text{as} \quad Z \to \infty,$$
$$\Omega(z) = \frac{\mu_1(1+2\kappa_2)-\kappa_1\mu_2}{\mu_2+\kappa_2\mu_1}\frac{X}{2\pi z} + \frac{\mu_1+\kappa_1\mu_2}{\mu_2+\kappa_2\mu_1}\left(Y + \frac{M}{z}\right)\frac{i}{2\pi z} \quad \text{as} \quad Z \to \infty; \tag{5.5.9}$$

in the lower half-plane

$$\Phi(z) = \frac{\mu_2-\mu_1(2+\kappa_2)}{\mu_1+\kappa_1\mu_2}\frac{X}{2\pi z} - \frac{\mu_2+\kappa_2\mu_1}{\mu_1+\kappa_1\mu_2}\left(Y + \frac{M}{z}\right)\frac{i}{2\pi z} \quad \text{as} \quad Z \to \infty,$$

$$\Omega(z) = \frac{X - iY}{2\pi z} - \frac{Mi}{2\pi z^2} \quad \text{as} \quad Z \to \infty. \tag{5.5.10}$$

Conditions at infinity when Eq. (5.5.3) *hold at infinity.* In this case, let us assume that $e(x)$ and $w(x)$ disappear at infinity so that:

$$\begin{aligned} \Phi(z) &= B_0 - \frac{\mu_1}{\mu_1 + \kappa_1\mu_2}\frac{X + iY}{2\pi z} \quad \text{as} \quad Z \to \infty, \\ \Omega(z) &= B_0 + B' - \frac{\kappa_2\mu_1}{\mu_2 + \kappa_2\mu_1}\frac{X + iY}{2\pi z} \quad \text{as} \quad Z \to \infty, \end{aligned} \tag{5.5.11}$$

where

$$\begin{aligned} &(\mu_2 - \mu_1)\left(\bar{B}_0 - B'\right) = B_0(\kappa_1\mu_2 - \kappa_2\mu_1), \\ &B_0 = \frac{1}{4}(N_1 + N_2), \quad B' = -\frac{1}{2}(N_1 - N_2)exp(-2i\omega). \end{aligned}$$

Here, N_1 and N_2 are the principal stresses at infinity; ω is the angle made by the direction of the N_1- axis and the $x-$ axis, measured from the latter; and (X, Y) is the equivalent force vector at infinity. In this case, functions $\Phi(z)$ and $\Omega(z)$ are continuous at infinity.

Boundary value problem for one function. Now, let us notice that according to Eqs. (5.5.1)–(5.5.3), we have $\left[\sigma_y - i\tau_{xy}\right]$ known on the whole $x-$ axis. From this, we can derive that

$$F^+(z) - F^-(z) = \mu_1(1 + \kappa_2)[p(x)] \quad \text{when} \quad y = 0. \tag{5.5.12}$$

Here,

$$\begin{aligned} &F(z) = (\mu_1 + \kappa_1\mu_2)\Phi(z) - (\mu_2 + \kappa_2\mu_1)\Omega(z); \\ &[p(x)] = p^+(x) - p^-(x) \quad \text{for} \quad x \in L_1; [p(x)] = 0 forx \in L_2 + L_3. \end{aligned} \tag{5.5.13}$$

Using the Sokhotski–Plemelj equation, we can write

$$\Omega(z) = \alpha\,\Phi(z) - \beta P(z). \tag{5.5.14}$$

Here,

$$\alpha = \frac{\mu_1 + \kappa_1\mu_2}{\mu_2 + \kappa_2\mu_1}, \quad \beta = \frac{\mu_1(1 + \kappa_2)}{\mu_2 + \kappa_2\mu_1}, \quad P(z) = \frac{1}{2\pi i}\int\limits_{L_1}\frac{[p(x)]}{x - z}dx.$$

Substituting function $\Omega(z)$ in Eqs. (5.5.4) and (5.5.5) by Eq. (5.5.14), we derive the general representation of stresses and displacements in terms of the single analytical function $\Phi(z)$ as follows:

in the upper half-plane

$$\begin{aligned}
&\sigma_x + \sigma_y = 4Re\,\Phi(z),\\
&\sigma_y - i\tau_{xy} = \Phi(z) + \alpha\Phi(\bar{z}) + (z-\bar{z})\overline{\Phi'(z)} - \beta P(z),\\
&2\mu_1\left(\frac{\partial u}{\partial x} + i\frac{\partial v}{\partial x}\right) = \kappa_1\Phi(z) - \alpha\Phi(\bar{z}) - (z-\bar{z})\overline{\Phi'(z)} + \beta P(\bar{z});
\end{aligned} \tag{5.5.15}$$

in the lower half-plane

$$\begin{aligned}
&\sigma_x + \sigma_y = 4\alpha Re\,\Phi(z) - 4\gamma Re P(z),\\
&\sigma_y - i\tau_{xy} = \alpha\Phi(z) + \Phi(\bar{z}) + \alpha(z-\bar{z})\overline{\Phi'(z)} - Q_1,\\
&2\mu_2\left(\frac{\partial u}{\partial x} + i\frac{\partial v}{\partial x}\right) = \alpha\kappa_2\Phi(z) - \Phi(\bar{z}) - \alpha(z-\bar{z})\overline{\Phi'(z)} - Q_2.
\end{aligned} \tag{5.5.16}$$

Here,

$$\begin{aligned}
&\gamma = \frac{\mu_1 - \mu_2}{\mu_2 + \kappa_2\mu_1}, \quad Q_1 = \gamma P(z) + P(\bar{z}) + \gamma(z-\bar{z})\overline{P'(z)},\\
&Q_2 = P(\bar{z}) + \gamma\kappa_2 P(z) + \gamma\overline{P'(z)}.
\end{aligned}$$

To determine analytic function $\Phi(z)$, we will use the following boundary conditions, which remain unsatisfied:

$$\left(\sigma_y - i\tau_{xy}\right)^+ = p^+(x) \quad \text{as} \quad y = 0, \quad x \in L_1; \tag{5.5.17}$$

$$\left[\frac{\partial v}{\partial x}\right] = 0, \quad \tau_{xy} + f\sigma_y = \tau_s \quad \text{as} \quad y = 0, \quad x \in L_2; \tag{5.5.18}$$

$$\left[\frac{\partial u}{\partial x} + i\frac{\partial v}{\partial x}\right] = e'(x) + iw'(x) \quad \text{as} \quad y = 0,\ x \in L_3. \tag{5.5.19}$$

Using Eqs. (5.5.15) and (5.5.16), these conditions can be written as follows:

$$\begin{aligned}
&\Phi^+ + \alpha\,\Phi^- = p(x) \quad as \quad x \in L_1;\\
&\mathrm{Im}\left(\Phi^+ - \Phi^-\right) = 0,
\end{aligned} \tag{5.5.20}$$

$$Re\left\{(f+i)\Phi^+ + \alpha(f+i)\Phi^-\right\} = \tau(x) \quad \text{as} \quad x \in L_2; \tag{5.5.21}$$

$$\Phi^+ - \Phi^- = d(x) \quad \text{as} \quad x \in L_3. \tag{5.5.22}$$

Here,

$$p(x) = p^{+}(x) + \beta P^{-}(x),$$
$$\tau(x) = \tau_s + \beta f\, Re P(x) - \beta \mathrm{Im} P(x), \quad \left(P^{+} = P^{-} on L_2 + L_3\right)$$
$$d(x) = \frac{2\mu_1\mu_2}{\mu_1 + \kappa_1\mu_2}\left\{e'(x) + iw'(x) - \frac{2}{\mu_2}P(x) - \frac{\gamma}{\mu_2}\overline{P(x)}\right\}.$$

Equations (5.5.20)–(5.5.22) form a boundary value problem for one analytic function.

Mixed problems. The boundary value problems with different conditions on different parts of the boundary are called mixed problems. Most of these problems are unexplored, and it is usually unclear whether a solution of such a problem exists, and if does whether it is unique. The study of these problems starts from the local points of discontinuity of boundary conditions which carry all basic information about the properties of the whole solution.

Infinitely remote point. The mathematical infinity arises always when we concentrate our study on some detail; it is a pay for extra-curiosity. In the old theory of elasticity, it is usually treated by St. Venant principle that says that only the resultant force and moment of loads play the role far from the place of application of loads. However, this principle is wrong for a great number of problems in which the resultant force and moment of local loads play almost no role far from their application; in these problems, the stress–strain distribution is determined by geometry and special conditions at infinity that provide the infinite resultant force. Certainly, the latter has no physical sense; nevertheless, the solution of these divergent problems provides an accurate information about the local stresses and strains (e.g., sufficient to solve the problem of fracturing or instability). It is appropriate to note that such divergent problems are more common in real life; the St. Venant principle is valid only for some classical problems; see [1–11] for more detail.

5.6 The Singular Integral Equation of the Problem

Proceeding to the solution of the boundary value problem in Eqs. (5.5.20)–(5.5.22), we put

$$U(x) = [\mathrm{Re}\,\Phi(z)] = \mathrm{Re}\left(\Phi^{+} - \Phi^{-}\right) \text{ when } y = 0,\ x \in L_2. \tag{5.6.1}$$

Function $U(x)$ is identical with $[\partial u/\partial x] = \partial u^{+}/\partial x - \partial u^{-}/\partial x$ on L_2, apart from a factor (plus a certain known function).

From Eqs. (5.6.1), (5.5.21), and (5.5.22), it follows that

$$\Phi^{+} - \Phi^{-} = U_1(x) \text{ when } y = 0, \quad x \in L_2 + L_3. \tag{5.6.2}$$

Here,

$$U_1 = U(x) \quad for \quad x \in L_2, \text{and} \quad U_1 = d(x) \quad for \quad x \in L_3. \tag{5.6.3}$$

Following Gakhov [12], we introduce the canonical function $X(z)$ of the boundary value problem of Eq. (5.5.20)

$$X^+(x) + \alpha X^-(x) = 0 \quad \text{when} \quad y = 0, \ x \in L_1. \tag{5.6.4}$$

Here, $X(z)$ is a function, analytic outside cuts along L_1 and having an integrable singularity at the ends of n line segments of L_1 along $a_{k-1} < x < a_k$ where $k =$ 1, 3, 5, ..., 2n−1. Then, we can write the solution to this problem as a product of n pairs, each corresponding to one segment of the $x-$ axis:

$$X(z) = \Pi(z - a_{k-1})^{\delta}(z - a_k)^{\bar{\delta}} \quad \text{where} \quad \delta = -\frac{1}{2} + i\frac{\ln \alpha}{2\pi}. \tag{5.6.5}$$

In the simplest case, we have only one segment when $n = 1$.
Using Eq. (5.6.4), we can write Eq. (5.5.20) as follows:

$$\left(\frac{\Phi}{X}\right)^+ - \left(\frac{\Phi}{X}\right)^- = \frac{p(x)}{X^+(x)} \quad \text{where} \quad y = 0, x \in L_1. \tag{5.6.6}$$

By Sokhotski's formula, the general solution of the boundary value problem of Eq. (5.6.6) bounded at infinity has the following form:

$$\Phi(z) = X(z)\{\Phi_1(z) + Q(z)\}, \tag{5.6.7}$$

$$\Phi_1^+ - \Phi_1^- = 0 \quad \text{when} \quad y = 0, \ x \in L_1, \tag{5.6.8}$$

$$Q(z) = \frac{1}{2\pi i}\int\limits_{L_1} \frac{p(x)dx}{X^+(x)(x - z)} + P_n(z). \tag{5.6.9}$$

Here, $P_n(z)$ is a polynomial of degree n.
Based on Eqs. (5.6.7), (5.6.2), and (5.5.22), function $\Phi_1(z)$, analytic on L_1 according to Eq. (5.6.8), has the following discontinuity on $L_2 + L_3$

$$\Phi_1^+ - \Phi_1^- = \frac{U(x)}{X(x)} \quad (x \in L_2); \quad \Phi_1^+ - \Phi_1^- = \frac{d(x)}{X(x)} \quad (x \in L_3). \tag{5.6.10}$$

The solution of the boundary value problem of Eq. (5.6.10), which vanishes at infinity, has the form

$$\Phi_1(z) = \frac{1}{2\pi i}\int\limits_{L_2} \frac{U(x)dx}{X(x)(x - z)} + \frac{1}{2\pi i}\int\limits_{L_3} \frac{d(x)dx}{X(x)(x - z)}. \tag{5.6.11}$$

Using Eqs. (5.6.7) and (5.6.11), the second boundary condition of Eq. (5.5.21) on the slip areas can be reduced to the following form:

$$(1-\alpha)a(x)U(x) + (1+\alpha)\int_{L_2} K(t,x)\frac{U(t)}{t-x}dt = b(x) \quad (t, x \in L_2). \tag{5.6.12}$$

Here,

$$a(x) = Re\{(f+i)X(x)\}, \quad K(t,x) = \frac{1}{\pi}\mathrm{Im}\left\{(f+i)\frac{X(x)}{X(t)}\right\},$$

$$b(x) = 2\tau(x) - \frac{1}{\pi}(1+\alpha)\int_{L_3} \mathrm{Im}\left\{(f+i)\frac{d(t)X(x)}{X(t)}\right\}\frac{dt}{t-x}.$$

This is a classical singular integral equation; numerical methods of its solutions have been well studied.

In Eq. (5.6.12), some additional boundary conditions are used at the ends of the slip areas on L_2. In the simplest cases, they lead to the requirement that function $U(x)$ should be bounded or integrable at these points.

5.7 The Closed Solution for Certain Pairs of Materials

In the case when $\alpha = 1$, the initial boundary value problem and the singular integral equation have a closed analytic solution. In this case, according to Eq. (5.5.14) we have for the plane strain

$$\frac{\mu_1}{\mu_2} = \frac{1-2\nu_1}{1-2\nu_2}, \quad \left(0 < \frac{\mu_1}{\mu_2} < \infty\right). \tag{5.7.1}$$

This equality holds not only for identical materials but for many other pairs. For example, it is sufficiently well satisfied for pairs of hardened rubber ($\nu_1 = 0.48$ and $\mu_1 = 7.5$ GPa) and structural steel ($\nu_2 = 0.29$ and $\mu_2 = 79$ GPa), and of silicate glass ($\nu_1 = 0.19$ and $\mu_1 = 32$ GPa) and the structural alloy Mg with 8.5%Al ($\nu_2 = 0.35$ and $\mu_2 = 16$ GPa), and for other pairs.

For the plane stress, the condition $\alpha = 1$ can be written as

$$\frac{\mu_1}{\mu_2} = \frac{1+\nu_2}{1+\nu_1}\frac{1-\nu_1}{1-\nu_2}, \quad \left(\frac{1}{3} < \frac{\mu_1}{\mu_2} < 3\right). \tag{5.7.2}$$

when $\alpha = 1$, at the end of interfacial cracks and also at the edge of the contact area of an elastic punch with another elastic body, the singularity $1/2$, typical for homogeneous materials, is preserved even under stick conditions.

For clarity, we provide the direct solution of the boundary value problem, Eqs. (5.5.20)–(5.5.22), when $\alpha = 1$. Using this solution, we can establish also the exact analytic solution of the corresponding class of singular integral equations; see Eq. (5.6.12).

Let us introduce a new function $\Gamma(z)$ which is analytic where the function $\Phi(z)$ is analytic:

$$\Gamma(z) = \overline{\Phi(\bar{z})} = \bar{\Phi}(z). \tag{5.7.3}$$

And so, on the $x-$ axis when $y = 0$, we have:

$$\Gamma^{+} = \overline{\Phi^{-}}, \qquad \Gamma^{-} = \overline{\Phi^{+}}. \tag{5.7.4}$$

Using this function, the boundary value problem, Eqs. (5.5.20)–(5.5.22), can be written as follows (when $\alpha = 1$):

$$\Phi^{+} + \Phi^{-} = p(x), \quad \Gamma^{+} + \Gamma^{-} = \overline{p(x)}; \quad (x \in L_1) \tag{5.7.5}$$

$$\Phi^{+} - \Phi^{-} + \Gamma^{+} - \Gamma^{-} = 0; \qquad (x \in L_2) \tag{5.7.6}$$

$$(f + i)\Phi^{+} + (f + i)\Phi^{-} + (f - i)\Gamma^{+} + (f - i)\Gamma^{-} = 2\tau(x); \, (x \in L_2) \tag{5.7.7}$$

$$\Phi^{+} - \Phi^{-} = d(x), \; \Gamma^{-} - \Gamma^{+} = \overline{d(x)}, \qquad (x \in L_3). \tag{5.7.8}$$

Hence, for the analytic functions $F(z)$ and $G(z)$ where

$$F(z) = \Phi(z) + \Gamma(z), \qquad G(z) = \Phi(z) + \frac{f - i}{f + i}\Gamma(z) \tag{5.7.9}$$

we obtain the following two Riemann boundary value problems for $F(z)$ and $G(z)$ that can be independently solved:

$$F^{+} + F^{-} = p(x) + \overline{p(x)}, \, G^{+} + G^{-} = \frac{f - i}{f + i}\overline{p(x)} + p(x); \, (x \in L_1) \tag{5.7.10}$$

$$F^{+} - F^{-} = 0, \qquad G^{+} + G^{-} = \frac{2}{f + i}\tau(x); \qquad (x \in L_2) \tag{5.7.11}$$

$$F^{+} - F^{-} = d(x) - \overline{d(x)}, \; G^{+} - G^{-} = d(x) - \frac{f - i}{f + i}\overline{d(x)}. \; (x \in L_3) \tag{5.7.12}$$

According to Eq. (5.7.9), the sought function $\Phi(z)$ is expressed in terms of the solution of these boundary value problems as follows:

$$\Phi(z) = \frac{1}{2}\{(1 + fi)F(z) + (1 - fi)G(z)\}. \tag{5.7.13}$$

The canonical functions for these boundary value problems are obviously different. The nodes of the Riemann problem for the function $F(z)$ are the points $z = a_k$ which are the ends of segments L_1 while the nodes of the Riemann problem for the function $G(z)$ are the points $z = b_k$ which are the ends of segments L_3. As a reminder, the $x-$ axis is the sum $L_1 + L_2 + L_3$.

The canonical functions $X_F(z)$ and $X_G(z)$ of the Riemann problems for $F(z)$ and $G(z)$ are the following products:

$$X_F(z) = (z - a_1)^{-1/2}(z - a_2)^{-1/2} \ldots (z - a_n)^{-1/2}; \tag{5.7.14}$$

$$X_G(z) = (z - b_1)^{-1/2}(z - b_2)^{-1/2} \ldots (z - b_m)^{-1/2}. \tag{5.7.15}$$

The number n in Eq. (5.7.14) is even, if the original boundary value problem of Eqs. (5.7.1)–(5.7.3) belongs to class S when its solution satisfies Saint-Venant's principle. The number n is odd, if the boundary value problem of Eqs. (5.7.1)–(5.7.3) belongs to class N in which Saint-Venant's principle is not valid.

The same assertion holds for Eq. (5.7.15). As a reminder, in the problems of class N the resultant force and moment at infinity are infinite. It has been proven [2] that the set of problems of class N in the theory of elasticity is equivalent to the set of problems of class S. In plain terms, the number of problems in which Saint-Venant's principle is not valid is not less than the number of problems in which this principle is valid.

Using the canonical functions $X_F(z)$ and $X_G(z)$, the closed solution of the Riemann problems of Eqs. (5.7.10)–(5.7.12) for $F(z)$ and $G(z)$ is given by Gakhov's equations. This is the method of a closed solution for the general boundary value problem of Eqs. (5.7.1)–(5.7.3) when $\alpha = 1$.

5.8 The Problem of a Flat Punch

To understand the development of slip areas and adhesion effect, let us start from the simplest problems. For this purpose, we will confine ourselves by the problem of one absolutely rigid wheel/punch which starts rolling over the free surface of the elastic half-plane $y \leq 0$ under plane-strain conditions when Poisson's ratio of the material is equal to ½. In this section, the curvature radius of the punch is assumed to be very large so that we will call it flat.

In this case, the stresses and displacements in the lower half-plane can be represented in terms of the single function $\Phi(z)$ which is analytic everywhere in the whole $z-$ plane outside the cut along the real axis representing the contact area:

$$\sigma_x + \sigma_y = 4Re\,\Phi(z), \quad \sigma_y - i\tau_{xy} = \Phi(z) - \Phi(\bar{z}) + (z - \bar{z})\overline{\Phi'(z)} \tag{5.8.1}$$

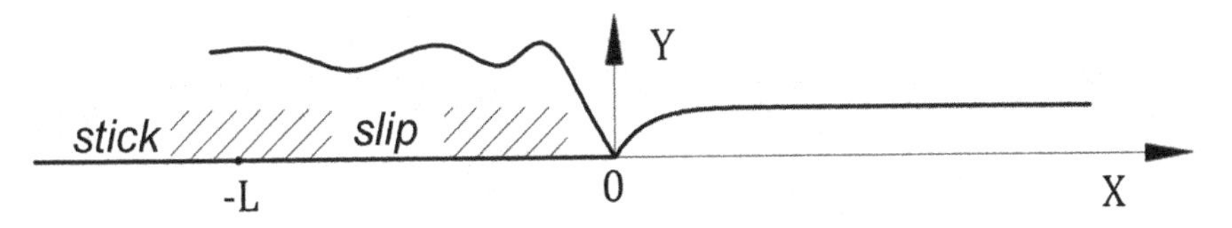

Fig. 5.3 A local slip zone at the edge of a punch

$$2\mu\left(\frac{\partial u}{\partial x}+i\frac{\partial v}{\partial x}\right)=\Phi(z)+\Phi(\bar{z})-(z-\bar{z})\overline{\Phi'(z)}. \tag{5.8.2}$$

The free boundary conditions outside the contact area are satisfied automatically. This function $\Phi(z)$ was introduced by N. I. Muskhelishvili; it differs from the author's function used in previous sections.

The corner edge of the punch. Suppose a semi-infinite flat punch is situated at $y=0,\ x<0$, while the remaining part of the surface of the lower half-plane is stress free. This problem describes the field of the stresses and strains in the small region of the corner edge of any punch. The corner point of the punch is at the coordinate origin, while the slip area is nearby at $y=0,\ 0>x>-l$. The stick-and-slip boundary conditions have the following form (Fig. 5.3):

$$\frac{\partial v}{\partial x}=0,\ \tau_{xy}+f\sigma_y=\tau_s,\quad (y=0, 0>x>-l) \tag{5.8.3}$$

$$\frac{\partial u}{\partial x}+i\frac{\partial v}{\partial x}=0,\qquad (y=0, x<-l) \tag{5.8.4}$$

And so, at $x<-l$ we get the stick and, at $0>x>-l$, slip conditions.

This boundary value problem belongs to class N where Saint-Venant's principle is not valid; the resultant force and moment of stresses at infinity are infinite. In this case the condition at infinity is written as follows:

$$\Phi(z)=-\frac{K_T+iK_N}{\sqrt{z}}+o\left(z^{-1/2}\right). \tag{5.8.5}$$

Here, $\sqrt{z}$ is analytic outside the cut along $y=0, 0>x>-\infty$ and positive when $y=0, x>0$; K_T and K_N are some parameters, which are determined from the solution of problems for punches of finite size (the sign of K_T and K_N is identical with the sign of the shear and normal stresses on the stick area as $x\to-\infty$).

According to Eqs. (5.8.1), (5.8.2), (5.7.3), and (5.7.4), the boundary value problem of Eqs. (5.8.3) and (5.8.4) can be written as

$$\begin{gathered}\Phi^{+}+\Phi^{-}-0,\quad \Gamma^{+}+\Gamma^{-}=0;\quad (y=0, x<-l)\\ (\Phi-\Gamma)^{+}=-(\Phi-\Gamma)^{-},\\ [\Phi(1-if)+\Gamma(1+if)]=2i\tau_s\cdot(y=0,-l<x<0)\end{gathered} \tag{5.8.6}$$

From here, it follows that $(\Phi - \Gamma)^+ = -(\Phi - \Gamma)^-$ on the whole semi-axis $x < 0$. Since $\Phi(z) = \bar{\Phi}(z)$, we get its solution using Eq. (5.8.5) at infinity:

$$\Phi(z) - \Gamma(z) = -\frac{2iK_N}{\sqrt{z}}. \tag{5.8.7}$$

Using Eqs. (5.8.6) and (5.8.7), we obtain the following boundary value problem:

$$\begin{aligned} &\Phi^+ = -\Phi^-, \quad (y = 0, x < -l); \\ &\Phi^+ - \Phi^- = i\tau_s - \frac{2K_N(1+if)}{\sqrt{|x|}}, \quad (y = 0, 0 > x > -l) \end{aligned} \tag{5.8.8}$$

The solution of this problem, which satisfies Eq. (5.8.5) at infinity, has the form

$$\begin{aligned} \Phi(z) &= -\frac{K_T + iK_N}{\sqrt{z+l}} - \frac{1}{2\pi i\sqrt{z+l}} \int_0^l \left(i\tau_s - 2K_N \frac{1+if}{\sqrt{r}} \right) \frac{\sqrt{l-r}}{r+z} dr \\ &= \frac{1}{\sqrt{z+l}} \left(-K_T - fK_N + \frac{1}{\pi}\tau_s\sqrt{l} \right) + K_N \frac{f-i}{\sqrt{z}} - \frac{\tau_s}{2\pi} ln \frac{\sqrt{l+z}+\sqrt{l}}{\sqrt{l+z}-\sqrt{l}}. \end{aligned} \tag{5.8.9}$$

Hence, this slip area along $-l < x < 0$ is a typical slip fracture, studied in the theory of adhesion of materials (see [2], pp. 363–368). The development of a slip fracture is determined by the shear stress intensity factor K_{II} on its growing edge $z = -l$; based on Eq. (5.8.9), it is equal to

$$K_{II} = 4\sqrt{2\pi} \left(K_T + fK_N - \frac{1}{\pi}\tau_s\sqrt{l} \right). \tag{5.8.10}$$

Since the open-mode stress intensity factor at the edge of any slip fracture is always equal to zero, the local slip condition has the form ([2], p. 366)

$$K_{II} = K_{IIC}. \tag{5.8.11}$$

Here, K_{IIC} is the slip toughness, characterizing the strength of adhesion of two materials (in this case the materials of the punch and the half-plane). It is related to their specific adhesion energy Γ_c for slip as follows ([2], p. 368) :

$$\Gamma_c = \frac{\kappa^3}{4\mu(1+\kappa)} K_{IIC}^2. \tag{5.8.12}$$

By Eqs. (5.8.10) and (5.8.11), the size of the slip area under the punch is given by the formula

$$\sqrt{l} = \frac{\pi}{\tau_s}\left(K_T + fK_N - \frac{1}{4\sqrt{2\pi}}K_{IIC}\right). \tag{5.8.13}$$

Suppose the punch starts moving in the direction of the $x-$ axis. A flat punch can be considered as a circular wheel of extremely large radius rolling over the elastic half-plane, since the stick of a roller at least over some part of the contact area is a necessary condition for rolling.

And so, the current problem can also describe the field near the edge of a roller so that the point $z = 0$ will be the leading edge of the roller, while the parameter K_T will be positive. Since K_N is always negative in contact problems, from Eq. (5.8.13) it follows that the slip area comes out when the parameter K_T reaches the critical value K_{TC}

$$K_{TC} = -fK_N + \frac{1}{4\sqrt{2\pi}}K_{IIC}. \tag{5.8.14}$$

After that, the size of the slip area increases proportionally to $(K_T - K_{TC})^2$.

Suppose now that the punch starts moving in a direction opposite to the direction of the $x-$ axis. In this case, the corner point $z = 0$ will be the trailing edge of the punch, and the parameter K_T will be negative. Since its sign is the same as the sign of the shear stress on the contact area, the slip condition on the slip area should be written in the form $\tau_{xy} - f\sigma_y = -\tau_s$ so that in Eqs. (5.8.9)–(5.8.13) we must change the sign in front of f, τ_s and K_{IIC}. As a result, Eqs. (5.8.13) and (5.8.14), and also the conclusion about the origination and growth of the slip area, are retained for the absolute value of K_T .

It should be noted that, when the punch moves in the direction of the $x-$ axis, a frontal resistance force acts on the leading edge of the punch, which is determined by the term $K_N(f - i)/\sqrt{z}$ in Eq. (5.8.9). From here, it follows that the stress intensity factors K_I and K_{II} on the leading edge $z = 0$ are equal to

$$K_I = 2\sqrt{2\pi}K_N, \quad K_{II} = 2f\sqrt{2\pi}K_N. \tag{5.8.15}$$

Hence, the frontal resistance force Γ_R acting on the leading edge of the punch is equal to

$$\Gamma_R = \frac{1}{8\mu}\left(K_I^2 + K_{II}^2\right) = \frac{\pi}{\mu}(1 + f^2)K_N^2. \tag{5.8.16}$$

This force is close to the energy dissipation spent per unit length of the path of the punch. The resultant force of normal stresses applied to the semi-infinite punch is infinite as it is common for the problems of class N .

As a result, the solution of the problem of the initial slip at the corner point of an arbitrary punch under loading conditions is determined by the following function:

$$\Phi(z) = \frac{K_T + iK_N}{\sqrt{z}} \quad \text{when} \quad 0 < K_T < K_{TC}; \tag{5.8.17}$$

$$\Phi(z) = -\frac{K_{IIC}}{4\sqrt{2\pi(z+l)}} + K_N\frac{f-i}{\sqrt{z}} - \frac{\tau_s}{2\pi}ln\frac{\sqrt{z+l}+\sqrt{l}}{\sqrt{z+l}-\sqrt{l}} \quad \text{when} \quad K_T > K_{TC}. \tag{5.8.18}$$

Here, l is determined by Eq. (5.8.13).

Now, let us assume that, when $K_T > K_{TC}$, beginning from a certain K_T, unloading occurs; i.e., K_T is reduced in value. The value of l initially remains the same as before, but the field is reorganized in accordance with Eq. (5.8.18), in which the quantity K_{IIC} is replaced by the decreasing quantity K_{II} given by Eq. (5.8.10). When K_{II} vanishes, l decreases, and the field is determined by Eqs. (5.8.13) and (5.8.18) at $K_{IIC} = 0$.

Hence, the condition of "finiteness of the stresses," usually assumed in contact problems, can hold only for some unloading conditions or provided that $K_{IIC} \ll \tau_s\sqrt{l}$. These conditions determine the limits of applicability of the solutions of the classical contact problem, in which the adhesion forces between the punch and the base are ignored.

This solution given for a semi-infinite punch describes asymptotically the beginning of the growth of the slip area from any corner edge of any punch.

A flat punch of finite dimensions. Let us study the classical problem of a rigid flat punch which starts moving on the boundary $y = 0$ of a lower elastic half-plane in the direction of the $x-$ axis. Suppose the edges of the punch at $x = \pm a$ are its corner points, and there are two slip areas at $a - l < |x| < a$ and a stick area at $|x| < a - l$ so that we have:

$$\begin{aligned} &\frac{\partial u}{\partial x} + i\frac{\partial v}{\partial x} = 0 \quad \text{when} \quad |x| < a - l,\, y = 0; \\ &\frac{\partial v}{\partial x} = 0,\, \tau_{xy} + f\sigma_y = \tau_s \quad \text{when} \quad a - l < |x| < a,\, y = 0. \end{aligned} \tag{5.8.19}$$

Suppose that (X, Y) is the principal vector of the external forces applied to the punch. In this section, we will assume that the moment of these forces is ignorably small and that $\partial v/\partial x = 0$ in the whole process.

First, we will consider the onset of the process by which the slip areas develop near the corner points. At the beginning, the size l of the slip areas will be the same at both edges of the punch and small compared with a so that the elastic field far from the slip areas can be represented by the following function:

$$\Phi(z) = -\frac{X + iY}{2\pi\sqrt{z^2 - a^2}}, \quad \left(\sqrt{z^2 - a^2} \to z \;\; as \;\; z \to \infty\right). \tag{5.8.20}$$

Comparing this field at $z \to \pm a$ with the beginning of the process described by Eq. (5.8.5), we obtain the parameters K_T and K_N which characterize the slip areas near the edges of the punch:

$$\begin{aligned} &\text{the leading edge } z = +a: \quad K_T = \frac{X}{2\pi\sqrt{2a}}, \quad K_N = \frac{Y}{2\pi\sqrt{2a}}; \\ &\text{the trailing edge } z = -a: l\ K_T = -\frac{X}{2\pi\sqrt{2a}}, \quad K_N = \frac{Y}{2\pi\sqrt{2a}}. \end{aligned} \tag{5.8.21}$$

When loading, the slip areas occur after the condition of Eq. (5.8.14) is satisfied, and they increase provided that

$$|X| > -fY + \frac{1}{2}K_{IIC}\sqrt{\pi a}. \tag{5.8.22}$$

Set of Eqs. (5.8.9), (5.8.17), (5.8.18), and (5.8.21) provides the asymptotic solution of the boundary value problem of Eq. (5.8.19) for small slip areas when $l \ll a$.

We now turn to the general case of a finite punch with slip areas of any dimensions. Using the method of Sect. 5.7, we derive the solution of the boundary value problem of Eq. (5.8.19); as a result, we obtain

$$\begin{aligned} \Phi(z) &= \frac{1}{2\pi\sqrt{z^2-b^2}}\left\{-X - iY + \int\limits_{b<|x|<a}\left(\tau_s - \frac{Y(1+if)}{\pi\left(\sqrt{x^2-a^2}\right)^+}\right)\frac{\sqrt{x^2-b^2}}{x-z}dx\right\} \\ &= \frac{Y(f-i)}{2\pi\sqrt{x^2-a^2}} + \frac{1}{2\pi\sqrt{x^2-b^2}}\left(-X - fY + 2\tau_s\sqrt{a^2-b^2}\right) \\ &+ \frac{\tau_s}{2\pi}ln\frac{\sqrt{z^2-b^2}-\sqrt{a^2-b^2}}{\sqrt{z^2-b^2}+\sqrt{a^2-b^2}}. \end{aligned} \tag{5.8.23}$$

Here, $b = a - l$; $\sqrt{z^2-a^2}$ and $\sqrt{z^2-b^2} \to z as z \to \infty$.

For very small slip areas when $l \ll a$, Eq. (5.8.9) follows from Eq. (5.8.23).

According to Eqs. (5.8.5) and (5.8.23), the parameters K_T and K_N on the leading edge of the punch are equal to

$$K_T = -\frac{fY}{2\pi\sqrt{2a}}, \quad K_N = \frac{Y}{2\pi\sqrt{2a}}. \tag{5.8.24}$$

From here, using Eqs. (5.8.15) and (5.8.16) we find the stress intensity factors and the front resistance at the leading edge of the punch:

$$K_I = \frac{Y}{\sqrt{\pi a}}, \quad K_{II} = -\frac{fY}{\sqrt{\pi a}}, \Gamma_R = \frac{1+f^2}{8\pi a\mu}Y^2. \tag{5.8.25}$$

It can be shown that the opposite force of the same value is applied at the trailing edge of the punch.

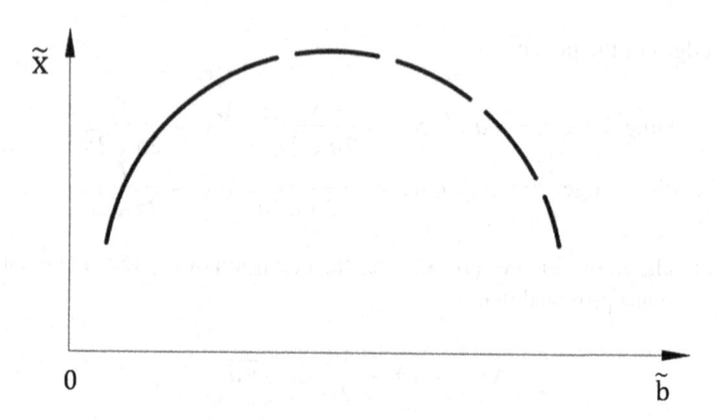

Fig. 5.4 Shear drag versus slip growth

Using Eq. (5.8.23) and the local slip condition of Eq. (5.8.11), we obtain the equation which determines the size of the slip area of any size

$$X + fY = 2\tau_s\sqrt{a^2 - b^2} + \frac{1}{2}K_{IIC}\sqrt{\pi b}. \tag{5.8.26}$$

The slip areas first occur when the condition of Eq. (5.8.22) is met.

Using dimensionless numbers, let us write Eq. (5.8.26) as follows (Fig. 5.4):

$$\breve{X} = -f\breve{Y} + \sqrt{1 - \breve{b}^2} + \lambda\sqrt{\breve{b}}. \tag{5.8.27}$$

Here,

$$\lambda = \frac{1}{4\tau_s}K_{IIC}\sqrt{\frac{\pi}{a}}, \quad \breve{X} = \frac{X}{2a\tau_s}, \quad \breve{Y} = \frac{Y}{2a\tau_s}, \quad \breve{b} = \frac{b}{a}. \tag{5.8.28}$$

On its physical meaning, the number λ can be called the brittleness number.

The function $\breve{X} = \breve{X}\left(\breve{b}\right)$ always has one maximum in the interval $(0, 1)$. It means that initially, as the shear force X increases, the slip areas grow stably. Then, on reaching a maximum, "pop-in" occurs and the slip areas rapidly grow in an unstable way, merging at the center. At this moment, the punch undergoes a jump in the direction of the $x-$ axis. And so, the present solution presents the theory of the "stick–slip" phenomenon, well-known from literary sources.

Let us designate the maximum value of the shear force X by X_m and the coordinate of the end of the right slip area at this instant by b_m . By Eq. (5.8.28), the brittleness number λ depends only on the ratio $\breve{b}_m = b_m/a$ and is equal to

$$\lambda = 2\breve{b}_m \sqrt{\frac{\breve{b}_m}{1-\left(\breve{b}_m\right)^2}}. \tag{5.8.29}$$

For example, for small values of b_m/a we have:

$$\lambda = 2(b_m/a)^{3/2} \tag{5.8.30}$$

In the opposite case, when b_m/a is close to 1, we have:

$$\lambda = \sqrt{2a/(a-b_m)}. \tag{5.8.31}$$

In the limiting cases, the maximum value of the shear force X is equal to X_p for the ideally plastic flow (yielding) and X_b for the ideally brittle fracturing so that:

$$\begin{aligned}
&\text{for } plastic\ flow \text{ as } \lambda \ll 1: X_p = -fY + 2a\tau_s, \breve{b}_m = (\lambda/2)^{2/3};\\
&\text{for } brittle\ \ fracture\ as\ \ \lambda \gg 1: X_b = -fY + \frac{1}{2}K_{IIC}\sqrt{\pi a}, \breve{b}_m = 1 - 2\lambda^{-2}.
\end{aligned} \tag{5.8.32}$$

In the case of an ideally brittle fracture, the slip cracks develop unstably from the very beginning, while in the case of an ideally plastic flow the slip lines from both ends develop stably until they merge at the center (Fig. 5.5).

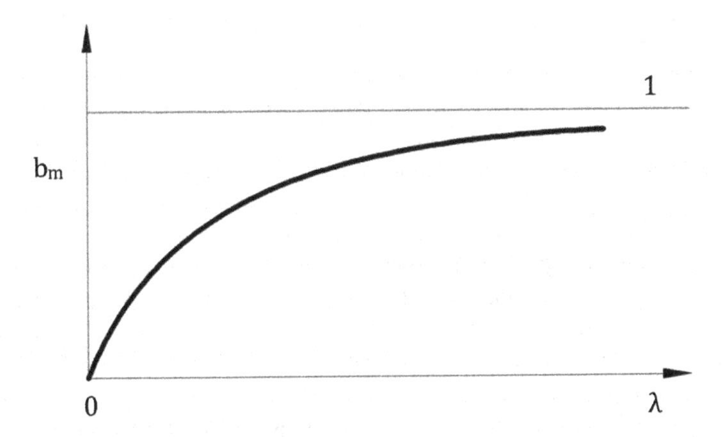

Fig. 5.5 From ideally brittle fracture to ideally plastic flow

We note also a simple relation:

$$\frac{X_b}{X_p} = \frac{\lambda - f\breve{Y}}{1 - f\breve{Y}}. \tag{5.8.33}$$

For any brittleness number, in order for the punch to shift and for merging to occur, the following force must be applied to the punch:

$$X_m = -fY + 2a\tau_s\left(\sqrt{1 - \left(\breve{b}_m\right)^2} + \lambda\sqrt{\breve{b}_m}\right) + \frac{1+f^2}{8\pi a\mu}Y^2. \tag{5.8.34}$$

Here, $\breve{b}_m$ is defined by Eq. (5.8.29) as a function of λ.

Formula of Eq. (5.8.34) provides the maximum resistance to the motion of the flat punch under "stick–slip" conditions.

5.9 Stick-and-Slip Effects While Rolling

In Chap. 3, we studied in detail the normal mode of rolling when no sliding, or stick, conditions hold over the whole contact area of a roller and a base. In this section, we consider other possible modes of rolling including the breakdown or slip mode, in which slip covers the whole contact area, and the most realistic stick-and-slip mode, in which slip covers some part of the contact area near its ends while stick conditions hold over the remaining contact area.

For more clarity, we will confine ourselves by the rolling of an absolutely rigid cylinder of radius R on an elastic half-plane under plane-strain conditions, and assume that Poisson's ratio of the material is equal to 0.5. In this section, we will again use Kolosov–Muskheishvili's function $\Phi(z)$ and Eqs. (5.8.1) and (5.8.2) for stresses and strains. As a reminder, $\Phi(z)$ is analytic in the whole z—plane outside the cut on the real axis corresponding to the contact area.

Rolling of a heavy cylinder of weight N occurs under the action of a tangential driving force T which is much less than N; both are applied to the center of the cylinder and counted per unit of its length. At rest, the vector (T, N) always eyes inside the contact area and at the edge of the contact area while rolling. We consider rolling as a result of monotonous increase of T at $N = \text{const}$.

Stick mode with account of adhesion and roughness of contact surfaces. Adhesion of the roller and base materials causes local tensile stresses near the trailing edge and increases local compressive stresses near the leading edge while the roughness of contact surfaces does an opposite effect. As a result, adhesion increases the size of contact area while roughness decreases it. These effects often can be ignored because adhesion and roughness act in opposite directions and often cancel one the other. It is only in the case of very smooth or very rough surfaces that they must be taken into account. Evidently, adhesion prevails for very smooth surfaces and roughness for very rough.

Let us estimate the effects of the adhesion and roughness using the solution of the contact problem singular at both leading and trailing edges of the contact area:

$$\Phi(z) = \frac{\mu i}{R}\left(z - \frac{2z^2 - a^2 - NR(\pi\mu)^{-1}}{2\sqrt{z^2 - a^2}}\right) + \frac{T}{2\pi\sqrt{z^2 - a^2}}. \tag{5.9.1}$$

According to Eq. (5.8.2), this function satisfies the boundary condition of the stick mode $\partial u/\partial x + i\partial v/\partial x = ix/R$ on the contact area $|x| < a$ and the condition at infinity $\Phi(z) = (T + iN)/(2\pi z)$ as $z \to \infty$.

The stresses on the contact area will be as follows:

$$\sigma_x = \sigma_y = \frac{\mu}{R}\frac{2x^2 - a^2 - NR(\pi\mu)^{-1}}{\sqrt{a^2 - x^2}}, \; \tau_{xy} = \frac{T}{\pi\sqrt{a^2 - x^2}}. \quad (y = 0, |x| < a) \tag{5.9.2}$$

when the rolling occurs, the vector (T, N) is directed to the leading end of the contact area and both ends of the contact area are in the limiting state so that

$$\sigma_y = \pm\sqrt{\frac{2\mu\Gamma_c}{\pi\varepsilon}} \quad \text{where} \quad \varepsilon = a - |x| \to 0, \; y = 0. \tag{5.9.3}$$

Here, Γ_c is the interfacial constant characterizing adhesion force between real rough surfaces of the cylinder and the foundation. The sign in Eq. (5.9.3) is positive when adhesion prevails over the resistance to roughness and negative when the effect of roughness is greater.

From Eqs. (5.9.2) and (5.9.3), it follows that

$$a^2 = \frac{R}{\pi\mu}\left(N \pm 2\sqrt{\pi a\mu\Gamma_c}\right). \tag{5.9.4}$$

Suppose the adhesion forces predominate. In this case, at the points $x = \pm x_*$ of the contact area where

$$x_* = \pm\frac{R}{\pi\mu}\left(N + \sqrt{\pi\mu a\Gamma_c}\right), \quad a^2 - x_*^2 = R\sqrt{a\Gamma_c/(\pi\mu)} \tag{5.9.5}$$

the stress σ_y is equal to zero. According to Eq. (5.9.5), we can neglect the adhesion forces and roughness only when the following condition is satisfied:

$$4R^2\Gamma_c \ll \pi\mu a^3. \tag{5.9.6}$$

Let us write Eq. (5.9.5) as follows:

$$\left(\frac{x_*}{R}\right)^2 = \left(\frac{a}{R}\right)^2 - \lambda^{1/2}\left(\frac{a}{R}\right)^{1/2} \quad \text{where} \quad \lambda = \frac{\Gamma_c}{\pi \mu R}. \tag{5.9.7}$$

From here, it follows that when $a^3 < \lambda R^3$ all contact area is under extension and tensile stresses, which describes the situation when a certain force balanced by the adhesion force and the weight of the cylinder is trying to pull it from the half-plane (Fig. 5.6). When $a^3 > \lambda R^3$, the x_* increases as a increases so that in the limit we have $x_* \to a$; see Eq. (5.9.6).

Let us determine the necessary condition for uniform rolling of a cylinder acted upon by a traction force T . By combining the equation of moments for rolling $TR = aN$ with Eq. (5.9.4) under the prevailing adhesion forces condition, we obtain

$$\bar{N} = m^2 - \sqrt{\lambda m}, \left(m = \frac{T}{N}, \bar{N} = \frac{N}{\pi \mu R}, \lambda = \frac{\Gamma_c}{\pi \mu R}\right) \tag{5.9.8}$$

The region to the left of this curve $\bar{N} = \bar{N}(m)$ in the $m\bar{N}$ plane corresponds to rest (no rolling), while the region to the right corresponds to accelerated motion of the cylinder by the traction force greater than the minimum value required to overcome the friction forces on the contact area (Fig. 5.7). When $m \le \lambda^{1/3}$ we will have $-\frac{3}{4}\left(\frac{\lambda}{2}\right)^{2/3} \le \bar{N} \le 0$; this part of the curve corresponds to inform rolling of a cylinder of weight N, suspended from an upper elastic half-plane and held to it by adhesion forces. In this region, two values of T correspond to each value of N, for the lesser of which equilibrium is unstable, while for the larger one it is stable. In this case, the maximum weight that can be held by the adhesion forces is

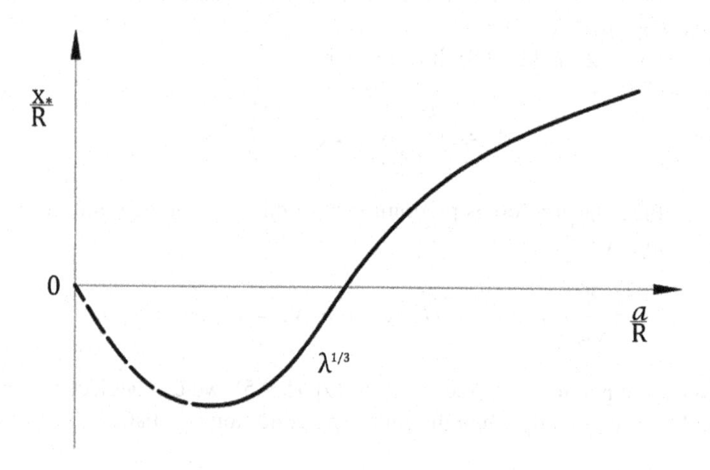

Fig. 5.6 Balance diagram of a cylinder on the elastic half-plane in the stick mode

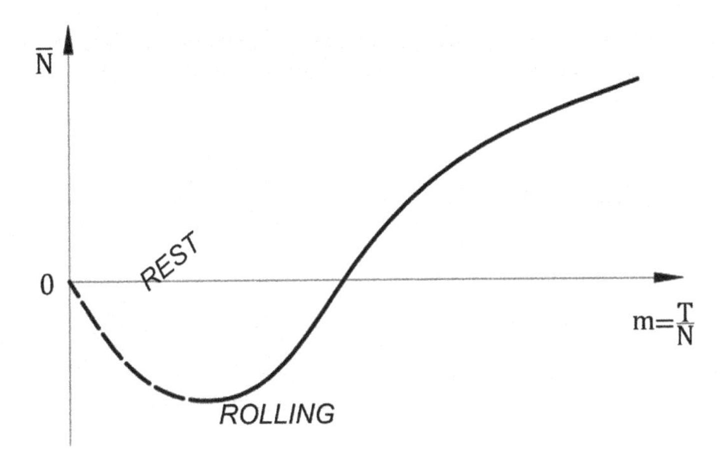

Fig. 5.7 Weight versus traction diagram of rolling in the stick mode

$N = \dfrac{3}{4}\pi\mu R(\lambda/2)^{2/3}$. In the region of positive $N > 0$, the value of N increases monotonically as m increases so that when $N \gg \pi\mu R(\lambda/2)^{2/3}$ the effect of the adhesion forces and the roughness can be neglected and we can assume that $\Gamma_c = 0$.

In a typical case of comparatively weak effects of adhesion and roughness, the solution of Eq. (5.9.4) can be written as

$$a^2 = \frac{R}{\pi\mu}\left(N \pm 2(\pi\mu RN)^{1/4}\Gamma_c^{1/2}\right). \tag{5.9.9}$$

In this case, the laws of rolling which take into account the effect of adhesion and roughness on both ends of the contact area are as follows:

$$\frac{M}{N} = \frac{TR}{N} = \sqrt{\frac{R}{\pi\mu}\left[N \pm 2(\pi\mu RN)^{1/4}\Gamma_c^{1/2}\right]}. \tag{5.9.10}$$

Also, we can conclude that the optimum conditions for rolling are the same conditions for which the effects of adhesion counteract those of roughness so that the equality $\Gamma_c = 0$ is valid. Besides, the effect of adhesion and roughness is less pronounced for the motion than in the state of rest.

Slip mode. Application and increase of the traction force are accompanied by the growth of shear stresses on the contact area and the development of local slip areas, until limits of the resistance forces on the contact area are reached, after which instantaneous development of slip areas occurs, which grip the whole contact area so that the cylinder skids and rolling friction sharply falls. Before turning to the general stick-and-slip mode, we will consider this limiting case of the skidding of a cylinder, when the slip covers the whole contact area.

Suppose Coulomb's Law of friction is valid on the whole contact area $|x| < a$, $y = 0$. It is easy to find the following classic solution of this problem for a heavy cylinder on the surface of an elastic half-plane:

$$\Phi(z) = \frac{\mu}{R}(i - f)\left(z - \sqrt{z^2 - a^2}\right) - \frac{\tau_s}{2\pi} ln \frac{z + a}{z - a}. \tag{5.9.11}$$

Here,

$$a^2 = \frac{NR}{\pi\mu}; \quad \text{when} \quad z \to \infty \sqrt{z^2 - a^2} \to z, \; ln \frac{z + a}{z - a} \to \frac{2a}{z}. \tag{5.9.12}$$

The stresses and displacements on the contact area are as follows:

$$\sigma_y = -\frac{\mu}{R}\sqrt{a^2 - x^2}, \quad \tau_{xy} = f\sigma_y + \tau_s, \quad \frac{\partial v}{\partial x} = \frac{x}{R}. \tag{5.9.13}$$

In this case, the laws of the uniform motion of a cylinder are simple:

$$T = fN + 2\tau_s\sqrt{\frac{NR}{\pi\mu}} \text{ for sliding;} \tag{5.9.14}$$

$$T = \frac{N^{3/2}}{\sqrt{\pi\mu R}} \text{ for rolling.} \tag{5.9.15}$$

These modes of motion are competing one with the other.

Evidently, in the process of increasing the traction force T the rolling starts first, if the following inequality is satisfied:

$$\sqrt{\frac{N}{\pi\mu R}} < f + 2\tau_s\sqrt{\frac{R}{\pi\mu N}}. \tag{5.9.16}$$

Otherwise, the sliding starts first which means the slip mode occurs characterizing the breakdown or emergency motion.

Let us write Eqs. (5.9.15) and (5.9.14) as follows (Fig. 5.8):

$$T_* = N_*^{3/2} \text{for rolling;} \quad T_* = fN_* + \tau_*\sqrt{N_*} \text{for sliding.} \tag{5.9.17}$$

Here,

$$T_* = \frac{T}{\pi\mu R}, \quad N_* = \frac{N}{\pi\mu R}, \quad \tau_* = \frac{2\tau_s}{\pi\mu}. \tag{5.9.18}$$

According to the graphs of Eq. (5.9.17) on the N_*T_* plane, the value of T_* for rolling is less than for sliding when $0 < N_* < N_T$ while it is greater than for sliding when $N_* > N_T$ where the threshold value of N_* is equal to

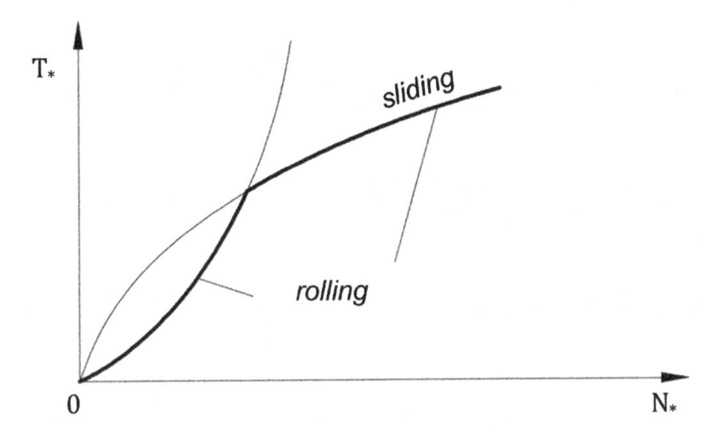

Fig. 5.8 Traction versus weight diagram of the slip mode of rolling

$$N_T = \frac{1}{4}\left(f + \sqrt{f^2 + 4\tau_*}\right)^2. \qquad (5.9.19)$$

From this, it follows that the breakdown or emergency state of sliding mode happens when the following inequality is satisfied

$$N_T < W(\pi n w \mu R)^{-1}. \qquad (5.9.20)$$

Here, W is the weight of a vehicle, n is the number of its wheels, and w is the width of the rut of the wheels.

In order to overcome the sliding mode, we should considerably increase the traction force T to shift up to the rolling mode on the $N_* T_*$ plane.

Stick-and-slip mode. The rolling in the stick-and-slip mode allows one to achieve the greatest traction force.

Let us first consider the development of small slip areas near the edges of the contact area. In the stick mode of the rolling, described in Chap. 3, the no-slip condition breaks down near the edges of the contact area, where slip areas start to develop from.

For small slip zones, matching the asymptotics of solutions in Eqs. (5.8.9) and (5.9.1) by Eqs. (5.8.5), (5.8.10)–(5.8.14), when $K_T > K_{TC}$, we obtain:

$$K_N = 0,\ K_T = \frac{T}{2\pi\sqrt{2a}},\ K_{TC} = \frac{K_{IIC}}{4\sqrt{2\pi}},\quad \sqrt{l} = \frac{1}{2\tau_s\sqrt{2\pi}}\left(\frac{T}{\sqrt{\pi a}} - \frac{K_{IIC}}{2}\right). \qquad (5.9.21)$$

Consequently, when $2T < K_{IIC}\sqrt{\pi a}$, slip areas will not occur near the edges of the contact area despite the infinitely high value of shear stresses (and the zero normal stresses), and when $2T > K_{IIC}\sqrt{\pi a}$, the slip areas will increase stably as T increases (Fig. 5.9).

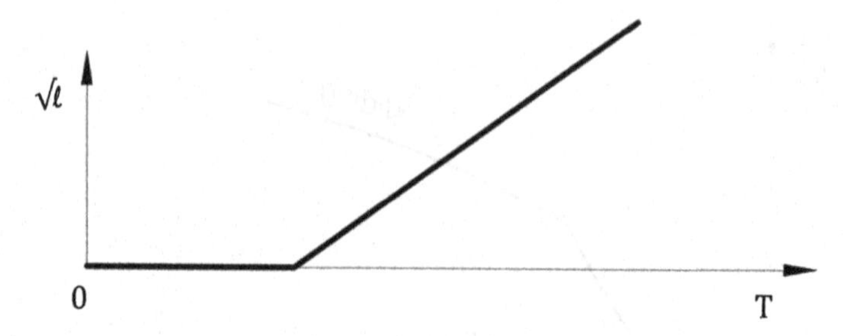

Fig. 5.9 Slip growth versus traction at the beginning of the slip

In the case of significant adhesion forces and roughness, according to Eqs. (5.9.1) and (5.9.9), we have

$$K_T = \frac{T}{2\pi\sqrt{2a}},\ K_N = \frac{\mu}{2R\sqrt{2a}}\left(a^2 - \frac{NR}{\pi\mu}\right),$$
$$\sqrt{l} = \frac{1}{2\tau_s\sqrt{2\pi}}\left(\frac{T}{\sqrt{\pi a}} + 2f\sqrt{\mu\Gamma_c} - \frac{1}{2}K_{IIC}\right). \tag{5.9.22}$$

We recall that in this case, Γ_c corresponds to the open mode while K_{IIC} corresponds to the shear mode. According to Eq. (5.9.22), no-slip areas occur so long as $2T < \sqrt{\pi a}\left(K_{IIC} - 4f\sqrt{\mu\Gamma_c}\right)$, and then they increase stably as T increases when $2T > \sqrt{\pi a}\left(K_{IIC} - 4f\sqrt{\mu\Gamma_c}\right)$.

As a reminder, all the previous equations correspond to a certain way of loading, namely initially by a normal force N and then by a shear force T for constant N. In the case of other loading ways, the corresponding equations can be different.

In the general case, we have the following boundary value problem:

$$\text{when } y = 0, |x| < b : \frac{\partial u}{\partial x} + i\frac{\partial v}{\partial x} = i\frac{x}{R}; \tag{5.9.23}$$

$$\text{when } y = 0, b < |x| < a : \quad \frac{\partial v}{\partial x} = \frac{x}{R},\ \tau_{xy} + f\sigma_y = \tau_s. \tag{5.9.24}$$

An external force $(T, -N)$ is applied to the cylinder center when $T > 0, N > 0$. It is required to determine the size of the stick area ($2b$) and the contact area ($2a$) and also the necessary condition for uniform rolling.

According to Eqs. (5.8.1) and (5.8.2), function $\Phi(z)$, analytic everywhere outside the cut $(-a, +a)$ of the $z-$ plane, by Eqs. (5.9.23) and (5.9.24), must be found from the following boundary value problem:

$$\text{when } y = 0, -b < x < b : \Phi^+ + \Phi^- = 2i\mu\frac{x}{R}; \tag{5.9.25}$$

when $y = 0, b < |x| < a :$

$$Im\left(\Phi^{+}+\Phi^{-}\right)=2\mu\frac{x}{R},\ Im\left(\Phi^{+}-\Phi^{-}\right)-fRe\left(\Phi^{+}-\Phi^{-}\right)=\tau_{s}. \tag{5.9.26}$$

This is a generalized Riemann boundary value problem. Using the method described above in Sect. 5.7, we arrive at the following solution of this problem in the class of integrable analytic functions:

$$\Phi(z)=i\frac{\mu}{R}z+\frac{\mu}{R}(f-i)\frac{z^{2}-\frac{1}{2}a^{2}-\frac{RN}{2\pi\mu}}{\sqrt{z^{2}-a^{2}}}+\frac{\tau_{s}}{2\pi}B(z)$$
$$-f\frac{\mu}{R}\frac{z^{2}-\frac{1}{2}b^{2}+\frac{R}{2\pi f\mu}(T-fN)-\frac{R\tau_{s}}{\pi f\mu}\sqrt{a^{2}-b^{2}}}{\sqrt{z^{2}-b^{2}}}. \tag{5.9.27}$$

Here,

$$B(z)=ln\frac{\sqrt{z^{2}-b^{2}}-\sqrt{a^{2}-b^{2}}}{\sqrt{z^{2}-b^{2}}+\sqrt{a^{2}-b^{2}}};\quad B(z)\to-2\frac{\sqrt{a^{2}-b^{2}}}{z}asz\to\infty;$$
$$\Phi(z)\to-\frac{T-iN}{2\pi z},\sqrt{z^{2}-a^{2}}\to z,\sqrt{z^{2}-b^{2}}\to zasz\to\infty.$$

The function $B(z)$ is analytic outside the cut $(-a,+a)$ of the $z-$ plane, where $B^{+}+B^{-}=0$ when $|x|<b$ and $\mathrm{Im}B(z)=\pm\pi i$ when $b<|x|<a$.

According to Eq. (5.9.27), the stresses on the contact area are as follows:

$$\sigma_{y}=\frac{\mu}{R}\frac{2x^{2}-a^{2}-\frac{RN}{\pi\mu}}{\sqrt{a^{2}-x^{2}}}\quad\text{when}\quad -a<x<+a; \tag{5.9.28}$$

$$\tau_{xy}=-\frac{f\mu}{R}\left(\frac{x^{2}-\frac{1}{2}a^{2}-\frac{RN}{2\pi\mu}}{\sqrt{a^{2}-x^{2}}}-\frac{x^{2}-\frac{1}{2}b^{2}+\frac{R}{2\pi f\mu}(T-fN)-\frac{R\tau_{s}}{\pi f\mu}\sqrt{a^{2}-b^{2}}}{\sqrt{b^{2}-x^{2}}}\right)$$
$$-\frac{\tau_{s}}{\pi}\theta\quad\text{when}\quad |x|<b;$$

$$\tau_{xy}=-\frac{f\mu}{R}\frac{x^{2}-\frac{1}{2}a^{2}-\frac{RN}{2\pi\mu}}{\sqrt{a^{2}-x^{2}}}+\tau_{s}\quad\text{when}\quad b<|x|<a;$$

$$\theta=\tan^{-1}\frac{2\sqrt{\left(a^{2}-b^{2}\right)\left(b^{2}-x^{2}\right)}}{x^{2}+a^{2}-2b^{2}}.$$

In accordance with this specified way of loading, when initially N increases at $T=0$, and then T increases at $N=$ const, the size of the contact area $2a$ is established in the first stage when $T=0$ and $b=a$. From the law of energy conservation, it follows that at this stage

$$\sigma_y = \pm 2\sqrt{\frac{\mu \Gamma_c}{2\pi \varepsilon}} \quad \text{when} \quad y = 0, x = a - \varepsilon, \ \varepsilon \ll a. \tag{5.9.29}$$

The plus sign corresponds to the prevailing effect of the adhesion forces, while the minus sign corresponds to the predominating effect of the roughness resistance. The value of Γ_c is equal to the elastic energy expended in forming unit square of the contact area.

Hence, using Eqs. (5.9.28) and (5.9.29), we obtain in the first stage

$$a^2 = \frac{RN}{\pi \mu} \pm 4R\sqrt{\frac{a\Gamma_c}{\pi \mu}}. \tag{5.9.30}$$

The application of force T, at the second stage, according to Eq. (5.9.28) gives rise to shear stresses on the contact area and, starting from a certain critical value of force T, to the development of a slip area, but the value of a remains unchanged in order to satisfy the non-penetration condition.

The following condition should be satisfied at the edge of the growing slip area

$$\tau_{xy} = \frac{K_{IIC}}{\sqrt{2\pi \varepsilon}} \quad \text{when} \quad y = 0, \ \ x = b - \varepsilon, \ \ \varepsilon \ll b. \tag{5.9.31}$$

Here, K_{IIC} is the slip toughness which is related to the specific adhesion energy by Eq. (5.8.12).

Hence, using Eqs. (5.9.28) and (5.9.31), we arrive at the following equation:

$$b^2 + \frac{R}{\pi f \mu}(T - fN) - \frac{2R\tau_s}{\pi f \mu}\sqrt{a^2 - b^2} = 2\frac{R}{f\mu}K_{IIC}\sqrt{\frac{b}{\mu}}. \tag{5.9.32}$$

Let us write down this equation in terms of six dimensionless variables:

$$T_* = fN_* - b_*^2 + \tau_*\sqrt{a_*^2 - b_*^2} + \lambda\sqrt{b_*}. \tag{5.9.33}$$

Here,

$$T_* = \frac{T}{\pi f \mu R}, N_* = \frac{N}{\pi f \mu R}, a_* = \frac{a}{R}, \lambda = \frac{2K_{IIC}}{f\mu\sqrt{\pi R}}, b_* = \frac{b}{R}, \tau_* = \frac{2\tau_s}{\pi f \mu}. \tag{5.9.34}$$

It can be shown that, for any positive λ and τ_*, there is always a single maximum $b_* = b_m$ of the function $T_* = T_*(b_*)$ in the range $0 < b_* < a_*$.

This maximum is the unique root $b_m = b_m(a_*, \tau_*, \lambda)$ of the following equation:

$$\frac{\lambda}{(b_m)^{3/2}} - \frac{2\tau_*}{\sqrt{(a_*)^2 - (b_m)^2}} = 4. \qquad (5.9.35)$$

Let us note two limiting cases when it is easy to find the roots of this equation:

$$b_m = \left(\frac{a_*}{4a_* + 2\tau_*}\right)^{2/3} \quad \text{when} \quad \tau_* \gg \lambda; \qquad (5.9.36)$$

$$b_m = a_* - 2\left(\frac{a_*\tau_*}{\lambda - 4(a_*)^{3/2}}\right)^2 \quad \text{when} \quad \lambda \gg \tau_*. \qquad (5.9.37)$$

The maximum possible value of the driving traction $T = T_m$ corresponding to the root of Eq. (5.9.35) is

$$T_m = \pi f \mu R\left[fN_* - (b_m)^2 + \tau_*\sqrt{(a_*)^2 - (b_m)^2} + \lambda(b_m)^{1/2}\right]. \qquad (5.9.38)$$

Hence, no-slip areas occur, i.e., $b = a$, when traction T increases from zero to $T = \pi f \mu R\left[fN_* - (a_*)^2 + \lambda\sqrt{a_*}\right]$. When T increases further, the slip areas develop stably, until the maximum is reached when $b_* = b_m$ and $T_* = T_m$. Then, the slip areas rapidly increase in the region of instability $0 < b_* < b_m$ and the loss of the driving force occurs when $T_* < T_m$. This is the same "stick–slip" phenomenon described in Sect. 5.8 for the plane punch. In the rolling problem considered here, this indicates skidding of the cylinder and almost no translational motion.

Literature

1. G.P. Cherepanov, The contact problem of the mathematical theory of elasticity with stick-and-slip areas. The theory of rolling and tribology. J. Appl. Math. Mech. (JAMM) **79**(1), 81–101 (2015)
2. G.P. Cherepanov, *Fracture Mechanics,* (Moscow-Izhevsk, IKI Publ., 2012), pp. 1–840
3. G.P. Cherepanov, Some new applications of the invariant integrals in mechanics. J. Appl. Math. Mech. (JAMM) **76**(5), 519–536 (2012)
4. G.P. Cherepanov, Singular solutions in the theory of elasticity. In: *Problems of Solid Mechanics,* (Leningrad, Sudpromgiz, 1970), pp 380–398
5. G. P. Cherepanov, Stresses in an inhomogeneous plate with cuts. *Izvestia Acad. Nauk SSSR, OTN Mekh. Mash.* **1,** 131–139 (1962)
6. G.P. Cherepanov, On propagation of cracks in compressed bodies. J. Appl. Math. Mech. (JAMM) **30**(1), 76–86 (1966)
7. G.P. Cherepanov, *Mechanics of Brittle Fracture* (McGraw-Hill, New York, 1978), pp. 1–952
8. G.P. Cherepanov (ed.), *Fracture. A Topical Encyclopedia of Current Knowledge,* (Malabar, Krieger, 1998), pp. 1–890
9. G.P. Cherepanov, *Methods of Fracture Mechanics: Solid Matter Physics* (Kluwer, Dordrecht, 1997), pp. 1–312

10. G.P. Cherepanov, Theory of rolling: Solution of the Coulomb problem. J. Appl. Mech. Tech. Phys. (JAMTP) **55**(1), 182–189 (2014)
11. G.P. Cherepanov, The laws of rolling. Phys. Mesomech. **21**(5), 435–451 (2018)
12. H. Hertz, Uber die Beruhrung Fester Elastischer Korper. Zeitschrift fur Reine Angew. Math. **92**, 156–180 (1882)
13. F.D. Gakhov, *Boundary Value Problems* (Dover, New York, 1990), pp. 1–490
14. L.A. Galin, *Contact Problems of Elasticity Theory* (Gostekhizdat, Moscow, 1953), pp. 1–280
15. N.I. Glagolev, The resistance to the rolling of cylindrical bodies. Prikl. Mat. Mekh. (PMM) **9**(4), 318–332 (1945)
16. N.I. Muskhelishvili, *Some Basic Problems of the Mathematical Theory of Elasticity* (Groningen, Amsterdam, 1958), pp. 1–720
17. M.A. Sadowsky, Zweidimensionale Probleme der Elastizitatstheorie. Zeitschrift fur Angewandte Math und Mech. **8**, 107–127 (1928)

Chapter 6
The Fracturing

Abstract This chapter concerns the phenomenon of leaks (gryphons) caused by the growth of undetected cavities and fissures filled up by oil/vapor, between different layers in laminated structures, acted upon by the pressure and temperature of the penetrated fluid. This unaccounted process caused many catastrophes, including the Chernobyl and Fukushima nuclear power stations' disasters, and the summer 2010 disaster in the Gulf of Mexico as a result of the breakthrough of gryphons in two boreholes. Therefore, a special attention is given to the development of casing string gryphons and cylindrical gryphon cracks. The problem solutions using invariant integrals are obtained in a closed, very effective shape. This chapter is for those who work or interested in oil and gas industry, and for those who design and operate nuclear power stations.

Application of invariant integrals to fracture mechanics began from this author's paper *Crack propagation in continua* published in *J. Appl. Math. Mech.* (*JAMM*), **31**(3), 1967. It enabled one to characterize the carrier of the fracture process which is the crack tip. In this chapter, some new, topical problems of fracture mechanics are studied using invariant integrals. They concern the problem of delamination, the problem of blisters, the problem of gryphons, and others.

6.1 The Delamination of a Film or Plate from a Substructure

At first, let us consider an inextensible, absolutely flexible film/membrane which is delaminated from a rigid foundation by a stretching force N. Let us assume conditions of plane deformation; see the cross section in Fig. 6.1, showing the delamination front. When the delamination front moves, the following equation is obviously valid:

$$N = \Gamma_c/(1 - \cos\beta) \quad \text{where } 0 < \beta \leq \pi. \qquad (6.1.1)$$

G. P. Cherepanov, *Invariant Integrals in Physics*,
https://doi.org/10.1007/978-3-030-28337-7_6

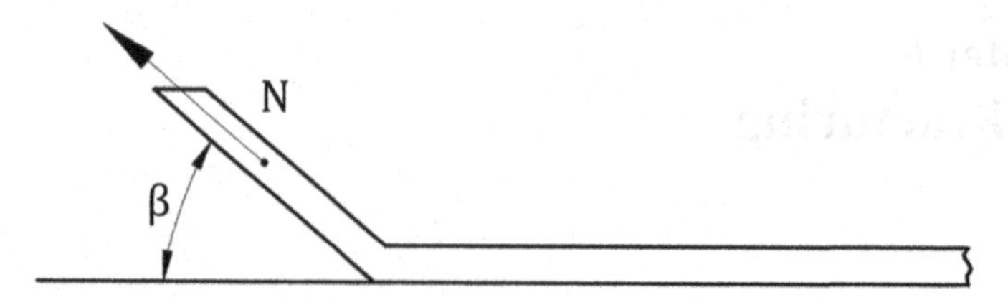

Fig. 6.1 Pulling-off a film from a solid

Here β is the angle between the foundation and the film, and Γ_c is the specific adhesion energy characterizing the strength of their bond. For $\beta \geq \pi/2$, this is mainly the open-mode fracturing.

Equation (6.1.1) follows directly from the law of the energy conservation for the inextensible, flexible film, since the length of the path traversed by the delamination front is $2\sin(\beta/2)$ times smaller than the path traversed by the end of the inextensible film to which force N is applied; this force deviates by angle $\pi/2 - \beta/2$ from the direction of the path traversed by this force.

Function $N = N(\beta)$ is monotonously decreasing when β increases so that:

$$N \to 2\Gamma_c/\beta^2 \text{ as } \beta \to 0; \quad N(\pi/2) = \Gamma_c; \quad N(\pi) = \Gamma_c/2. \tag{6.1.2}$$

The problem of delamination of thin films is very important for people and industries because everything we use is covered by either protective or decorative films, from skin on human bodies to special coatings on stealth jets, and from the glaze of porcelain to anticorrosive coatings of metals. Everything exists only because it is protected by something thin and almost imperceptible. The strength and durability of coatings determine the lifetime of almost any enclosure.

The latter two equations in Eq. (6.1.2) provide a simple but persuasive criterion for the estimate of the strength of any coating; this criterion can serve the force N necessary to tear off a coating from a substrate.

Now, let us discuss the effect of elasticity of a thin film that is delaminated by a force from a rigid foundation (Fig. 6.2). We confine ourselves by small deformations

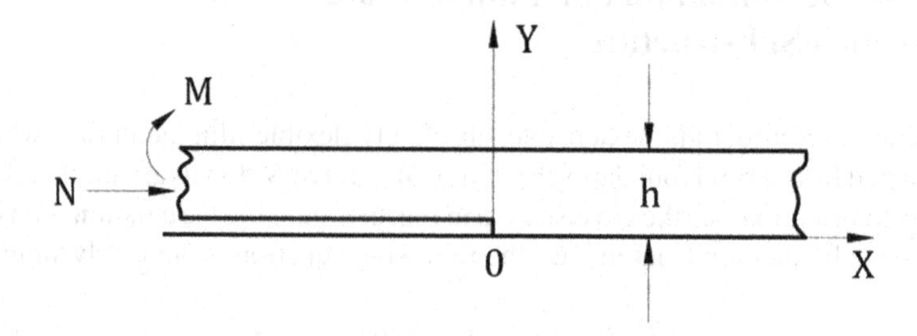

Fig. 6.2 Tearing-off a plate from a substrate

so that, as distinct from the previous problem, the film is an elastic beam or strip $0 < y < h$ of thickness h bonded to a rigid substrate at $y = 0, x > 0$. Boundary at $y = 0, x < 0$ and $y = h$ is free from tractions. At infinity, when $x \to -\infty$, force N and the bending moment M are applied (evidently, this force can have only the x-component).

And so, we have the following boundary value problem for the strip:

$$\sigma_y = \tau_{xy} = 0 \quad \text{when } y = 0, x < 0 \text{ and } y = h; \tag{6.1.3}$$

$$u = v = 0 \text{ or } v = 0, \ \tau_{xy} = 0 \quad \text{when } y = 0, \ x > 0. \tag{6.1.4}$$

When

$$x \to -\infty\colon \ \sigma_x = \frac{N}{h} - 12\frac{M}{h^3}\left(y - \frac{h}{2}\right), \ \sigma_y = \tau_{xy} = 0. \tag{6.1.5}$$

Evidently, in this case, force N produces the cracking of shear mode while the bending moment M brings about the open-mode crack.

This problem can be strictly solved by the Wiener–Hopf method, but we can elementarily find the main result using the following invariant integral:

$$\Gamma_x = \oint \left(Un_x - \sigma_{ij}n_j u_{i,x}\right)\mathrm{d}S. \quad (i, j = 1, 2) \tag{6.1.6}$$

Here the closed contour of integration is an infinitesimal circle encompassing the origin of coordinates $x = y = 0$.

Using the invariance of this integral, let us deform the contour of integration into the boundary of the strip $y = 0$ and $y = h$, and two segments at $x \to \pm\infty, \ 0 < y < h$. Because of the boundary conditions, Eqs. (6.1.3)–(6.1.5), the integral in Eq. (6.1.6) is not equal to zero only along the segment $x = \text{const} \to -\infty, 0 < y < h$.

And so, Eq. (6.1.6) is reduced to the following equation:

$$\Gamma_x = \int_0^h (\sigma_x \varepsilon_x - U)\mathrm{d}y = \frac{1}{2E}\int_0^h \sigma_x^2 \mathrm{d}y \quad \text{where } x \to -\infty. \tag{6.1.7}$$

Here $n_x = n_1 = -1, \ \sigma_x = \sigma_{11} = E\varepsilon_x, \ \varepsilon_x = u_{1,1}, \ 2U = \sigma_x\varepsilon_x$.

Using Eq. (6.1.5) for the asymptotic value of σ_x in Eq. (6.1.7), we easily derive:

$$\Gamma_x = \frac{N^2}{2Eh} + \frac{6M^2}{Eh^3}. \tag{6.1.8}$$

From this equation, we can also derive the values of the stress intensity factors of fracture mechanics describing the open mode (K_I) and the shear mode (K_{II}):

$$K_I = \frac{2\sqrt{3}}{h\sqrt{h}} M, \quad K_{II} = \frac{N}{\sqrt{2h}}. \tag{6.1.9}$$

Equation (6.1.8) was derived for the plane-stress condition, but it will be also valid for the plane-strain condition if we multiply by $1 - \nu^2$ the right-hand part of this equation. The equation for the shear mode, the second Eq. (6.1.9), found an application to the cutting of rocks and to the design of drilling bits in mining [1–3].

Evidently, the properties of a substructure/foundation can be essential, only if its compliancy is comparable to the compliancy of a film/coating.

Also, it is important to keep in mind that Eqs. (6.1.1) and (6.1.8) embrace different areas, namely Eq. (6.1.1) describes large deformations of very flexible, inextensible plates, while Eq. (6.1.8) relates only to small deformations of elastic but very solid plates.

Summary Generalizing the results obtained by Eqs. (6.1.1) and (6.1.8), we can conclude that, based on the principle of superposition and the principle of microscope, the basic invariant integral for small deformations of thin plates and shells obeys the following law:

$$\Gamma_x = \frac{1}{2} N\beta^2 + \frac{1}{2} k_e \varepsilon_e^2 + \frac{1}{2} k_b R^{-2}. \tag{6.1.10}$$

Here:

$$k_e = \frac{E}{h}, \quad \varepsilon_e = \frac{N}{E}, \quad k_b = \frac{1}{12} E h^3, \quad R^{-1} = 12M/\left(Eh^3\right). \tag{6.1.11}$$

The quantities k_e and k_b are the tensile and flexural stiffness, and ε_e and R^{-1} are the extension and curvature of the middle line of the plate near the crack front. The first and third terms in Eq. (6.1.10) describe the open mode while the second term the tearing mode. The first term is essential only when $N < \beta h E$ or $N \sim \beta h E$. Equation (6.1.11) is written for plane stress; in the case of plane strain, the value of E in these equations should be replaced by $E/\left(1 - \nu^2\right)$.

The first term in Eq. (6.1.10) describes the work of the external force spent to increase the free surface of the body per unit area, while the second and third terms describe the loss of the elastic energy due to the same increase of the free surface. Some applications of Eqs. (6.1.10) and (6.1.11) are brought about in next sections.

6.2 Two Example Problems of Delamination

Let us consider two elastic films 1 and 2 that have been glued together and are separated by forces N_1 and N_2 (Fig. 6.3). From the equilibrium condition, we have

$$N_1 \cos \beta_1 + N_2 \cos \beta_2 = N, \quad N_1 \sin \beta_1 = N_2 \sin \beta_2. \tag{6.2.1}$$

The notation is shown in Fig. 6.3. It is supposed that $\beta_1 + \beta_2 \leq \pi$.

This equilibrium is possible only until the limiting state of delamination is achieved which is, based on Eq. (6.1.1), characterized by the following equation:

$$\Gamma_c = N_1(1 - \cos \beta_1) + N_2(1 - \cos \beta_2). \tag{6.2.2}$$

Here the specific energy Γ_c describes the strength of adhesion of both films.

Solving the equation system, Eqs. (6.2.1) and (6.2.2), provides the limiting forces of delamination:

$$N_1 = \frac{\Gamma_c \cos \dfrac{\beta_2}{2}}{2 \sin \dfrac{\beta_1}{2} \sin \dfrac{\beta_1 + \beta_2}{2}}, \; N_2 = \frac{\Gamma_c \cos \dfrac{\beta_1}{2}}{2 \sin \dfrac{\beta_2}{2} \sin \dfrac{\beta_1 + \beta_2}{2}}, \quad N = N_1 + N_2 - \Gamma_c. \tag{6.2.3}$$

In the case of the symmetric splitting, when $\beta_1 = \beta_2 = \beta$, we have

$$N_1 = N_2 = \Gamma_c \left(2 \sin \frac{\beta}{2} \right)^{-2}, \quad N = 2N_1 - \Gamma_c. \tag{6.2.4}$$

This solution determines the local forces of delamination near the edge of a bubble of any shape between two glued, momentless membranes/films.

The Problem of Bubble Between Films Let us consider the problem of axisymmetric cavity/bubble between two glued films that lie in a horizontal plane and are

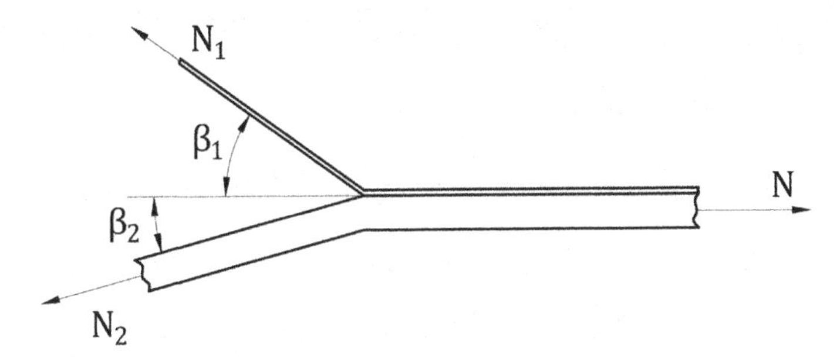

Fig. 6.3 Glued films stretched by forces

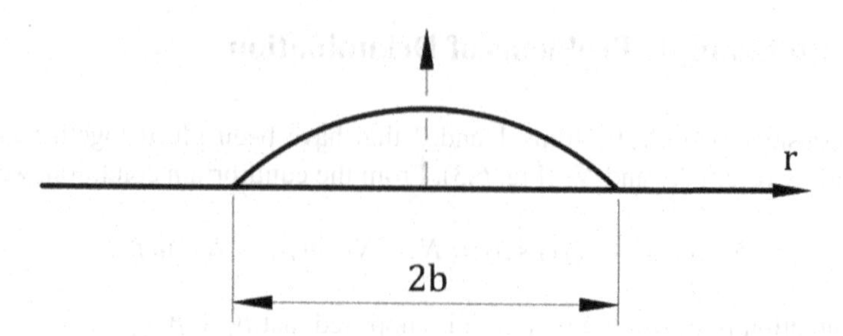

Fig. 6.4 A blister on the surface of a body

stretched by a force N in all horizontal directions. The bubble is assumed to be filled with a gas under a pressure p. This problem is of interest for some medical and biological applications.

The bubble is produced by the splitting of the film thickness. Because of the adhesion, the bubble takes the shape of two spherical segments characterized by the maximum height h, the curvature radius R, and the volume V (the segments are different in the case of asymmetric splitting of the film).

And so, for the upper, with subscript 1, and lower, with subscript 2, segments of the bubble, we have (Fig. 6.4):

$$R_1 = \frac{1}{2}h_1\left(1 + \frac{r^2}{h_1^2}\right), \quad V_1 = \frac{\pi}{6}h_1^3\left(1 + 3\frac{r^2}{h_1^2}\right), \quad N_1 = \frac{1}{2}pR_1; \tag{6.2.5}$$

$$R_2 = \frac{1}{2}h_2\left(1 + \frac{r^2}{h_2^2}\right), \quad V_2 = \frac{\pi}{6}h_2^3\left(1 + 3\frac{r^2}{h_2^2}\right), \quad N_2 = \frac{1}{2}pR_2. \tag{6.2.6}$$

$$p(V_1 + V_2) = R_G MT. \tag{6.2.7}$$

Here r is the radius of the common foundations of the upper and lower segments, M and T is the mass and absolute temperature of the gas in the bubble, and R_G is the gas constant.

Equations (6.2.3)–(6.2.7) allow us to predict the development of the bubble in terms of pressure, temperature, and mass of the gas injected into the cavity. For example, in the case of symmetric splitting when $h = h_1 = h_2$, $V_1 = V_2 = 0.5V$, $R_1 = R_2 = R$, and $N_1 = N_2 = N_0$, we have:

$$2N_0 = N + \Gamma_c = pR, \quad pV = R_G MT, \quad R = \frac{1}{2}h\left(1 + \frac{r^2}{h^2}\right), \quad V = \frac{\pi}{3}h^3\left(1 + 3\frac{r^2}{h^2}\right). \tag{6.2.8}$$

$$2\sin\frac{\beta}{2} = (\Gamma_c/N_0)^{1/2}. \quad (0 < \beta < \pi/2) \tag{6.2.9}$$

At the beginning of the development, when the initial cavity satisfies condition $r \gg h$, from Eqs. (6.2.8) and (6.2.9), it follows that:

$$r = \left(\frac{4}{\pi} R_G\right)^{1/4} \left(\frac{MTN_0}{p^2}\right)^{1/4}, \quad 2\sin\frac{\beta}{2} = \sqrt{\frac{\Gamma_c}{N_0}}, \quad 2N_0 = N + \Gamma_c. \qquad (6.2.10)$$

And so, the radius of the delamination cavity grows directly proportional to the one-fourth degree of MTN_0p^{-2} which is the only parameter it depends on. Evidently, if this law is valid at the beginning, it will be valid afterward while r increasing.

A Problem of Local Buckling Suppose a body covered by a coating is subject to some compression along its surface so that a portion of a delaminated coating can loose stability and be torn off. Let us model this problem assuming the plane-stress condition.

A coating layer of thickness h is split off from a half-plane on the length L, where $L \gg h$. And so, we come to the following boundary value problem:

$$EI\frac{\mathrm{d}^2 v}{\mathrm{d}x^2} = -Nv \quad \text{where } I = \frac{h^3}{12}; \qquad (6.2.11)$$

$$v = 0 \quad \text{when } x = \pm L/2. \qquad (6.2.12)$$

Here v is the lateral deflection of the elastic beam supported at the ends, and N is the value of the compressive force equal to h times the compressive stress in the coating.

The end part of the beam is subjected to all three components of the loadings considered in Sect. 6.1; see Eq. (6.1.11). The beam near its ends is tied to the body by some adhesion forces which breakage can increase the beam length. In this modified Euler's problem, it is required to also take into account the splitting process.

Let us write down the solution of this boundary value problem as follows:

$$v = \frac{M}{N}\cos\left(x\sqrt{\frac{N}{EI}}\right), \quad \beta = \frac{M}{N}\sqrt{\frac{N}{EI}}\sin\left(\frac{L}{2}\sqrt{\frac{N}{EI}}\right). \qquad (6.2.13)$$

Here M is the bending moment at $x = \pm L/2$ after the buckling occurs.

At the moment of buckling, Euler's relation is valid:

$$NL^2 = \pi^2 EI. \qquad (6.2.14)$$

According to Eqs. (6.1.11) and (6.2.13), the values of β, M and N determine the process of fracturing and delamination of the beam when

$$\Gamma_c = \frac{6M^2}{Eh^3}\sin^2\left(\frac{L}{2}\sqrt{\frac{N}{EI}}\right) + \frac{N^2}{2Eh} + \frac{6M^2}{Eh^3}. \qquad (6.2.15)$$

However, at the beginning the deflection of the beam was equal to zero, i.e., $M = 0$, so that summarizing Eqs. (6.2.14) and (6.2.15), we derive the following value of the critical force $N = N_*$

$$N_* = \min\left\{\sqrt{2hE\Gamma_c}, \pi^2 EIL^{-2}\right\}. \tag{6.2.16}$$

The buckling starts, only if $288L^4\Gamma_c > \pi^4 Eh^5$. Otherwise, the process starts from the delamination of the beam so that the value of L will increase until the Euler's critical force is achieved, after which the process of fracturing and delamination becomes dynamical.

6.3 A Blister on the Body Surface

The Delamination of a Film from a Half-Space Let us discuss the problem of an absolutely flexible elastic film/coating that is delaminated from a solid half-space with the formation of an axisymmetric cavity by injecting gas into the cavity/blister (Fig. 6.4). The displacement of the film under the action of a pressure p will be as follows:

$$w = \frac{p}{4N}\left(b^2 - r^2\right) \quad \text{where } r < b. \tag{6.3.1}$$

Here b is the blister radius, and N is the tension of the film caused by the gas injection. It is required to find p, b, and N in terms of mass of the gas injected. This problem can model some skin illnesses in medicine, some accidental problems in gas/oil industry, etc.

The film tension in the cavity is to be determined from the energy conservation law $E_p = W_p$ and Hooke's law for the coating extension:

$$E_p = \pi h b^2 \cdot \frac{1-\nu}{E}\left(\frac{N}{h}\right)^2, \quad W_p = 2\pi p \int_0^b r w(r) \mathrm{d}r = \frac{\pi p^2 b^4}{8N}. \tag{6.3.2}$$

It is assumed that all work W_p of pressure inside the cavity turns into the energy E_p of the elastic film. Then, from Eqs. (6.3.2), we get:

$$8(1-\nu)N^3 = hEb^2p^2. \tag{6.3.3}$$

According to Eq. (6.3.1), the angle β of the film on the delamination front is equal to $pb/(2N)$. From here, using Eq. (6.1.11) for the limiting state on this front, we obtain:

$$\Gamma_c = \frac{b^2 p^2}{8N} + \frac{(1-\nu^2)N^2}{2hE}. \tag{6.3.4}$$

Since the volume of the cavity is equal to $\pi p b^4/(8N)$, the state equation of the gas in the cavity can be written as follows:

$$\pi p^2 b^4 = 8 R_G N T M. \tag{6.3.5}$$

Here T and M are the absolute temperature and mass of the gas in the cavity, respectively.

From Eqs. (6.3.3) to (6.3.5), we find the required values of p, b and N:

$$b^2 = \frac{(3-\nu)R_G}{2\pi \Gamma_c} TM, \quad p^2 = \frac{8(1-\nu)^2}{hE^2} \frac{N^5}{R_G TM}, \quad N^2 = \frac{2hE\Gamma_c}{(1-\nu)(3+\nu)}. \tag{6.3.6}$$

The cavity radius increases in a stable manner in direct proportion to the square root of the mass and absolute temperature of gas injected into the cavity. The tension in the elastic film is found to be independent of mass of gas injected, and it is also independent of the pressure and the cavity dimensions.

The Delamination of a Plate from a Rigid Half-Space Suppose an elastic plate of thickness h, bonded to a rigid half-space, delaminates from it forming an axisymmetric cavity between them, into which a gas is pumped under a pressure p. The elastic displacement w of the plate is governed by the biharmonic equation $\Delta^2 w = p/k_b$ which solution can be written as follows:

$$w = \frac{p}{64k_b}\left(b^2 - r^2\right)^2, \quad k_b = \frac{Eh^3}{12\left(1-\nu^2\right)}. \tag{6.3.7}$$

Here it was assumed that the plate is clamped at the edge so that $w = \mathrm{d}w/\mathrm{d}r = 0$ at $r = b$. The values of b and p have to be found. This problem is characteristic for the delamination of rigid coatings.

In this case, based on Eq. (6.1.11), the limiting condition at the cavity edge has the form

$$\Gamma_c = k_b/\left(2R^2\right). \tag{6.3.8}$$

Here R is the radius of curvature in a section of the middle surface of the plate perpendicular to the delamination front at the corresponding point of the front. According to Eq. (6.3.7), we have $1/R = pb^2/(8k_b)$ when $r = b$. Then, from Eq. (6.3.8), it follows that

$$128 k_b \Gamma_c = p^2 b^4. \tag{6.3.9}$$

Calculating the cavity volume using Eq. (6.3.7), the state equation of gas in the cavity can be written as

$$\pi p^2 b^6 = 192 k_b R_G T M. \tag{6.3.10}$$

From Eqs. (6.3.9) and (6.3.10), we determine the required quantities:

$$b^2 = \frac{3R_G}{2\pi \Gamma_c} TM, \quad p = \frac{16\pi}{3R_G TM}\sqrt{2k_b \Gamma_c^3}. \tag{6.3.11}$$

Based on the Eqs. (6.3.6) and (6.3.11), the cavity radius for a film is only $\sqrt{1-(\nu/3)}$ times, i.e., practically by about 5%, less than the cavity radius for a plate, if the adhesion energy is the same in both cases.

6.4 A Crack Between Layers in Laminated Composites

Let us discuss the general case of an interface crack in a laminated composite of a bimetal type when both the longitudinal stretching of a composite plate and its transverse bending are important. In this case, from Eq. (6.1.11), it follows that the limiting condition on the front of an axisymmetric flat crack between two arbitrary plates can be written in the form:

$$\Gamma_c = \frac{1}{2}\sum_{i=1}^{2}\left[N_i\left(\frac{\mathrm{d}w_i}{\mathrm{d}r}\right)^2 + k_{ei}N_i^2 + k_{bi}\left(\frac{\mathrm{d}^2 w_i}{\mathrm{d}r^2}\right)^2\right], \tag{6.4.1}$$

$$k_{ei} = \frac{1-\nu_i^2}{h_i E_i}, \quad k_{bi} = \frac{E_i h_i^3}{12\left(1-\nu_i^2\right)}. \tag{6.4.2}$$

Here all quantities characterize the flat crack front neighborhood, and $i = 1$ and 2 is related to the upper and lower layers inside the crack. Equation (6.4.1) is valid also for curvilinear-in-plan cracks between layers, with r meaning the normal direction to the crack front in its plane.

Let us consider the axisymmetric problem of the cracked, two-layer composite plate stretched by a force N from all sides at infinity and subjected to the constant pressure p inside the interface crack. This problem can characterize the main emergency situation sometimes arising in the work of nuclear power stations as a result of the development of interface cracks in the multilayered protective walls of nuclear reactors due to some leaks of coolant. The catastrophes, which happened in Chernobyl, Ukraine, in 1986, and in Fukushima, Japan in 2012, drew the attention of scientists and engineers to this problem.

In this case, the value of the transverse displacement of each plate about a neutral plane of zero bending stresses obeys the following equation which is valid for the layers inside the interface crack

$$k_{bi}\Delta^2 w_i + N_i \Delta w_i = (-1)^i p, \quad \Delta = \frac{\mathrm{d}^2}{\mathrm{d}r^2} + \frac{1}{r}\frac{\mathrm{d}}{\mathrm{d}r}, \quad r < b, \; i = 1, 2. \tag{6.4.3}$$

Here the sign before the constant load p is positive for the lower bank of the crack, and negative for the upper bank. The value of p represents the pressure of gas inside the crack.

The stretching forces N_i in the composite plate are determined from the compatibility condition of a common tensile deformation of layers and the equilibrium equation. We have:

$$\frac{1-\nu_1}{E_1}\frac{N_1}{h_1} = \frac{1-\nu_2}{E_2}\frac{N_2}{h_2}, \quad N_1 + N_2 = N. \tag{6.4.4}$$

From here, we find:

$$N_1 = \frac{h_1(1-\nu_2)E_1}{(1-\nu_2)h_1E_1 + (1-\nu_1)h_2E_2}N, \quad N_2 = \frac{h_2(1-\nu_1)E_2}{(1-\nu_2)h_1E_1 + (1-\nu_1)h_2E_2}N. \tag{6.4.5}$$

Because of the structural asymmetry, these stretching forces give rise to a certain bending moment M_r which is, about the interface border, equal to

$$M_r = \frac{1}{2}\frac{h_1^2(1-\nu_2)E_1 - h_2^2(1-\nu_1)E_2}{(1-\nu_2)h_1E_1 + (1-\nu_1)h_2E_2}N. \tag{6.4.6}$$

It leads to a bending of the bimetal plate, which takes the form of a paraboloid of revolution close to a spherical segment with the curvature radius of the order of R_B, where

$$R_B = \eta\sqrt{\frac{k_b}{|N_1 - N_2|}}, \quad M_r = \frac{1}{\eta}\sqrt{k_b|N_1 - N_2|}, \quad k_b = \sum_{i=1}^{2}\frac{E_i h_i^3}{12(1-\nu_i^2)} \tag{6.4.7}$$

Here η is a constant. In the case of symmetry, we have $N_1 = N_2, \; M_r = 0, \; R_B \to \infty$.

We assume that the bending moment of the same value but of the opposite sign is applied at infinity in order to remove this bending effect. Anyway, the effect of this moment on the transverse displacement of the plates inside the crack, when $r < b$, is negligibly small, and it can be ignored. Therefore, we can accept that the transverse displacement of each layer at the edge of the crack is equal to zero.

And so, the edge of the plates inside the interface crack can be assumed clamped:

$$w_i = 0, \quad \mathrm{d}w_i/\mathrm{d}r = 0 \quad \text{when } r = b. \tag{6.4.8}$$

Also, the function dw_i/dr should be equal to zero when $r = 0$. This requirement and Eq. (6.4.8) complete the boundary value problem for Eq. (6.4.3).

Let us solve Eq. (6.4.3) in the following simplified designations:

$$k\left(\frac{d^2 f}{dr^2} + \frac{1}{r}\frac{df}{dr}\right) + Nf = p, \quad f = \Delta w = \frac{1}{r}\frac{d}{dr}\left(r\frac{dw}{dr}\right). \tag{6.4.9}$$

The general solution of the first Eq. (6.4.9), limited at the coordinate origin, is:

$$f = \frac{p}{N} + C\frac{1}{N}J_0\left(r\sqrt{\frac{N}{k}}\right). \tag{6.4.10}$$

Here C is an arbitrary constant, and $J_0(t)$ is the Bessel function of the first kind of zero order which is the following entire function:

$$J_0(t) = \sum_{n=0}^{\infty}\frac{(-1)^n}{(n!)^2}\left(\frac{t}{2}\right)^{2n} = 1 - \frac{1}{4}t^2 + \frac{1}{64}t^4 - \frac{1}{2304}t^6 + \ldots \tag{6.4.11}$$

With the account of Eq. (6.4.10), the general solution of the second Eq. (6.4.9) can be written as follows:

$$w = C_0 + C_1 lnr + \int_r^b \frac{dr}{r}\int_0^r \left[\frac{p}{N}x + C\frac{x}{N}J_0\left(x\sqrt{\frac{N}{k}}\right)\right]dx. \tag{6.4.12}$$

Here, to satisfy the boundary conditions, we must put:

$$C_0 = C_1 = 0, \quad \int_0^b \left[px + CxJ_0\left(x\sqrt{N/k}\right)\right]dx = 0. \tag{6.4.13}$$

The solution of Eq. (6.4.3) can be written now as follows:

$$w = \frac{pb^2}{4N}\left\{1 - \frac{r^2}{b^2} - 2\left[\int_0^{\lambda} tJ_0(t)dt\right]^{-1}\int_r^b \frac{dr}{r}\int_0^{\lambda r/b} tJ_0(t)dt\right\} \quad \text{where } \lambda = b\sqrt{\frac{N}{k}}. \tag{6.4.14}$$

By adding a corresponding subscript 1 or 2 to w, N, and k in this expression, we come to the solution of the original boundary value problem, Eqs. (6.4.3) and (6.4.8).

The Bessel functions of zero order can be expressed in terms of the Bessel function of the first order as follows:

$$xJ_0(x) = \frac{\mathrm{d}}{\mathrm{d}x}[xJ_1(x)]. \tag{6.4.15}$$

By the help of this equation, Eq. (6.4.14) is simplified:

$$w = \frac{pb^2}{4N}\left\{1 - \frac{r^2}{b^2} - 2[\lambda J_1(\lambda)]^{-1}\int\limits_{\lambda r/b}^{\lambda} J_1(t)\mathrm{d}t\right\}. \tag{6.4.16}$$

Let us use another property of the Bessel functions

$$J_1(x) = -\frac{\mathrm{d}}{\mathrm{d}x}J_0(x). \tag{6.4.17}$$

The sought solution acquires the final shape

$$w = \frac{pb^2}{4N}\left[1 - \frac{r^2}{b^2} - 2\frac{J_0(\lambda r/b) - J_0(\lambda)}{\lambda J_1(\lambda)}\right] \quad \text{where } \lambda = b\sqrt{\frac{N}{k}}. \tag{6.4.18}$$

Using Eqs. (6.4.15) and (6.4.18), we calculate the volume of the cavity from either one side

$$V = 2\pi\int\limits_0^b rw(r)\mathrm{d}r = \frac{\pi pb^4}{2N}\left[\frac{3}{4} - \frac{2}{\lambda^2} + \frac{J_0(\lambda)}{\lambda J_1(\lambda)}\right]. \tag{6.4.19}$$

According to Eq. (6.4.8), we have $f = \mathrm{d}^2w/\mathrm{d}r^2$ when $r = b$, and since $\mathrm{d}w/\mathrm{d}r$ is equal to zero at the edge of the crack, the limiting condition in Eq. (6.4.1) can be written as:

$$k_{b1}f_1^2 + k_{b2}f_2^2 = A \quad \text{where } A = 2\Gamma_c - \frac{1-\nu_1^2}{h_1E_1}N_1^2 - \frac{1-\nu_2^2}{h_2E_2}N_2^2 \quad \text{when } r = b. \tag{6.4.20}$$

From Eq. (6.4.18), when $r = b$, we can find

$$f = -\frac{p}{2N}\left[1 + \lambda J_0''(\lambda)/J_1(\lambda)\right] \quad \text{where } \lambda = b\sqrt{\frac{N}{k}}. \tag{6.4.21}$$

Substituting the quantities f_1, f_2, N_1 and N_2 in Eq. (6.4.20) by Eqs. (6.4.5) and (6.4.21) provides the function $p = p(b, N)$ in the general case:

$$p^2\left\{N_1^{-2}k_{b1}\left[1 + \lambda_1J_0''(\lambda_1)/J_1(\lambda_1)\right]^2 + N_2^{-2}k_{b2}\left[1 + \lambda_2J_0''(\lambda_2)/J_1(\lambda_2)\right]^2\right\} = 4A. \tag{6.4.22}$$

Using Eq. (6.4.22), the state equation for gas $pV = MTR_G$ and Eq. (6.4.19) for the volume of the cavity, we can determine the crack growth $b = b(M)$ in terms of mass M of gas in the cavity. Generally, this solution depends on seven dimensionless parameters.

In most important particular case when $\lambda < 2$, for the Bessel functions, we can use the first approximation:

$$J_0(\lambda) = 1 - \frac{1}{4}\lambda^2, \quad J_0''(\lambda) = -\frac{1}{2}\left(1 - \frac{3}{8}\lambda^2\right), \quad J_1(\lambda) = \frac{1}{2}\lambda\left(1 - \frac{1}{8}\lambda^2\right). \tag{6.4.23}$$

Let us illustrate the solution results by a particular case when $h_1 = h_2 = h,\ \nu_1 = \nu_2 = \nu,\ E_1 = E_2 = E,\ k_{b1} = k_{b2} = k_b,$ and $N_1 = N_2 = N$. In this case, $\lambda_1 = \lambda_2 = \lambda$. The values of $V,\ f,\ p,\ M$ and $\lambda = b\sqrt{N/k_b}$ are determined by Eqs. (6.4.19)–(6.4.21) and by the state equation for gas in the cavity. Solving this equation system provides:

$$M_* = \pi\lambda^4\left[\frac{3}{4} - \frac{2}{\lambda^2} + \frac{J_0(\lambda)}{\lambda J_1(\lambda)}\right]\left[1 + \frac{\lambda J_0''(\lambda)}{J_1(\lambda)}\right]^{-1} \quad \text{where } M_* = MTR_G\frac{N^2}{k_b\sqrt{2k_bA}}. \tag{6.4.24}$$

This is the implicit function $b = b(M)$ in the dimensionless variables. When $\lambda < 2$, using Eqs. (6.4.23) and (6.4.24) provides the following simple result:

$$M_* = 2\pi\lambda^2\left(1 - \frac{3}{16}\lambda^2\right) \quad \text{where } \lambda^2 = b^2N/k_b < 4. \tag{6.4.25}$$

And so, when gas under a pressure leaks into a crack, the crack radius b increases as follows:

$$b^2 = \frac{k_b}{N}\left(1 - \sqrt{1 - \frac{3}{8\pi}M_*}\right) \quad \text{when } M_* < 2\pi,\ Nb^2 < 4k_b. \tag{6.4.26}$$

As distinct from Griffith's crack, any even, however, small, original crack in a structure will stably grow if a pressurized gas or steam goes into the crack.

6.5 Delamination of a Graphene Film from a Plate

The case when one of the plate layers is a momentless film cannot be obtained as a limit from the preceding solution since, unlike the case of a plate, the angle β on the delamination front of the film is nonzero. However, this problem is interesting in the limiting case when the momentless film is inextensible, which is a good approximation to graphene films having a very great Young's modulus. The thickness of these films is of the order of one or several interatomic spacings of the diamond crystal. Therefore, they can be considered momentless while their ultimate tensile strength is about hundred times greater than for structural steel. Any force stretching a plate with such a film will be entirely transferred into the film at a distance of two to three plate thicknesses from the point of the force application.

When there is a cavity between a graphene film and a plate, the load from a force N applied to a plate with such a film from all sides will be entirely transferred onto the film in the cavity so that all stresses in the plate far from the point of the force application will be equal to zero. The pressure p in the cavity can give rise to a vertical deflection of an inextensible film only due to a horizontal displacement of the film edge $r = b$, directed toward the center. The relative deflection w/b of the film is of the order of $pb/(2N)$, and the horizontal displacement of the film edge toward the center is equal to $b(pb/N)^2/8$, that is, by an order of magnitude less than the deflection. Such a small change in the boundary as well as its effect on the elastic field of the plate can obviously be neglected.

Hence, when a plate covered on one side with an inextensible film is stretched at infinity and a cavity under the film is formed by pumping gas in, the plate is only subject to bending in the cavity (the plate extension is equal to zero). The film deflection, when $r < b$, occurs under a constant tension which is the same as at infinity. In this case, the bending moment at infinity is equal to zero in spite of the structure asymmetry, the plate does not undergo bending or stretching outside the cavity, and the horizontal displacement of the film edge on the delamination front can be neglected. In the cavity, hence, Eq. (6.3.7) can be used for the deflection of the plate and Eq. (6.3.1) can be used for the deflection of the film.

According to Eq. (6.4.1), in this case the limiting condition on the delamination front has the following form

$$\Gamma_c = \frac{1}{2}N\left(\frac{\mathrm{d}w_1}{\mathrm{d}r}\right)^2 + \frac{1}{2}k_b\left(\frac{\mathrm{d}^2w_2}{\mathrm{d}r^2}\right)^2 \quad \text{when } r = b. \tag{6.5.1}$$

Here the first term on the right-hand side of Eq. (6.5.1) refers to the film, and the second term to the plate.

The deflection of the film and plate is given by Eqs. (6.3.1) and (6.3.7), correspondingly. Using these equations and Eq. (6.5.1), we can easily calculate the cavity volume and find both the condition of the isothermal expansion of the gas in the cavity and the limiting condition at its front. As a result, in terms of mass and temperature of gas in the cavity, we get:

$$\frac{\pi}{8}p^2b^4\left(\frac{1}{N} + \frac{b^2}{24k_b}\right) = MTR_G; \tag{6.5.2}$$

$$p^2b^2\left(\frac{1}{N} + \frac{b^2}{16k_b}\right) = 8\Gamma_c. \tag{6.5.3}$$

From here, we find the cavity radius and the pressure in terms of the mass of gas in the cavity:

$$b_*^2 = \frac{1}{2}\left(\frac{3}{2}M_* - 1 + \sqrt{1 + M_* + \frac{9}{4}M_*^2}\right); \tag{6.5.4}$$

$$p_*^2 = b_*^{-2}\left(1 + \frac{3}{2}b_*^2\right)^{=1}. \tag{6.5.5}$$

Here:

$$b_*^2 = \frac{Nb^2}{24k_b}, \quad p_*^2 = \frac{3p^2k_b}{N^2\Gamma_c}, \quad M_* = MT\frac{NR_G}{24\pi k_b \Gamma_c}. \tag{6.5.6}$$

The cavity radius increases in a stable manner as the gas mass in the cavity becomes greater, and the radius is directly proportional to $\sqrt{M_*}$ for small values of M_* and to $\sqrt{1.5M_*}$ for large values of M_*.

It is important to emphasize that the common analysis of fracture mechanics based on loadings (here, pressure) would lead to a confusing result because, according to Eq. (6.5.5), the pressure is less for a greater radius, which would mean the crack instability. The point is that the pressure is produced by gas, which pressure depends on the cavity itself. And so, the physical nature of loadings should be taken into account in any analysis and prediction of fracture mechanics.

The above results on the development of cavities can be recalculated for other physical processes such as, for example, the polytropic expansion/compression of a gas or a process in which the mass and temperature of the gas supply vary in a certain way. Then, the development of cavities will occur differently: There can be halts in the growth, an alteration of stable and unstable development, and so on. In chemical and biochemical systems, the load-transferring gas can be produced from solid walls of cavities as a result of some reaction.

6.6 A Crack at the Interface Between Dissimilar Half-Spaces

Let us consider an infinite elastic space consisting of two media with dissimilar elastic properties when the interface between the media is a certain plane. The space is compressed at infinity by a stress q, normal to the boundary plane. Gas or fluid under a greater pressure p is pumped at a certain point of the boundary and an axisymmetric crack of radius b arises and develops at the interface between the two media under the action of this pressure. It is necessary to find the crack radius in terms of the mass of gas or fluid pumped and in terms of physical properties of the media. This problem is encountered in gas/oil industry using the method of fracking for extraction of these minerals. Fracking is the hydraulic fracturing of a well by pumping a drilling mud into it with the aim of fracturing the rock and increasing the output of a borehole (see next Chap. 7 of this book).

Using the linear superposition principle, we will now reduce this elastic problem to that about a constant normal load $p - q$ applied to the sides of the axisymmetric crack, with the space at infinity being load-free. The energy flow density at the crack front is determined by the invariant integral of Eq. (1.9.1) which in this case is equal to [1]

$$\Gamma = Db(p-q)^2/E_1 \quad \text{where } p > q; \tag{6.6.1}$$

$$D = \frac{\eta_0\left(1+\eta_0^2\right)\left[\eta_2 E_1^2 + \eta_1 E_2^2 + 2\eta_1 E_1 E_2 (1+\nu_1)(1+\nu_2)\right]}{E_2(1-2\nu_2)[E_1(1+\nu_2) - E_2(1+\nu_1)]}. \tag{6.6.2}$$

Here:

$$\eta_0 = \frac{1}{2\pi} ln \frac{E_1(1+\nu_2)(3-4\nu_2) + E_2(1+\nu_1)}{E_1(1+\nu_2) + E_2(1+\nu_1)(3-4\nu_1)}, \tag{6.6.3}$$

$$\eta_i = (1+\nu_i)^2(3-4\nu_i) \quad \text{where} \quad i = 1, 2, \tag{6.6.4}$$

$$\eta_3 = (1-2\nu_1)(1-2\nu_2) + 4(1-\nu_1)(1-\nu_2). \tag{6.6.5}$$

In this case, the volume of the crack cavity is equal to [1]

$$V = \frac{4\pi D}{3E_1}(p-q)b^3 \quad \text{where } p > q. \tag{6.6.6}$$

The condition $\Gamma = \Gamma_c$ is satisfied in the limiting state on the moving crack front. In the case of a homogeneous space, the following relations are valid:

$$E = E_1 = E_2, \quad \nu = \nu_1 = \nu_2, \quad D = (4/\pi)\left(1-\nu^2\right).$$

Further, we consider the cases of a gas or the drilling mud pumped into the crack cavity:

(i) *The drilling mud is an incompressible fluid*

In this case, we can assume that $M = \rho V$ where ρ is a constant density of the drilling mud.

(ii) *Gas obeys the state equation for the perfect gas*

In this case, the state equation $pV = MTR_G$ is valid. This case is more relevant for leaks in nuclear reactors.

Incompressible Fluid In this case, Eqs. (6.6.1) and (6.6.6) provide the following equation system:

$$Db(p-q)^2 = \Gamma_c E_1, \quad 4\pi\rho D(p-q)b^3 = 3ME_1. \tag{6.6.7}$$

From this system, we find:

$$b = \left(\frac{9E_1}{16\pi^2\rho^2 D}\right)^{1/5} M^{2/5}, \quad p-q = \left(\frac{\Gamma_c E_1}{D}\right)^{1/2}\left(\frac{16\pi^2\rho^2 D}{9E_1}\right)^{1/5} M^{-2/5}. \tag{6.6.8}$$

And so, the pumping of the drilling mud increases the crack radius directly proportionally to $M^{2/5}$, where M is the mud mass in the cavity.

Perfect Gas In this case, Eqs. (6.6.1) and (6.6.6) give us the following equations:

$$b(p-q)^2 = \frac{\Gamma_c E_1}{D}, \quad pb^3(p-q) = \frac{3E_1}{4\pi D} MTR_G. \tag{6.6.9}$$

These equations can be reduced in a dimensionless shape as follows:

$$b = \frac{\Gamma_c E_1}{Dq^2y^2}, \quad y = \frac{p}{q} - 1 \quad \text{where } y > 0; \tag{6.6.10}$$

$$y^5 - dy - d = 0. \tag{6.6.11}$$

Here:

$$d = \frac{4\pi\Gamma_c^3 E_1^2}{3q^4 B^2 MTR_G}. \tag{6.6.12}$$

Equation (6.6.11) has one real root $y = y(d)$ which monotonically increases as d increases so that when $d \to 0$ then $y \to d^{1/5}$ and when $d \to \infty$, then $y \to d^{1/4}$. The following inequalities are valid: $y(d) > d^{1/5} > d^{1/4}$ when $d < 1$, and $y(d) > d^{1/4} > d^{1/5}$ when $d > 1$.

These are some roots of Eq. (6.6.11):

d	0	1/48	1/2	32/3	1024/5
y	0	0.5	1	2	4

As the mass M of gas pumped into the crack cavity increases, the cavity radius also increases. For small M, it is proportional to $M^{1/2}$ and, for large values of M, it is proportional to $M^{2/5}$. The gas pressure decreases as M increases. A small mass of a high pressure is necessary to create an initial crack.

It should be noticed that, in all the problems considered, the crack begins to develop from zero and increases in a stable manner as a greater mass of gas or fluid is pumped into the cavity.

6.7 Development of a Casing String Gryphon

The growth of cylindrical cracks as gryphons outside a borehole casing string due to the high pressure of gas, which is usually formed from a liquid gas condensate, is characterized by an initial stable imperceptible growth which, under certain conditions, changes to the extremely rapid unstable growth.

The uncontrolled growth of a gryphon can lead to an ecological catastrophe. For example, in 2010 such a catastrophe occurred in the Gulf of Mexico as a result of the breakthrough of gryphons at two boreholes.

Let us consider the development of a gryphon along a vertical string casing which is a cylindrical tube of radius R, outside of which there is a homogeneous elastic body (rock) saturated by a gas condensate. Let us assume that the rock and the tube material are bonded along a common surface with the exception of a certain part where there is a thin cavity (a gryphon) filled with gas that enters from the deposit by seepage through the porous rock.

Let us make the following assumptions: (1) The problem is axisymmetric, (2) the radial distance between the opposite sides of the gryphon, that is, its thickness, is small as compared with the tube radius, (3) the gryphon length along the axis is large compared with the tube radius, (4) the pressure gradient of the gas in the cavity of the gryphon is small compared with the pressure gradient in the porous body, and (5) the gryphon length along the axis is small compared with the characteristic size of the porous body.

These assumptions justify the model of the active process zone of a gryphon as a semi-infinite cylindrical cavity, the sides of which are subjected to a constant pressure p_0. A gryphon can develop: (i) due to erosion, that is, the washing out and removal of rock particles by the gas, or (ii) as a brittle crack due to the local fracturing of the rock by gas pressure. These are completely different physical mechanisms. An erosion gryphon develops downward driven by the gas seepage flow while a gryphon crack develops in the direction of the gas flow, that is, upward.

Below we calculate the driving force of the slow erosion gryphon. In the next section, the fast gryphon crack will be considered. In this and next section, we use the cylindrical coordinates rz.

Erosion Gryphon Suppose a cylindrical gryphon located at $r = R,\ z > 0$ develops in the direction of the gas flow, that is, upward in the unperturbed field. Here z is the axis of symmetry, which is directed upward. A steady flow of a polytropic gas in a porous medium can be described using the following invariant integral [1]

$$\Gamma_i = k \oint_S (-\varphi_{,j}\varphi_{,j}n_i + 2\varphi_{,j}n_j\varphi_{,i})\mathrm{d}S \quad \text{where } i, j = 1, 2, 3. \tag{6.7.1}$$

Here φ is the gas flow potential, k is the coefficient of permeability (seepage intensity), S is the integration surface, and n_i are the components of the unit outward normal to this surface.

The gas velocity components v_i are expressed in terms of the potential φ as follows:

$$\rho v_i = q_i = -\varphi_{,i}, \quad v_i = -kp_{,i} \quad \text{where } i = 1, 2, 3; \tag{6.7.2}$$

$$\varphi = kAp^\gamma/\gamma, \quad \rho = Ap^{\gamma-1}. \tag{6.7.3}$$

Here p and ρ are the pressure and density of the gas, q_i are the components of the mass velocity vector of the gas, and A and γ are the constants of the polytropic process of the gas expansion, respectively.

The integrals Γ_i over any closed surface S encompassing a volume, in which there are no field singularities, are equal to zero which, in this case, manifest themselves as the laws of conservation of mass and momentum. From Eq. (6.7.1), it follows that Laplace's equation $\varphi_{,ii} = 0$ holds at all regular points of the field; it expresses the same conservation laws locally.

In our problem about the gryphon, all the points of the field when $r \geq R$ will be regular, with the exception of the singular gryphon front at $r = R,\ z = 0$, where the flow velocity becomes infinite. The last fact is modeled by the following boundary value conditions:

$$\partial\varphi/\partial r = 0 \quad \text{when } r = R,\ z < 0; \tag{6.7.4}$$

$$\varphi = \varphi_0 \quad \text{when } r = R,\ z > 0; \ \varphi_0 = kAp_0^\gamma/\gamma. \tag{6.7.5}$$

The first condition forbids the gas flow into the tube, while the second one describes the local thin layer of liberated gas escaped from the rock into free cavity on the wall.

The unperturbed flow far from the gryphon has the potential $\varphi = -qz$, where q is the mass velocity of the unperturbed flow; and the flow perturbed by the gryphon obviously has another potential at infinity:

$$\varphi = \varphi_0 R/r - qz \quad \text{when } r > R,\ z \to +\infty; \tag{6.7.6}$$

$$\varphi = -qz \quad \text{when } r > R,\ z \to -\infty. \tag{6.7.7}$$

The potential of the perturbed flow when $z \to -\infty$ is obviously equal to zero. The process zone under consideration is practically about two to three diameters of the casing string/borehole; therefore, the pressure variation inside the cavity can be ignored.

Let us consider the invariant integral Γ_z of this flow. As the closed surface of integration, let us take the cylinder surface $r = R$ plus the cylinder surface when $r \to \infty$ plus the cylinder end faces perpendicular to the axis when $z \to -\infty$ and $z \to +\infty$, and plus the surface of a half of a torus formed by the rotation of a circle, which radius is much less than R, around the singular front of the gryphon $r = R,\ z = 0$. The integral over this closed surface is equal to zero. By the boundary conditions and the conditions at infinity, the integrals over both cylindrical surfaces are also equal to zero.

The integral over the toroidal surface is equal to $2\pi R\Gamma$, where Γ is the density of the gas flow momentum on the gryphon front, which is the driving force of the erosion gryphon. According to Eqs. (6.7.6) and (6.7.7), the summary integral over the upper and lower end faces, when $z \to +\infty$ and $z \to -\infty$, is equal to $(\pi/2)k\varphi_0^2$. As a result, the driving force of the erosion gryphon is equal to

$$\Gamma = \frac{k\varphi_0^2}{4R} = \left(k^3A^2p_0^{2\gamma}\right)/\left(4\gamma^2R\right). \tag{6.7.8}$$

As seen, the driving force of the erosion gryphon is directly proportional to the third degree of the rock permeability and the gas pressure in the cavity to degree 2γ.

The development of an erosion gryphon can be characterized by some critical values of the driving force describing the start of intense erosion, some acceleration of the front movement, etc. It is noteworthy that unlike the usual dimension of force per length for driving force of cracks and fractures, the driving force of the erosion gryphon has dimension of momentum per length.

Proper Potential Field of Cylinder The problem of cylindrical potential has some peculiarities which are worth of a discussion. Let us consider the following simplest boundary value problem of the potential for $r \geq R,\ -\infty < z < +\infty$ which is an infinite space having a cylindrical cavity:

$$\Delta\varphi = 0 \quad \text{where } \Delta = \frac{1}{r}\frac{\partial}{\partial r}\left(r\frac{\partial}{\partial r}\right) + \frac{\partial^2}{\partial z^2}; \tag{6.7.9}$$

$$\frac{\partial\varphi}{\partial r} + a\varphi = 0 \quad \text{when } r = R,\ a = \text{const}. \tag{6.7.10}$$

$$\varphi \to 0 \quad \text{when } r \to \infty. \tag{6.7.11}$$

Let us find eigenfunctions of this homogeneous boundary value problem which is a certain canonic problem. We apply the calculus of groups to solve it.

Let the function $\varphi(r, z)$ be the general solution of this problem. Then, the function $C_1\varphi(r, z + C_2)$ will, evidently, be the general solution, too (C_1 and C_2 are arbitrary complex constants). Hence, the set of sought solutions makes a group, and the following functional equation is valid

$$\varphi(r, z) = C_1\varphi(r, z + C_2). \tag{6.7.12}$$

Let us solve this equation. Suppose the solution is a certain differentiable function. Substituting it in Eq. (6.7.12), we find that C_1 is a function of C_2, that is $C_1 = C_1(C_2)$. Let us put it in Eq. (6.7.12), which becomes an identity. Let us differentiate this identity over C_2. As a result, we obtain

$$\frac{dC_1}{dC_2}\varphi(r, z + C_2) + C_1(C_2)\frac{\partial\varphi}{\partial(z + C_2)} = 0. \tag{6.7.13}$$

Let us designate

$$\lambda = \frac{R}{C_1(C_2)}\frac{dC_1}{dC_2}. \tag{6.7.14}$$

Here λ is the proper number of this problem (generally, a complex one).
Equation (6.7.13) turns into the following equation

$$\frac{\partial\varphi}{\partial(z + C_2)} + \frac{\lambda}{R}\varphi(r, z + C_2) = 0. \tag{6.7.15}$$

The general solution of this equation has the form

$$\varphi(r, z) = F(r)e^{-\lambda z/R}. \tag{6.7.16}$$

Here $F(r)$ is an arbitrary function of r.
Using Eqs. (6.7.9) and (6.7.16), we find

$$\frac{d^2F}{dr^2} + \frac{1}{r}\frac{dF}{dr} + \left(\frac{\lambda}{R}\right)^2 F = 0. \tag{6.7.17}$$

From here, it follows that in Eq. (6.7.16), we can replace λ by $-\lambda$.
The general solution of Eq. (6.7.17) is:

$$F = B_1 J_0(\lambda r/R) + B_2 Y_0(\lambda r/R). \tag{6.7.18}$$

Here $J_0(\lambda r/R)$ is the Bessel function of zero order, $Y_0(\lambda r/R)$ is the Weber function of zero order, and B_1 and B_2 are some constants.

From the condition at infinity, Eq. (6.7.11), it follows that $B_2 = 0$ because the Weber function tends to infinity when $r \to \infty$ (while the Bessel function tends to zero). Now, from Eqs. (6.7.10) and (6.7.18), we derive

$$\lambda J_1(\lambda) = a_* J_0(\lambda), \tag{6.7.19}$$

where

$$\mathrm{d}J_0(\lambda r/R)/\mathrm{d}r = (\lambda/R)J_1(\lambda r/R), \quad a_* = aR. \tag{6.7.20}$$

Here $J_1(\lambda r/R)$ is the Bessel function of the first order.

Equation (6.7.19) is the characteristic equation, each root of which $\lambda = \lambda_n$ gives a particular solution, $\varphi_n = J_0(\lambda_n r/R)e^{-\lambda_n z/R}$ called an eigenfunction of this problem.

It can be shown that the characteristic equation, Eq. (6.7.19), has an infinite number of roots; all roots are real and positive. Hence, the general solution of the canonic problem stated by Eqs. (6.7.9)–(6.7.11) is given by the superposition of all particular solutions

$$\varphi = \sum_{n=0}^{\infty} c_n J_0(\lambda_n r/R)\exp(-\lambda_n z/R). \tag{6.7.21}$$

Here c_n are arbitrary constants.

For our problem of a slow erosion gryphon, Eqs. (6.7.4)–(6.7.7), we obtain:

When $z \to -\infty,\ a_* = 0,\ \lambda J_1(\lambda) = 0,\ (\varphi = -qz)$:

$$\varphi = -qz + c_0 J_0\left(\lambda_0 \frac{r}{R}\right)\exp\left(\lambda_0 \frac{z}{R}\right) + \dots \quad (\lambda_0 \cong 3.82); \tag{6.7.22}$$

When $z \to +\infty,\ a_* \to \infty,\ J_0(\lambda) = 0,\ \left(\varphi = \varphi_0 \frac{R}{r} - qz\right)$:

$$\varphi = \varphi_0 \frac{R}{r} - qz + c_1 J_0\left(\lambda_1 \frac{r}{R}\right)\exp\left(-\lambda_1 \frac{z}{R}\right) + \dots \quad (\lambda_1 \cong 2.4). \tag{6.7.23}$$

As seen, the pressure in front of a gryphon varies much steeper than behind.

The values of $c_0,\ c_1,\ c_2\ \dots$ can be obtained using the solution of the mixed boundary value problem, Eqs. (6.7.4)–(6.7.7). Its exact solution can be found using the Wiener–Hopf method, which is outside the subject of this book.

6.8 A Cylindrical Gryphon Crack

Suppose a semi-infinite cylindrical gryphon crack situated along $r = R,\ z < 0$ is propagating upward under the action of a high gas pressure $p = p_0$ in the gryphon cavity located between the rock at $r > R$ and the casing string (the tube) at $r = R$. As distinct from the slow, downward-growing erosion cavity, it is a fast, fracturing crack.

Let us assume that the rock is a homogeneous isotropic elastic body with a shear modulus μ_r. The tube with a wall thickness t is assumed to be elastic with a Young's modulus E and Poisson's ratio ν. Away from the tube, the rock is compressed by a lateral rock pressure p_b.

Using the linear superposition principle, let us study the axisymmetric problem of the development of the semi-infinite cylindrical crack between the tube and the rock under the action of a constant pressure $p_0 - p_b$ in the crack cavity and zero stresses at infinity. The boundary conditions of this problem have the form:

$$\sigma_r = -(p_0 - p_b),\ \tau_{rz} = 0 \quad \text{when } r = R,\ z < 0; \tag{6.8.1}$$

$$u_r = u_z = 0 \quad \text{when } r = R,\ z > 0; \tag{6.8.2}$$

$$\sigma_r = \tau_{rz} = 0 \quad \text{when } r \to \infty. \tag{6.8.3}$$

Here σ_r, σ_θ and τ_{rz} are stresses, and u_r, u_θ and u_z are displacements.

Let us use the following invariant integral of fracture mechanics in cylindrical coordinates

$$\Gamma_z = \int_S \left(U n_z - T_r \frac{\partial u_r}{\partial z} - T_z \frac{\partial u_z}{\partial z} \right) \mathrm{d}S. \tag{6.8.4}$$

Here:

$$T_r = \sigma_r n_r + \tau_{rz} n_z, \quad T_z = \sigma_z n_z + \tau_{rz} n_z. \tag{6.8.5}$$

When $z \to -\infty$, the elastic field of the rock is well known. Based on Eq. (6.8.1), it has the form:

$$u_r = (p_0 - p_b)\frac{R^2}{2\mu_r r}, \quad \sigma_r = -\sigma_\theta = -(p_0 - p_b)\frac{R^2}{r^2}, \tag{6.8.6}$$

$$u_\theta = u_z = 0, \quad \sigma_z = 0 \quad \text{when } R < r < \infty. \tag{6.8.7}$$

The latter equation $\sigma_z = 0$ is a remarkable property of this specific plane-strain problem, due to the equality: $\sigma_r + \sigma_\theta = 0$.

It should be noted that the process zone of this problem has an order of about two diameters of the tube near the fracture front; therefore, within this range, both the rock pressure and gas pressure can be assumed some constants (which depend on the depth of the fracture front, as well).

As the integration surface S, let us consider the same closed surface as in the case of an erosion gryphon, taking account of the opposite position of the gryphon crack. The integral Γ_z over this closed surface is equal to zero. The integral over the upper end face of the cylinder is equal to zero since, when $z \to +\infty$, the elastic stresses decrease exponentially. By virtue of Eq. (6.8.2), the integral over the surface $r = R,\ z > 0$ is also equal to zero. According to Eq. (6.8.6), the integrand is of the order of r^{-4} when $r \to \infty$; hence, the integral over the infinitely distant cylindrical surface is equal to zero.

The integral over the toroidal surface around the gryphon crack front is equal to $2\pi R\Gamma$, where Γ is the driving force density at the gryphon crack front. It is the main parameter characterizing the propagating and destructive abilities of the gryphon.

According to Eq. (6.8.6), the integral over the lower end face of the cylinder, where $n_z = -1,\ n_r = 0$ when $z \to -\infty$, is equal to

$$\int_R^\infty \int_0^{2\pi} U(r) r dr d\theta = -(p_0 - p_b)^2 \frac{R^2}{2\mu_b}. \tag{6.8.8}$$

The integral over the crack surface $r = R, z < 0$, where $n_z = 0$ and $n_r = -1$, has the following value

$$2\pi R \int_{-\infty}^{0} \sigma_r \frac{\partial u_r}{\partial z} \mathrm{d}z = -2\pi R(p_0 - p_b)\delta. \tag{6.8.9}$$

Here δ is the crack opening displacement at infinity when $z \to -\infty$. It is equal to the displacement of the elastic medium $(p_0 - p_b)R/(2\mu_b)$ plus the displacement of the elastic tube that is equal to $\left(1 - \nu^2\right)(p_0 - p_b)R^2/(tE)$. The latter equation follows from the well-known equations of deformation of thin-walled elastic tubes under pressure:

$$u_r = R\varepsilon_\theta = \frac{1-\nu^2}{E} R\sigma_\theta, \quad u_z = u_\theta = 0, \quad \sigma_\theta = \frac{1}{\nu}\sigma_z = -\frac{(p_0 - p_b)R}{t}. \tag{6.8.10}$$

Finally, collecting all the terms, we obtain the following expression for the force driving the gryphon crack

$$\Gamma = \frac{R(p_0 - p_b)^2}{4\mu_b}\left(1 + \frac{4\left(1 - \nu^2\right)\mu_b R}{tE}\right). \tag{6.8.11}$$

If $(R/t) \gg (E/\mu_b)$, this equation can be simplified as follows:

$$\Gamma = \left(1 - \nu^2\right)(p_0 - p_b)^2 R^2 (tE)^{-1}. \tag{6.8.12}$$

The limiting condition $\Gamma = \Gamma_c$ is the criterion of the catastrophic gryphon crack propagation because the driving force does not depend on the distance run by the gryphon. In this case, the value of Γ_c characterizes the least of either the specific work spent for fracturing the contact of rock and casing string, i.e., the adhesion energy, or the effective surface energy of the rock.

The speed of the gryphon propagation is determined by the process of gas transport in the cavity of the gryphon. However, it cannot exceed the speed of gas discharge into vacuum, which is about an order less than the velocity of the Rayleigh surface wave in the rock.

The knowledge of the crack driving force Γ helps one characterize also the slow subcritical crack growth because this force controls any crack development.

Literature

1. G.P. Cherepanov, *Fracture Mechanics* (IKI Publisher, Moscow-Izhevsk, 2012), pp. 1–870
2. G.P. Cherepanov, *Mechanics of Brittle Fracture* (McGraw-Hill, New York, 1978), pp. 1–950
3. G.P. Cherepanov, *Methods of Fracture Mechanics: Solid Matter Physics* (Kluwer Publ., Dordrecht, 1997), pp 1–300
4. G.P. Cherepanov (ed.), *Fracture. A Topical Encyclopedia of Current Knowledge* (Krieger Publ., Malabar, 1998), pp. 1–870
5. G.P. Cherepanov, Crack propagation in continuos media. J. Appl. Math. Mech. (JAMM) **31**(3), 503–512 (1967)
6. G.P. Cherepanov, Some new applications of the invariant integrals in mechanics. J. Appl. Math. Mech. (JAMM) **76**(5), 519–536 (2012)
7. M. Abramowitz, I.S. Stegun (eds.), *Handbook of Mathematical Functions* (Dover Publ., New York, 1972)

Chapter 7
Theory of Fracking

Abstract The mathematical problems of hydraulic fracturing (fracking) for the oil-and-gas extraction have been just recently solved by the author using invariant integral, complex variables, boundary layers, and his method of functional equations (published in the *Royal Proceedings*). In the chapter, the shape of the destructed rock volume and the gas/fluid output of the borehole are determined in terms of the geometrical, physical, and instrumental parameters of fracking process for horizontal drilling using thick muds and proppants. Three basic regimes of fracking are studied, including the permeation, non-permeation, and mixed regimes. Also, some new related problems of gas/fluid flow in porous media and of heat flow in cracked materials are effectively solved. This chapter is a "must-to-learn" for those who are engaged in the fracking technology or business.

Hydraulic fracturing called shortly fracking is an extraction method of oil and gas using horizontal drilling and special thick muds with proppants that keep fractures open. This method made a revolution in oil/gas industry of the twenty-first century by solving the problem of energy supply of the man for the next two hundred years. Invented still in the 1940th it has only recently become cost-effective after many improvements.

Our task is to derive the physical laws and equations governing the fluid flow in the porous rock being fractured by the fluid itself. The theory treated below is based on this author's paper published in *Philosophical Transactions of the Royal Society,* **A373**, 2014, 0119.

7.1 Temperature Track Behind a Moving Crack or Dislocation

To understand any fracturing behavior, we need, first, to look into the crack-tip where it happens. Temperature and heat flow stay behind all non-equilibrium processes including fracturing. What happens if all irreversible work of fracturing turns into

G. P. Cherepanov, *Invariant Integrals in Physics*,
https://doi.org/10.1007/978-3-030-28337-7_7

heat at the crack-tip? It is a fundamental and simplest question; besides, the heat flow in solids is very similar to the fluid flow in porous materials.

Therefore, let us study, first, the local stationary temperature field near the front of an arbitrary crack or dislocation in a solid. In this case, the field does not depend on x_3 so that a crack or dislocation in the moving coordinate frame Ox_1x_2 is at $x_2 = 0,\ x_1 < 0$. This front moves along the x_1-axis with respect to the elastic solid.

The crack driving Γ force Γ_1, see Eq. (1.5.1) in Chap. 1, can be calculated from the local elastic field of stresses $\sigma_{ij}(x_1, x_2)$ and displacements $u_j(x_1, x_2)$ near the crack front using the following equation [1]:

$$\Gamma_1 = \frac{\pi}{2} \lim\left[\sigma_{2j}(+\varepsilon, 0)u_j(-\varepsilon, 0)\right] \quad \text{where } \varepsilon \to 0,\ j = 1, 2, 3. \tag{7.1.1}$$

This equation is valid for dynamic and static cracks in arbitrary anisotropic linearly elastic materials and for interface cracks. Besides, it is valid for nonlinear power-law hardening incompressible materials, with replacing coefficient $\pi/2$ by another coefficient dependent of power [1]. For dislocations, equation $\Gamma_1 = \sigma^0_{1j} B_j$ is valid, where B_j is the displacement discontinuity and σ^0_{1j} are the stresses at the front location without the dislocation.

Let us assume that this specific energy Γ_1 of the Γ force driving a moving crack or dislocation turns into heat because the losses of energy for acoustic and electromagnetic radiation, for latent residual stresses and for surface energy, are negligibly small. The work spent on local plastic deformations turns into heat. Hence, the moving front of a crack or dislocation is mainly a heat source. Griffith's theory about surface energy is certainly a historic anachronism.

And so, we come to the problem of a linear source of heat Γ_1 at the crack-tip moving at constant speed $v_1 = v$ along the x_1-axis. Let us assume that a steady-state temperature field is set up in the crack-tip neighborhood. In this case, we can use the following invariant integral of the temperature field taken on a closed contour over the crack-tip [1, 2]

$$\Gamma_1 = \oint \left(\rho_m c_H T n_1 + \frac{k_T}{v^2} T_{,i} n_i v_1\right) \mathrm{d}S, \quad i = 1, 2, 3. \tag{7.1.2}$$

Here $T(x_1, x_2)$ is the temperature increase owing to the heat source at the beginning of coordinates which is point O, Γ_1 is the value of the force driving the crack or dislocation which is equal to the specific work spent by the stresses to move this crack or dislocation, ρ_m, c_H and k_T are the mass density, specific heat and thermal conductivity of the matter. As a matter of fact, Eq. (7.1.2) represents the energy conservation law.

If no heat sources are inside the integration surface, then $\Gamma_1 = 0$, so that applying the divergence theorem to Eq. (7.1.2) provides the following partial differential equation which is valid at all regular points of the stationary temperature field

$$k_T\left(T_{,11} + T_{,22}\right) = -\rho_m c_H v T_{,1}. \tag{7.1.3}$$

A similar equation emerged in the problem of convective heat/mass transfer in a fluid flow past a cylinder of an arbitrary cross section for low Prandtl and any Peclet numbers [3]. Using the result of this work, we can write the easily verifiable solution of Eq. (7.1.3), which is singular at point O, vanishes at infinity, and is an even function of x_2

$$T = A\lambda^{-2} e^{\lambda x_1} K_0(\lambda r) \quad \text{where } r = \sqrt{x_1^2 + x_2^2},\ \lambda = \frac{v\rho_m c_H}{2k_T}. \tag{7.1.4}$$

Here A is a constant to be found, and $K_0(\lambda r)$ is the modified Bessel function which has the following asymptotes

$$K_0(\lambda r) \to -\ln\frac{\lambda r}{2} \text{ when } \lambda r \to 0 \text{ and } K_0(\lambda r) \to \sqrt{\frac{\pi}{2\lambda r}} e^{-\lambda r} \text{ when } \lambda r \to \infty. \tag{7.1.5}$$

Let us substitute function $T(x_1, x_2)$ in Eq. (7.1.2) by its asymptotic value for small λr; see Eqs. (7.1.4) and (7.1.5). By taking a circle of infinitely small radius as the integration contour in Eq. (7.1.2) and by calculating the integral, we get

$$A = \frac{v\lambda^2}{2\pi k_T}\Gamma_1. \tag{7.1.6}$$

And so, the local temperature field produced by a moving crack or dislocation is

$$T = \frac{v\Gamma_1}{2\pi k_T} e^{\lambda x_1} K_0(\lambda r) \quad \text{where } r^2 = x_1^2 + x_2^2,\ \lambda = \frac{\rho_m v c_H}{2k_T}. \tag{7.1.7}$$

The temperature has a logarithmic singularity (a heat source) at the moving front of a crack or dislocation, with intensity being directly proportional to the square of fracture toughness in the case of open-mode cracks.

In light of this new approach, Griffith's view on the fracturing as a reversible exchange of elastic and surface energies seems very poor and delusive. The irreversible exchange of elastic and heat energies is a much more reasonable theory. This implies also that the crack can grow owing to other energy-consuming mechanisms, for example, by the vaporization or fluidization of an infinitely small amount of the material at the crack-tip.

7.2 Drainage Crack Growth in Porous Materials Filled with a Fluid

Let us study the small neighborhood of the end of a crack in a porous body filled with a fluid. Let us show, first, that this problem can be reduced to a mathematical problem which is very similar to the problem of the previous section. The fluid flow in a horizontal layer, which can be an oil/gas bearing bed, is governed by the following equations:

$$\frac{\partial p}{\partial t} = a_* \left(\frac{\partial^2 p}{\partial x^2} + \frac{\partial^2 p}{\partial y^2} \right), \; v_x = -\frac{k}{\rho g} \frac{\partial p}{\partial x}, \; v_y = -\frac{k}{\rho g} \frac{\partial p}{\partial y} \quad \text{where } a_* = \frac{k}{\beta \rho g}. \tag{7.2.1}$$

Here t is time, $p(x, y, t)$ is the fluid pressure, $\vec{v}(v_x, v_y)$ is the fluid/gas flow rate (real velocity equals $\vec{v}/\varepsilon_p$), k is the filtration coefficient, ρg is the specific weight of fluid, ε_p is the rock effective porosity, and β is the compressibility coefficient of the saturated rock. It is assumed that $p/\rho g$ is much greater than the bed thickness.

Other constants commonly used are:

$$k_p = \frac{k \eta_f}{\rho g}, \; \eta_f = \nu \rho. \tag{7.2.2}$$

Here k_p is the permeability of the porous rock, and ν and η_f are the kinematic and dynamic viscosities of gas/fluid. There are some empirical relations, for example, the Kozeny equation

$$k_p = C_K \frac{d^2}{(1 - \varepsilon_p)^2} \varepsilon_p^3. \tag{7.2.3}$$

Here d is the effective size of rock grains, and C_K is a constant.

Suppose a vertical crack cuts through the horizontal oil bed at $x_2 = 0$, $x_1 < 0$ so that its front moves along the x-axis where

$$x_1 = x - vt, \; x_2 = y. \tag{7.2.4}$$

Here v is the constant velocity of the crack front.

Let us study the local steady-state fluid flow near the crack front. Equation (7.2.1) in the new variables x_1 and x_2 turns to the following one:

$$a_* (p_{,11} + p_{,22}) = -v p_{,1}. \tag{7.2.5}$$

This equation can also be derived using the following invariant integral of the stationary fluid flow in a porous medium

$$\Gamma_1 = \oint \left(p n_1 + \frac{a_*}{v} p_{,i} n_i \right) \mathrm{d}S \quad \text{where } i = 1, 2. \tag{7.2.6}$$

Here $p(x_1, x_2)$ is the local pressure depression caused by the moving sink at the crack front, and Γ_1 is the drainage force driving the local depression field in the solid bed when the integration contour in the plane Ox_1x_2 embraces the crack front at $x_1 = x_2 = 0$.

It is easy to verify that the solution of Eq. (7.2.5), which is singular at $x_1 = x_2 = 0$ and which vanishes at infinity, has the following form

$$p = B\left(\frac{2a_*}{v}\right)^2 K_0\left(\frac{vr}{2a_*}\right)\exp\frac{vx_1}{2a_*} \quad \text{where } r^2 = x_1^2 + x_2^2. \tag{7.2.7}$$

Here B is a constant to be found, and $K_0\left(\frac{vr}{2a_*}\right)$ is the modified Bessel function of zero order.

Let us substitute function $p(x_1, x_2)$ in Eq. (7.2.6) by its asymptotic value for small r; see Eqs. (7.1.5) and (7.2.7). Using a circle of infinitely small radius as the integration contour in Eq. (7.2.6) and by calculating the integral, we find

$$B = \frac{1}{8\pi}\left(\frac{v}{a_*}\right)^3 \Gamma_1. \tag{7.2.8}$$

And so, the pressure field of a local depression near the front of a moving crack in a bed is given by the following equation:

$$p = \frac{v}{\pi a_*}\Gamma_1 K_0\left(\frac{vr}{2a_*}\right)\exp\frac{vx_1}{2a_*}. \tag{7.2.9}$$

The drop of the local fluid pressure near the moving front of a crack in a bed has a logarithmic singularity of a sink which intensity is directly proportional to $v\Gamma_1/a_*$.

It should be noticed that the total force for the fracturing of a solid skeleton saturated by a fluid is equal to the drainage driving force plus the crack driving force if all other forces are negligibly small. From this equation, it follows that the greater is the drainage driving force Γ_1, the less is the crack driving force fracturing the skeleton. And so, in the limit the drainage driving force, Γ_1 can fracture the rock skeleton by itself.

7.3 Fluid/Gas Flow in Porous Materials: Binary Continuum

Let us model a stationary process of the flow of a viscous fluid or gas in a porous material by a binary continuum so that two different continua are assumed to be at each point of space. One of these continua is an elastic solid characterized by the following invariant integrals [1]:

$$\Gamma_k = \oint \left(Un_k - \sigma_{ij}n_j u_{i,k}\right)\mathrm{d}S, \tag{7.3.1}$$

$$\oint (\sigma_{ij} n_j + \varepsilon_p p n_i) \mathrm{d}S = 0. \quad (7.3.2)$$

Here U is the elastic potential, u_i and σ_{ij} are the displacements and stresses, and $\varepsilon_p p n_i$ are the components of the volume force acted by the other continuum upon this one.

The other continuum is a fluid or gas flow characterized by the following invariant integrals [1]:

$$\Gamma_k = \oint (f_{ki} n_i - \rho_f v_i n_i v_k) \mathrm{d}S, \quad (7.3.3)$$

$$\oint \rho_f v_i n_i \mathrm{d}S = 0, \quad (7.3.4)$$

$$\oint (f_{ki} n_i - \varepsilon_p p n_k) \mathrm{d}S = 0, \quad (7.3.5)$$

and by the following state equations for fluid or gas:

$$f_{ij} = -p\delta_{ij} + \eta_f (v_{i,j} + v_{j,i}) \text{ for fluid;} \quad p\rho_f^\chi = C_p \text{ for gas.} \quad (7.3.6)$$

Here ε_p is the effective porosity, η_f is the dynamic fluid viscosity, v_i is the fluid/gas flow rate (the real velocity equals v_i/ε_p), ρ_f, p, and f_{ij} are the density, pressure, and stresses in the fluid/gas continuum, and χ and C_p are constants characterizing the polytropic process for gas.

In Eqs. (7.3.1) and (7.3.3), the value of Γ_k is the same due to the interconnection of the both continua. The value $\varepsilon_p p n_i$ of the volume force follows from the tubular model of porous materials. The effective porosity accounts only for the volume of interconnected pores where the fluid flows. At regular points, the equations of the theory of elasticity and fluid or gas dynamics can be obtained from Eqs. (7.3.1)–(7.3.6) by means of the divergence theorem.

Below we study some problems arising from the horizontal drilling of porous rocks, especially the theory of fracking which concerns the fracturing of the bed by a drill mud.

7.4 Horizontal Borehole: Stationary Extraction Regime

Let us study, at first, the fluid flow in horizontal boreholes. Let the x_3-axis coincides with the axis of a vertical borehole so that plane $x_1 x_2$ is parallel to the day surface. Suppose there is also a horizontal cylindrical borehole with axis $x_1 = x$ issuing from the vertical borehole. Let us use the cylindrical coordinate frame Oxr, where $r^2 = x_2^2 + x_3^2$, and point O is the issue of the horizontal borehole. We assume that a horizontal borehole of radius r_0 is embedded inside a horizontal fluid deposit which size is much greater than length l_H of this borehole and which thickness is much greater than radius r_0.

The porous rock is subject to stress $\sigma_{33} = -w_r$ which is equal to the weight of higher-placed rocks per unit square and to stresses $\sigma_{11} = \sigma_{22} = -\delta_T w_r$ where the lateral thrust coefficient δ_T is equal to $\delta/(1-\delta)$ in terms of Poisson's ratio, in the plane-strain model of the rock structure. Since $1 \geq \delta_T \geq 0$, the fracking wins the best advantage from the horizontal drilling because fractures in the rocks tend to grow along vertical planes which are perpendicular to the day surface.

Let us find the fluid flow field of a horizontal borehole ignoring elasticity of the porous medium. Luckily, in this 3D problem, there are two small dimensionless parameters λ_1 and λ_2

$$\lambda_1 = \frac{r_0}{l_H}, \quad \lambda_2 = \frac{k_p}{r_0^2}. \quad (\lambda_1 \ll 1, \ \lambda_2 \ll 1) \tag{7.4.1}$$

Here k_p is the permeability of the porous medium. Parameter λ_2 is vey small so that the fluid transport inside a borehole is much faster than through the porous rock.

These small parameters signal that there is a boundary layer in the domain $0 < x < l_H, r_0 < r < r_*$, where $r_* - r_0$ is the thickness of the boundary layer. Let us use the invariant integral of mass conservation, Eq. (7.3.4). By applying the divergence theorem, we derive equation $v_{i,i} = 0$. Using the empirical Darcy law for oil flow, we substitute $v_i = (k_p/\eta_f) p_{,i}$ and derive Laplace's equation for pressure p, which has the following form in the cylindrical coordinate frame Oxr

$$\frac{\partial^2 p}{\partial r^2} + \frac{1}{r}\frac{\partial p}{\partial r} + \frac{\partial^2 p}{\partial x^2} = 0. \tag{7.4.2}$$

Under the conditions of Eq. (7.4.1), we have $p_{,xx} \ll p_{,rr}$ in the narrow boundary layer of the most active flow process so that the solution of Eq. (7.4.2) in this layer can be written as follows:

$$p = P_B(x) + \frac{p_\infty - P_B(x)}{\ln \dfrac{r_*}{r_0}} \ln \frac{r}{r_0}; \tag{7.4.3}$$

$$v_r = -\frac{k_p}{\eta_f}\frac{\partial p}{\partial r}, \quad v_x = -\frac{k_p}{\eta_f}\frac{\partial p}{\partial x}. \tag{7.4.4}$$

Here v_r and v_x are the fluid flow rates, p_∞ is the initial pressure in the deposit and on the border $r = r_*$ of the boundary layer, and $P_B(x)$ is the pressure in the horizontal borehole.

From here, we arrive at the following system of ordinary differential equations:

$$\frac{\mathrm{d}V_B}{\mathrm{d}x} = 2\pi r_0 q_B, \quad V_B(x) = \frac{\pi r_0^4}{8\eta_f}\frac{\mathrm{d}P_B}{\mathrm{d}x}, \quad q_B(x) = \frac{k_p}{\eta_f} \cdot \frac{p_\infty - P_B(x)}{r_0 \ln \dfrac{r_*}{r_0}}. \tag{7.4.5}$$

Here $V_B(x)$ is the fluid flow rate through the borehole cross section, and $q_B(x)$ is the inflow rate of fluid into the borehole. The second equation follows from the famous Poiseuille equation.

The solution of Eq. (7.4.5) can be written in the following form:

$$P_B = p_\infty - \frac{p_\infty - p_b}{\mathrm{sh}(l_H/l)} \mathrm{sh}\frac{l_H - x}{l}, \quad q_B = \frac{k_p}{r_0 \eta_f} \cdot \frac{p_\infty - p_b}{\ln(r_*/r_0)} \cdot \frac{1}{\mathrm{sh}(l_H/l)} \mathrm{sh}\frac{l_H - x}{l} \tag{7.4.6}$$

$$V_B = \frac{\pi r_0^4}{8 l \eta_f} \cdot \frac{p_\infty - p_b}{\mathrm{sh}(l_H/l)} \mathrm{ch}\frac{l_H - x}{l} \quad \text{where } l = \frac{1}{4} r_0^2 \sqrt{\frac{1}{k_p} \ln\frac{r_*}{r_0}}. \tag{7.4.7}$$

Here p_b is the pressure at the issue of the borehole, where $x = 0$.

And so, the fluid output of the horizontal borehole without fractures per unit time is equal to V_B at $x = 0$, that is, to

$$Q_b = \frac{\pi}{2} \frac{r_0^2}{\eta_f} \sqrt{k_p} \left(\ln\frac{r_*}{r_0} \right)^{-1/2} (p_\infty - p_b) \mathrm{cth}\frac{l_H}{l}. \tag{7.4.8}$$

To find r_*/r_0, it is convenient to use equality $r_*/r_0 = (l_H/r_0)^\alpha$, where α is a fitting constant that can be found from one numerical solution of the problem [1]. It is equal to about 0.7.

7.5 Penny-Shaped Fracture: Stationary Extraction Regime

Let a penny-shaped fracture of radius R_0 be issuing from the horizontal borehole at $x = x_0$ so that its center is on the x-axis and its plane is perpendicular to this axis. We move the frame Oxr along the x-axis to the center of the fracture and designate it as $O\xi r$, where $\xi = x - x_0$.

The distance between opposite banks of the fracture at $r_0 < r < R$ where $R < R_0$ can be taken constant equal to d_p, where d_p is the diameter of hard particles (proppants) in the drill mud used to make the fracture by fracking. The particles remain inside the open fracture after the mud is removed when the rock pressure is closing the opening. These particles keep the fracture open like a wedge does.

The value of R is determined by the fracking process while the difference $R_0 - R$ can be found from the corresponding plane-strain problem of fracture mechanics when $R_0 - R \ll R$. We provide the result of the solution concerning this problem of wedging [1]

$$\frac{E d_p}{2(1-\delta^2)\sqrt{2\pi(R_0 - R)}} - \delta_T w_r \sqrt{\frac{1}{2}\pi(R_0 - R)} = K_{IC}. \tag{7.5.1}$$

Here K_{IC} is the rock fracture toughness.

From Eq. (7.5.1), it follows that $R_0 - R$ is less than $Ed_p\left[2\pi\delta_T w_r(1-\delta^2)\right]^{-1}$, that is, less than about 0.1 m for sandstones at the depth 1 km and $d_p \approx 0.5\,\text{cm}$. Thus, $R_0 - R \ll R$ indeed.

In what follows we also assume that $R \gg r_*$; otherwise, the fracking has no advantage.

The fluid flow near the fracture has the structure of a boundary layer along $|\xi| < x_*$, $r_0 < r < R$, where $\lambda_3 = x_*/R \ll 1$ and $d_p^2 \gg k_p$ so that $V_F \gg d_p v_r$. Here V_F is the flow rate through the fracture cross section per unit length.

In the boundary layer, the approach of Sect. 7.4 provides the following basic equations

$$\frac{\mathrm{d}}{\mathrm{d}r}(rV_F) = rq_F, \quad V_F = -\frac{d_p^3}{12\eta_f}\cdot\frac{\mathrm{d}P_F}{\mathrm{d}r}, \quad q_F = 2\frac{k_p}{\eta_f}\cdot\frac{p_\infty - P_F(r)}{x_*}. \tag{7.5.2}$$

Here $P_F(r)$ and $q_F(r)$ are the pressure in, and inflow rate into, the fracture. From Eq. (7.5.2), it follows that

$$\frac{1}{r}\frac{\mathrm{d}}{\mathrm{d}r}\left(r\frac{\mathrm{d}P_F}{\mathrm{d}r}\right) = -\frac{1}{b^2}[p_\infty - P_F(r)] \quad \left(\text{where } b = \frac{d_p}{2}\sqrt{\frac{x_* d_p}{6k_p}}\right)$$
$$(P_F = p_\infty \quad \text{when } r = R, \text{ and } P_F = p_b \quad \text{when } r = r_0) \tag{7.5.3}$$

Here p_b is the pressure at the issue of the fracture on the horizontal borehole.

The solution of the boundary value problem of Eq. (7.5.3) can be written as follows:

$$P_F = p_\infty - D(p_\infty - p_b)\left[I_0\left(\frac{R}{b}\right)K_0\left(\frac{r}{b}\right) - K_0\left(\frac{R}{b}\right)I_0\left(\frac{r}{b}\right)\right], \tag{7.5.4}$$

where $D = \left[K_0\left(\frac{r_0}{b}\right)I_0\left(\frac{R}{b}\right) - K_0\left(\frac{R}{b}\right)I_0\left(\frac{r_0}{b}\right)\right]^{-1}$.

Here $I_0(r/b)$ is the modified Bessel function of zero order so that:

$$I_0\left(\frac{r}{b}\right) \to \exp\frac{r}{b}\left(2\pi\frac{r}{b}\right)^{-1/2} \quad \text{when } \frac{r}{b} \to \infty; \; I_0\left(\frac{r}{b}\right) \to 1 \quad \text{when } \frac{r}{b} \to 0. \tag{7.5.5}$$

According to Eqs. (7.5.2) and (7.5.4), the fluid output from the fracture into the horizontal borehole per unit time is equal to

$$Q_F = \frac{\pi r_0 d_p^3}{6b\eta_f}D(p_\infty - p_b)\left[I_0\left(\frac{R}{b}\right)K_1\left(\frac{r_0}{b}\right) + K_0\left(\frac{R}{b}\right)I_1\left(\frac{r_0}{b}\right)\right]. \tag{7.5.6}$$

Here $K_1(r_0/b)$ and $I_1(r_0/b)$ are the corresponding modified Bessel functions of the first order.

For very large fractures when $R \gg b \gg r_0$, Eq. (7.5.6) is reduced to the simpler equation

$$Q_F = \frac{\pi d_p^3 (p_\infty - p_b)}{6\eta_f \ln(b/r_0)}. \tag{7.5.7}$$

And so, the output of a very large fracture significantly depends on d_p, $p_\infty - p_b$, and η_f, and much less on the fitting parameter b/r_0. The value of the latter parameter has to be found from one numerical solution of the problem under some typical conditions, or from a modeling experiment.

Evidently, this extraction process can be productive only for large fractures in rocks of high permeability. In the case of several fractures, when the distance between any two neighboring fractures is greater than x_*, the total output is given by summation of Eqs. (7.4.8) and (7.5.7). For a closer distance, more study is required.

7.6 Hydraulic Fracturing (Fracking)

Let us consider the hydraulic fracturing process used in oil/gas industry for the drilling of boreholes, especially, in shale gas reservoirs. Shales are characterized by high porosity, low permeability, and low fracture toughness. They are fractured by minor tensile stresses so that the shale destruction opens the way to extract gas stored in closed pores. Because of low permeability, the fluid flow in rock beyond the fractured zone can be ignored.

Horizontal boreholes in shales can be as long as 2 km. High pressure of the drill mud upon the rock surface and chemicals dissolving links between the rock fragments at the front of fractures produce a well-fractured volume in the local neighborhood of horizontal boreholes. Practically, all gas can be extracted from this volume. And so, the capacity of the borehole depends on the volume of fractured rock. The fractures keep open using proppants embedded by the drill mud inside fractures during fracking.

When the drill mud pressure is low, the horizontal cylindrical channel in an elastic bed of rock is subject to the following stresses in the surface layer of the channel

$$\sigma_r = -p_m, \ \sigma_\theta = p_m - w_r[1 + \delta_T + 2(1 - \delta_T)\cos 2\theta] \tag{7.6.1}$$

Here p_m is the drill mud pressure; w_r and δ_T are the weight of higher-situated rock per unit area and the lateral thrust coefficient; and θ is the angle between the horizontal plane and radius in the polar system of coordinates $Or\theta$ in the cylinder cross section with the center at the axis.

The stresses far from this channel are equal to

$$\sigma_{11} = \sigma_{22} = -\delta_T w_r, \ \sigma_{33} = -w_r. \tag{7.6.2}$$

The fracturing starts at the top point $\theta = \pi/2$ when $p_m \geq (3\delta_T - 1)w_r$.

Below we study three basic regimes of fracking. In the permeation regime, the drill mud penetrates everywhere inside fractures, while in the non-permeation regime, it penetrates nowhere in the rock. The most practical regime is that of partial permeation. The zone of fractured rock is assumed to enclose the horizontal borehole.

We assume that many fractures issue from the borehole, all being radial, that is, propagating along planes $\theta = \text{const}$. The border of the domain of fractured rock in the cylinder cross section represents an oval more extended in the vertical direction. The exact shape of the fractured domain is found using the method of functional equations and the theory of functions of a complex variable.

This method was elaborated by this author in 1963 in the paper *A method of solving elastic-plastic problems*, J. Appl. Math. Mech. (JAMM), vol. 27, see [4]. Based on the principle of analytic continuation, a boundary value problem is replaced by a functional equation which is valid in the whole plane of a complex variable. Used by many authors later, this method brought exact solutions of some nonlinear problems in the theory of plasticity, cavitation, and local buckling.

7.7 The Permeation Regime of Fracking

The friable shale is fractured by minor tensile stresses caused by the pressure of the drill mud that permeates multiple fractures. As a result, the hydrostatic pressure p_m is setting in everywhere in the well-fractured rock so that

$$\sigma_{11} = \sigma_{22} = \sigma_{33} = -p_m \quad (\text{inside } Z_F) \tag{7.7.1}$$

Here Z_F is the closed contour of the boundary of fractured rock in the normal cross section Ox_2x_3 of the horizontal borehole. This stress state is similar to a specific fluidized state [5]. The rock outside Z_F is intact and elastic, and is in a prefractured state on Z_F so that

$$\sigma_n = -p_m, \ \sigma_t = -\sigma_s \quad \text{if } \sigma_t \leq 0, \ \sigma_{nt} = 0 \text{ on } Z_F. \tag{7.7.2}$$

Here σ_n, σ_{nt}, and σ_t are the normal, shear, and tangential stresses on Z_F satisfying a failure criterion, for example, the von Mises criterion where k_s is a constant

$$(\sigma_n - \sigma_t)^2 + (\sigma_n + \delta_T w_r)^2 + (\sigma_t + \delta_T w_r)^2 + 6\sigma_{nt}^2 = 3k_s^2. \tag{7.7.3}$$

In the extreme case, we get $\sigma_t = 0$ on Z_F due to the effect of special chemicals so that we can neglect the tensile strength of the rock.

It is required to find contour Z_F and stresses outside Z_F meeting these boundary conditions. This is an inverse problem of the theory of elasticity. Let us solve it. We apply the invariant integral of Eq. (1.6.1) to an arbitrary elastic domain outside Z_F and use the divergence theorem; as a result, we have

$$\sigma_{ij,j} = 0, \quad \sigma_{ij} = \frac{\partial U}{\partial \varepsilon_{ij}}, \quad \varepsilon_{ij} = \frac{1}{2}\left(u_{i,j} + u_{j,i}\right). \tag{7.7.4}$$

This is the equation system of the theory of elasticity.

In the plane-strain case of a linearly elastic homogenous isotropic rock, the following representation is valid for stresses σ_{22}, σ_{33}, and σ_{23} outside contour Z_F [1]

$$\sigma_{22} + \sigma_{33} = 4\mathrm{Re}\Phi(z), \quad (z = x_2 + ix_3) \tag{7.7.5}$$

$$\sigma_{33} - \sigma_{22} + 2i\sigma_{23} = 2\left[\bar{z}\Phi'(z) + \Psi(z)\right]. \tag{7.7.6}$$

Here the Kolosov–Muskhelishvili potentials $\Phi(z)$ and $\Psi(z)$ are analytic functions outside contour Z_F that is unknown beforehand and has to be found.

From Eqs. (7.6.2), (7.7.2), and (7.7.5), it follows that based on Liouville's theorem

$$4\Phi(z) = -(1 + \delta_T)w_r = -p_m - \sigma_s. \tag{7.7.7}$$

Hence, the pressure of the drill mud necessary for this regime of fracking is

$$p_m = (1 + \delta_T)w_r - \sigma_s. \tag{7.7.8}$$

Using Eqs. (7.7.2), (7.7.6), and (7.7.7), and the following equation

$$\sigma_t - \sigma_n + 2i\sigma_{tn} = \mathrm{e}^{2i\alpha}(\sigma_{33} - \sigma_{22} + 2i\sigma_{23}), \tag{7.7.9}$$

we can formulate our boundary value problem as follows:

$$2\mathrm{e}^{2i\alpha}\Psi(z) = p_m - \sigma_s \quad (x_2 + ix_3 = z \in Z_F). \tag{7.7.10}$$

Here α is the angle between the external normal to Z_F and axis x_2 being counted from the axis to the normal.

Let the conformal mapping of domain $|\zeta| \geq 1$ onto the domain outside contour Z_F be provided by function $z = \omega(\zeta)$, where ζ is a new parametric complex variable. Since $\overline{\omega'(\zeta)\mathrm{e}^{2i\alpha}} = -\zeta^2\omega'(\zeta)$ on $|\zeta| = 1$, the boundary condition (7.7.10) can be written as

$$(p_m - \sigma_s)\overline{\omega'(\zeta)} + 2\zeta^2\omega'(\zeta)\Psi(\omega(\zeta)) = 0 \quad (\zeta = 1). \tag{7.7.11}$$

Using the method of functional equations and the principle of analytic continuation, the boundary value problem of Eq. (7.7.11) is reduced to the following functional equation which is valid in the whole plane of the complex variable ζ

$$(p_m - \sigma_s)\overline{\omega'}\left(\frac{1}{\zeta}\right) + 2\zeta^2\omega'(\zeta)\Psi(\omega(\zeta)) = 0. \tag{7.7.12}$$

For more detail and for many other nonlinear problems solved by this method, we can refer to the book [6]. As can be directly verified, the solution to the functional equation of Eq. (7.7.12) can be written in the following shape:

$$\omega(\zeta) = A\left(\zeta - \frac{\kappa}{\zeta}\right), \quad \Psi(\omega(\zeta)) = -\frac{1}{2}(p_m - \sigma_s)\frac{1 + \kappa\zeta^2}{\kappa + \zeta^2}. \tag{7.7.13}$$

Here A is an arbitrary constant, and κ is equal to

$$\kappa = \frac{1 - \delta_T}{1 + \delta_T}. \quad (0 < \kappa < 1, 0 < \delta_T < 1) \tag{7.7.14}$$

And so the boundary of the fractured rock has a shape of ellipse which diameters in the vertical and horizontal directions are:

$$D_V = 2A(1 + \kappa), \quad D_H = 2A(1 - \kappa). \tag{7.7.15}$$

The output of shale gas in this regime is directly proportional to $\pi\left(1 - \kappa^2\right)A^2 l_B$.

For $A \gg r_0$, the value of A is directly proportional to the square root of the volume of the drill mud pumped into the horizontal borehole. In this regime of fracking, the shale gas output is directly proportional to the drill mud volume pumped into the borehole.

7.8 The Non-permeation Regime of Fracking

Let us study the other extreme case when the permeation of the drill mud into the fractured rock can be ignored. In the continuum approximation, for many radial fractures inside contour Z_F, we get

$$\sigma_r = -p_m\frac{r_0}{r}, \ \sigma_\theta = 0, \ \sigma_{r\theta} = 0. \quad (r \geq r_0 \text{ and inside } Z_F) \tag{7.8.1}$$

At infinity, the stresses determined by Eq. (7.6.2) are valid.

Similarly to the previous problem of Sect. 7.7, we use the conformal mapping of domain $|\zeta| \geq 1$ onto the domain outside Z_F by function $z = \omega(\zeta)$ and from Eq. (7.8.1) arrive at the following boundary value problem when $|\zeta| = 1$:

$$4\mathrm{Re}\Phi(\omega(\zeta)) = -\frac{r_0 p_m}{|\omega(\zeta)|}, \tag{7.8.2}$$

$$\frac{\overline{\omega(\zeta)}}{\omega'(\zeta)}\frac{\mathrm{d}}{\mathrm{d}\zeta}\Phi(\omega(\zeta)) + \Psi(\omega(\zeta)) = \frac{r_0 p_m \overline{\omega(\zeta)}}{2\omega(\zeta)|\omega(\zeta)|}. \tag{7.8.3}$$

Using the principle of analytic continuation, let us continue Eqs. (7.8.2) and (7.8.3) onto the whole plane of the complex variable ζ

$$\Phi(\omega(\zeta)) + \bar{\Phi}\left(\omega\left(\frac{1}{\zeta}\right)\right) = -\frac{1}{2} r_0 p_m \left(\omega(\zeta)\bar{\omega}\left(\frac{1}{\zeta}\right)\right)^{-1/2}, \tag{7.8.4}$$

$$\bar{\omega}\left(\frac{1}{\zeta}\right)\left(\omega'(\zeta)\right)^{-1}\frac{\mathrm{d}}{\mathrm{d}\zeta}\Phi(\omega(\zeta)) + \Psi(\omega(\zeta)) = \frac{1}{2} r_0 p_m (\omega(\zeta))^{-3/2}\left(\bar{\omega}\left(\frac{1}{\zeta}\right)\right)^{1/2} \tag{7.8.5}$$

These functional equations are valid everywhere in the ζ-plane. Their exact solution can be written as follows [4, 7]:

$$\omega(\zeta) = B\left(2\zeta^2 - \lambda\right)^2 \zeta^{-3}, \quad \Phi(\omega(\zeta)) = w_r(1+\delta_T)\left[\frac{1}{4} - \frac{\zeta^2}{2\zeta^2 - \lambda}\right] \tag{7.8.6}$$

$$\Psi(\omega(\zeta)) = -\frac{1}{2}\lambda w_r (1+\delta_T)\frac{\zeta^4(1+\zeta^2)\left[2(4+\lambda^2)\zeta^2 + \lambda(4-3\lambda^2)\right]}{\left(2\zeta^2-\lambda\right)^3\left(2\zeta^2+3\lambda\right)}. \tag{7.8.7}$$

Here:

$$B = \frac{r_0 p_m}{(1+\delta_T)\left(4-\lambda^2\right)w_r}, \quad \lambda^3 + 4\lambda = 8\kappa, \quad \kappa = \frac{1-\delta_T}{1+\delta_T}. \tag{7.8.8}$$

Notice of erratum: The denominator of the second Eq. (5.2.23) in book [6] should be equal to $\sigma_x^\infty + \sigma_y^\infty - 2\sigma_s$ instead of $\sigma_s - p$. In Sects. 7.7 and 7.8 of the present text, it is assumed that $\sigma_s = 0$.

According to Eq. (7.8.6) for $\omega(\zeta)$, the oval contour Z_F is described by the following parametric equations

$$\begin{aligned} x_2 &= B\left[4(1-\lambda)\cos\beta + \lambda^2\cos 3\beta\right], \quad (2\pi \geq \beta \geq 0) \\ x_3 &= B\left[4(1+\lambda)\sin\beta - \lambda^2 \sin 3\beta\right]. \end{aligned} \tag{7.8.9}$$

The vertical and horizontal diameters of the fractured domain are as follows:

$$D_V = 2B(2+\lambda)^2, \quad D_H = 2B(2-\lambda)^2. \tag{7.8.10}$$

Contour Z_F encloses the horizontal borehole when

$$\frac{p_m}{w_r} \geq (1+\delta_T)\frac{2-\lambda}{2+\lambda}. \tag{7.8.11}$$

It can be shown that the solution given by Eq. (7.8.9) satisfies the condition of Eq. (7.8.11) when

$$\frac{2}{3} \geq \lambda \geq 0 \quad \text{i.e} \quad \text{when} \ \frac{17}{37} \leq \delta_T \leq 1. \tag{7.8.12}$$

when $\lambda = 2/3$, cusps appear at points $x_3 = \pm\frac{1}{2}D_V$ of contour Z_F. At this state, the vertical diameter of the cross section of the fractured domain is four times greater than the horizontal diameter, i.e., $D_V = 4D_H$. This means that for $\lambda > 2/3$, some fractures grow in the intact rock issuing from the cusps, along the vertical plane x_1x_3, because of the square-root singularity of tensile elastic stresses at the cusps.

According to Eq. (7.8.9), the area of the cross section of the fractured zone is equal to

$$\pi B^2(16 - 16\lambda^2 - 3\lambda^4) - \pi r_0^2.$$

And so, in the non-permeation regime under study, the parameters δ_t and p_m/w_r are the main parameters of the fracking process. When $p_m \gg w_r$, the volume of fractured rock in this regime is equal approximately to $l_H r_0^2 (p_m/w_r)^2$.

7.9 The General Regime of Fracking

The general regime of partial permeation occurs when the drill mud penetrates into some part of fractures at $r_0 \leq r \leq r_*$ while gas liberated from fractured pores of the rock permeates all the remaining part of fractures. This regime is of the most practical importance. The stresses in the rock between any two neighboring radial fractures meet the following equation

$$\frac{\partial \sigma_r}{\partial r} + \frac{\sigma_r - \sigma_\theta}{r} = 0, \ \sigma_{r\theta} = 0 \quad (\theta_0 \leq \theta \leq \theta_0 + \Delta\theta) \tag{7.9.1}$$

Since $\Delta\theta \ll 1$, we can put $\sigma_\theta = -p_m$ in the area where the drill mud wets the fracture surface, i.e., when $r_0 \leq r \leq r_*$. This is an axisymmetric area because it is determined by the axisymmetric conditions in the neighborhood of the horizontal borehole. In the remaining part inside Z_F when $r \geq r_*$, we can put $\sigma_\theta = -p_G$ where p_G is the pressure of shale gas liberated from fractured pores.

And so, from here and Eq. (7.9.1), in the continuum approximation for all fractured area inside contour Z_F, we get:

$$\sigma_r = \sigma_\theta = -p_m, \ \sigma_{r\theta} = 0 \quad (r_0 \leq r \leq r_*); \tag{7.9.2}$$

$$\sigma_r = -p_G + (p_G - p_m)\frac{r_*}{r}, \ \sigma_\theta = -p_G, \ \sigma_{r\theta} = 0 \quad (r \geq r_* \text{ in } Z_F) \tag{7.9.3}$$

Evidently, r_* increases if $p_m > p_G$, and r_* decreases if $p_m < p_G$, but the boundary velocity is much less than c_T.

In this case, using the method of functional equations, we come to the following solution [4]

$$z = \omega(\zeta) = D\left(2\zeta^2 - \mu\right)^2\zeta^{-3} \quad (|\zeta| \geq 1); \tag{7.9.4}$$

$$\Phi(\omega(\zeta)) = -p_G + \frac{1}{4}(1+\delta_T)w_r - \frac{(w_r + \delta_t w_r - 2p_G)\zeta^2}{2\zeta^2 - \mu}. \tag{7.9.5}$$

$$D = \frac{(p_m - p_G)r_*}{(4-\mu^2)(1+\delta_T - 2\overline{p_G})w_r}, \quad \mu^3 + 4\mu = \frac{8(1-\delta_T)}{1+\delta_T - 2\overline{p_G}}, \quad \overline{p_G} = \frac{p_G}{w_r}. \tag{7.9.6}$$

Function $\Psi(\omega(\zeta))$ coincides with that in Eq. (7.8.7) if, in Eq. (7.8.7), factor $w_r(1+\delta_t)$ is replaced by $(1+\delta_T)w_r - 2p_G$ and λ is replaced by μ.

In this case, contour Z_F of the fractured zone and its vertical and horizontal diameters are provided by Eqs. (7.8.9) and (7.8.10), where λ has to be replaced by μ, and B by D.

It can be shown that the solution given by Eqs. (7.9.4)–(7.9.6) exists when $2/3 \geq \mu \geq 0$. When $\mu = 2/3$, some cusps appear at points $x_3 = \pm D(2+\mu)^2$ so that for $\mu > 2/3$, two fractures grow in the intact rock along plane $x_1 x_3$ issuing from those points.

In this general case, the area of fractured rock is equal to

$$\pi D^2\left(16 - 16\mu^2 - 3\mu^4\right) - \pi r_0^2. \tag{7.9.7}$$

And so this regime of fracking is determined by the following four dimensionless parameters δ_t, p_m/w_r, p_G/w_r and r_*/r_0 which have to meet the following conditions

$$37\delta_T \geq 17 + 20\overline{p_G}, \quad r_*/r_0 \geq 1, \quad D(2-\mu)^2 \geq r_0. \tag{7.9.8}$$

The fractured area grows when $p_m > p_G$. When $p_G > p_m$, the gas pushes out the drill mud and fractures close up to the level supported by proppants.

Literature

1. G.P. Cherepanov, *Fracture Mechanics* (IKI Publ., Moscow-Izhevsk, 2012), pp. 1–872
2. G.P. Cherepanov, The mechanics and physics of fracturing: application to thermal aspects of crack propagation and to fracking. Philos. Trans. R. Soc. **A373**, 0119 (2014)
3. G.P. Cherepanov, Two-dimensional convective heat/mass transfer for low Prandtl and any Peclet numbers. SIAM J. Appl. Math. **58**, 942–960 (1998)
4. G.P. Cherepanov, A method of solving elastic-plastic problems. J. Appl. Math. Mech. (JAMM) **27**(3), 428–438 (1963)
5. G.P. Cherepanov, *Rock Fracture Mechanics in Drilling* (Nedra, Moscow, 1987), pp. 1–320
6. G.P. Cherepanov, B.D. Annin, *Elastic-Plastic Problems* (ASME Press, New York, 1988), pp. 1–250

7. G.P. Cherepanov, The theory of fluidization, Parts 1 and 2. Ind. Eng. Chem. Fundam. (IECF) **11**, 1–20 (1972)
8. G.P. Cherepanov, Some new applications of the invariant integrals in mechanics. J. Appl. Math. Mech. (JAMM) **76**(5), 823–849 (2012)
9. G.P. Cherepanov, Invariant integrals in continuum mechanics. Sov. Appl. Mech. **26**(1), 3–16 (1990)
10. G.P. Cherepanov, The invariant integral: some news, in *Proceedings of ICF-13, Sir Alan Cottrell Symposium*, Beijing, China, 2013
11. G.P. Cherepanov, Cracks in solids. Solids Struct. **4**, 811–831 (1968)
12. G.P. Cherepanov, On the opening-up of oil and gas boreholes. Dokl. USSR Acad. Sci. (Geophys.) **284**(4), 745–748 (1985)
13. G.P. Cherepanov, *Fracture Mechanics of Composite Materials* (Nauka, Moscow, 1983), pp. 1–250
14. G.P. Cherepanov, *Mechanics of Brittle Fracture* (McGraw Hill, New York, 1978), pp. 1–950
15. G.P. Cherepanov, Invariant Γ-integrals and some of their applications in mechanics. J. Appl. Math. Mech. (JAMM) **41**, 397–410 (1977)
16. G.P. Cherepanov, Point defects in solids, in *Deformation and Fracture in Solids*, J. D. Eshelby memorial volume, ed. by B. Bilby (Cambridge University Press, 1984)

Chapter 8
Fatigue and Superplasicity

Abstract This chapter deals with the birth and growth of dislocations and cracks in nanoscale, when the atomistic structure of a material being taken into account. Invariant integral is used as a basic variable. The fatigue crack threshold is determined by one atomic spacing of the crack growth per cycle. Superplastic flow of superfine grains is studied using the modified Arrhenius equation, and the theory of superplasticity is advanced. As a result, the neck-free elongation to failure, the maximum possible elongation, the critical size of ultrafine grains necessary to stop the growth of microcracks, characteristic dimensionless number, and other values of superplasticity are calculated in terms of stress, strain rate, and temperature. The theory was compared with, and supported by, the test data for the superplastic Pb–62%Sn eutectic alloy and Zn–22%Al eutectoid alloy, obtained by famous Japanese scholars. This chapter is designed for those who are interested in materials science.

Fatigue is a material property to fail under the action of cyclic or variable loads in which magnitude is much less than ultimate strength. Fatigue is caused by the slow, subcritical growth of microcracks. Superplasticity is a property of some materials with ultrafine grains to neck-free elongations exceeding three and more times original length. Both are a subject of fracture nanomechanics drawn up for the scales from some nanometers to some hundred micrometers. Invariant integral plays the role of the basic parameter characterizing the growth and arrest of micro- and nanocracks.

The chapter is based on this author's paper *Theory of superplasticity and fatigue of polycrystalline materials based on nanomechanics of fracturing* published in *Journal of Physical Mesomechanics*, **21**(6), 2018.

8.1 Introduction

Fracture nanomechanics is concerned with birth, growth, and arrest of cracks and dislocations in the nanoscale. Below, it is used to build the theory of superplasticity and fatigue of metals describing the fatigue threshold and enormous neck-free superplastic elongation in terms of strain rate, stress, and temperature. The optimum strain

G. P. Cherepanov, *Invariant Integrals in Physics*,
https://doi.org/10.1007/978-3-030-28337-7_8

rate of the maximum superplastic elongation is determined in terms of temperature, creep index, and other material constants. Further, we estimate the critical size of ultrafine grains necessary to stop the growth of microcracks and open way to the superplastic flow, and find the superplastic deformation of grains, their maximum possible elongation and the activation energy of superplastic state.

A dimensionless number is introduced in order to characterize the capability of different materials in yielding the superplastic flow. At a very high elongation, the alloying boundary of grains proves to be broken by a periodical system of dead fractures of some definite period. It is shown that experimental results of the testing of the Pb–62%Sn eutectic alloy and Zn–22%Al eutectoid alloy have supported this theory of superplasticity.

Fatigue Theory Fracturing from fatigue is caused by the slow growth of a crack or microcrack due to the accumulation of local plastic deformations near the crack front. The analytical theory of fatigue crack growth used an exact analysis of these plastic deformations. According to this theory, the rate of the fatigue crack growth for cyclic loadings in terms of invariant integrals is equal to [1–6]

$$\frac{\mathrm{d}L}{\mathrm{d}n} = -\beta\left(\frac{\Gamma_{\max} - \Gamma_{\min}}{\Gamma_{\mathrm{IC}}} + \ln\frac{\Gamma_{\mathrm{IC}} - \Gamma_{\max}}{\Gamma_{\mathrm{IC}} - \Gamma_{\min}}\right). \tag{8.1.1}$$

here n is the number of loading cycles; $\mathrm{d}L/\mathrm{d}n$ is the crack growth rate; $\Gamma_{\max}$ and $\Gamma_{\min}$ are the maximum and minimum values of the main invariant integral of fracture mechanics during the loading cycle designated as Γ_1 in Eq. (1.5.2) above; Γ_{IC} is the maximum value of the main invariant integral corresponding to the gross fracturing at the front of fast-moving crack; and β is a material constant of the length dimension. In the case of compression, when $K_{\mathrm{I}\min} < 0$, we should put $\Gamma_{\min} = 0$ in Eq. (8.1.1).

Let us remind how the main invariant integral of fracture mechanics Γ is expressed in terms of stress intensity factors K_{I}, K_{II} and K_{III} [1–6]:

(i) *Homogeneous, isotropic, and linearly elastic material*

$$\Gamma = \frac{1}{2\mu}\left[(1-\nu)\left(K_{\mathrm{I}}^2 + K_{\mathrm{II}}^2\right) + K_{\mathrm{III}}^2\right]; \tag{8.1.2}$$

(ii) *Interface crack on the border of two dissimilar elastic materials*

$$\Gamma = \frac{1}{4}\frac{(\mu_1 + \mu_2\kappa_1)(\mu_2 + \mu_1\kappa_2)}{\mu_1\mu_2[\mu_2(\kappa_1 + 1) + \mu_1(\kappa_2 + 1)]}\left(K_{\mathrm{I}}^2 + K_{\mathrm{II}}^2\right) + \frac{1}{4}\left(\frac{1}{\mu_1} + \frac{1}{\mu_2}\right)K_{\mathrm{III}}^2. \tag{8.1.3}$$

Here, μ is the shear modulus, and $\kappa = 3 - 4\nu$ where ν is Poisson's ratio.

In most practical cases, the fatigue crack grows as an open mode, tensile crack so that $K_{\mathrm{II}} = K_{\mathrm{III}} = 0$.

Equations (8.1.2) and (8.1.3) are valid only for the small-scale yielding when the size of the region of plastic deformation near the crack-tip is much less than the crack length. For more complicated cases, see the special literature [1–6]. As seen, the main invariant integral is the governing parameter of subcritical fracture mechanics.

This simple analytical theory satisfactorily describes the tensile crack growth in metals and alloys caused by cyclic loading, except for the cases of extremely low loads and high subcritical loads [1–6]. Similar equations were derived for random and any variable loads [5, 6].

Particularly, when $\Gamma_{max} \ll \Gamma_{Ic}$ and $\Gamma_{min} = 0$, Eq. (8.1.1) is reduced to the following one

$$\frac{\mathrm{d}L}{\mathrm{d}n} = \frac{1}{2}\beta\left(\frac{\Gamma_{max}}{\Gamma_{Ic}}\right)^2 \qquad (8.1.4)$$

This is the empirical Paul Paris' law. Extensive tests showed that it works well in most metals and alloys when the number of cycles to failure is not too large.

Both Eqs. (8.1.1) and (8.1.4) are not valid for extremely low loads when cracks do not propagate, at all. This property is controlled by the threshold value Γ_F of the invariant integral such that, if the invariant integral Γ_{max} is less than Γ_F, the fatigue crack does not grow. Evidently, in this case, it is necessary to use the methods of fracture nanomechanics.

The value of Γ_F is of paramount practical importance because the fatigue strength of structures, especially made from metals or alloys, is determined by their resistance to the subcritical crack growth. It is difficult to find this value experimentally because of enormous number of cycles necessary for adequate testing. The problem of Γ_F is addressed below in Sect. 8.2.

Superplasticity discovered just about half a century ago is poorly studied, and all basic features of this phenomenon have been so far unexplained, although well documented in experiments. The following questions are in need of answers:

(i) What ultrafine grain size is necessary for superplasticity, and why?
(ii) What is the maximum elongation to failure, and why is it so enormous?
(iii) How to characterize the superplastic property of a material in terms of its physical microstructure and temperature?

These and other relevant problems are studied in Sects. 8.3–8.9 below from the position of the nanomechanics of fracturing.

8.2 The Rule of Thumb: Fatigue Threshold

To study the start of the crack growth, it is necessary to take into account the atomic structure of the material and emission of individual dislocations from the front of the crack. This account has been provided by the nanomechanics of fracturing, see [7–14]. According to its basic concept, the minimum value of the crack-tip growth corresponding to the stable settlement of a first single pair of elementary edge dislocations emitted from the crack-tip is equal to the interatomic spacing.

This concept allows us to also evaluate the threshold value of the invariant integral using Eqs. (8.1.1) or (8.1.4). According to these equations, the crack growth for one cycle of loading, ΔL, is equal to

$$\Delta L = \frac{1}{2}\beta\left(\frac{\Gamma_{\max}}{\Gamma_{IC}}\right)^2. \tag{8.2.1}$$

Evidently, when the crack growth is equal to the interatomic spacing, the maximum stress intensity factor in any loading cycle is equal to the threshold stress intensity factor because the crack cannot grow for an amount less than the interatomic spacing. From here and from Eq. (8.2.1), it follows that

$$\frac{\Gamma_F}{\Gamma_{Ic}} = \left(\frac{2a}{\beta}\right)^{1/2} \tag{8.2.2}$$

Here, a is the interatomic spacing.

This equation requires some significant experimental work with measurements of the controlled fatigue crack growth in order to determine the fitting constant β in Eqs. (8.1.1) or (8.1.2). Then, Eq. (8.2.2) allows one to evaluate the threshold value of the invariant integral Γ_F for any specific material.

The analysis of many experimental works on the growth of fatigue cracks in various metals and alloys has shown that the value of β varies in the range from 0.001 to 1 mm [1–6, 15]. Since $a \approx 10^{-7}$ mm, from here and from Eq. (8.2.2), it follows that

$$\Gamma_F = (0.001 \div 0.01)\Gamma_{Ic}. \tag{8.2.3}$$

From this estimate, we can derive that when $\Gamma_{\max} < 0.001\Gamma_{Ic}$ no crack growth occurs and when $\Gamma_{\max} > 0.01\Gamma_{Ic}$ the fatigue crack growth is unstoppable like that of an avalanche. In the latter case, the most important parameter to watch and control is the time of loading. This is the simple rule of thumb.

The intermediate range $0.001\Gamma_{Ic} < \Gamma_{\max} < 0.01\Gamma_{IC}$ is more complicated for study. In this range, a more accurate estimate can be given by using the model of two rows of edge dislocation pileups on each side of the tensile crack [10].

8.3 Characterization of Polycrystalline Materials

Both local plastic deformations near a crack-tip and a crack growth are some interconnected effects taking place in polycrystalline materials simultaneously during the loading part of a loading–unloading cycle. Both of them proceed in primary acts such that the stable settlement of a pair of elementary edge dislocations emitted from the tip of a tensile crack is always accompanied by the one interatomic spacing crack growth. Plastic deformation is a continuum description of the strain produced by a very large number of dislocations. Everywhere below, we assume that the polycrystalline materials under study are homogenous and isotropic, except for some designated cases. For the modeling purpose, we use the case of the cubic lattice of atoms.

According to fracture nanomechanics, at the beginning of the process of loading, the crack does not grow and no dislocations emanate from the crack-tip, until the main invariant integral achieves some critical value Γ_1 and the stress intensity factor achieves some critical value of k_1. At that moment, the crack grows one interatomic spacing and simultaneously the first pair of elementary edge dislocations emanates from the crack-tip and settles down at a certain distance whereby [7–14]

$$\Gamma_1 = 13.25 a\tau_0, \quad k_1 = 3.64\sqrt{2a\mu\tau_0/(1-\nu)}. \tag{8.3.1}$$

Here, μ is the shear modulus of the polycrystalline material; τ_0 is its Schmid's constant of friction on gliding planes of the crystal lattice; ν is Poisson's ratio; a is the interatomic spacing; and Γ_1 and k_1 is the specific value of the main invariant integral and the open-mode stress intensity factor K_I for the cubic lattice of atoms. It should be noticed that all criteria coincide, in terms of either the invariant integral or the stress intensity factor, only in the case of the small-scale yielding and open-mode fracturing.

This first pair of settled dislocations finds a stable position at distance ρ_1 from the crack-tip on the gliding planes issuing from the crack front under angles $\pm 45°$ to the crack plane in the cubic lattice of atoms so that [7–14]

$$\rho_1 = \frac{\mu b_e}{4\pi\tau_0(1-\nu)}. \tag{8.3.2}$$

Here, b_e is the absolute value of the Burgers vector of the elementary edge dislocation depending on the parameters of the crystal lattice ($b_e = a\sqrt{2}$ for the cubic lattice).

If the invariant integral is less than Γ_1 or, the stress intensity factor is less than k_1, there are no stable-set dislocations and no crack growth so that Eq. (8.3.1) describes the lowest threshold of the dislocation emission from the crack-tip and of the crack growth. When $\Gamma < \Gamma_1$, or $K_I < k_1$, no dislocations settle down and the crack does not grow because, despite the infinite stress at the crack-tip, all dislocations emitted from the crack-tip by thermal fluctuations are unstable and come back to the crack-tip.

For example, using Eqs. (8.3.1) and (8.3.2) for aluminum crystals [16, 17], we get:

$$a = 2.85 \times 10^{-7}\ \text{mm}, \quad \mu = 25\ \text{GPa}, \quad \nu = 0.35, \quad \tau_0 = 0.75\ \text{MPa},$$
$$\Gamma_1 = 0.028\ \frac{\text{N}}{\text{m}}, \quad k_1 = 0.38\ \text{MPa}\sqrt{\text{mm}}, \quad \rho_1 = 2.3\ \mu\text{m}. \tag{8.3.3}$$

To compare: fracture toughness of structural aluminum, K_{Ic}, is of the order of 1 GPa$\sqrt{mm}$ while k_{Ic} is of the order of 10 MPa$\sqrt{mm}$ where k_{Ic} is the stress intensity factor of the aluminum lattice for the brittle crack in the cleavage plane corresponding to the dislocation-free growth of the crack.

Brittle Versus Ductile Behavior of Crystals Equations (8.3.1) and (8.3.2) allow us to also characterize brittle versus ductile properties of crystals. The crack-tip can advance not only by the ductile mechanism with emission of dislocations, but also by a brittle mechanism. If the brittle mechanism is realized, the atomic bond ahead of the crack-tip is ruptured without emission of dislocations. It occurs when the value of Γ_1 is greater than the surface energy of the dislocation-free crack growth or the superfine scale stress intensity factor k_1 at the crack-tip exceeds the critical value k_{Ic} corresponding to this surface energy.

The superfine scale stress intensity factor k_I is the total stress intensity factor at the crack-tip which is equal to the sum of the stress intensity factors due to the external load, K_I, and to the elastic field of dislocations, k_{ID}. The dislocations generated by the crack-tip during the loading process create a compression in the superfine scale region of the crack-tip and, hence, cause the local relaxation of the stresses induced by the external load so that [7–14]

$$k_I = K_I + k_{ID} \quad (k_{ID} < 0). \tag{8.3.4}$$

In other words, the generated dislocations play the role of a screen or shield protecting the crack-tip from external loads.

And so, a crystal is ideally brittle, if the crack can grow along a cleavage plane without to settle down any dislocations emitted from the crack front, that is, if the following inequality is met

$$\Gamma_1 > 2\gamma \quad \text{or} \quad k_1 > k_{Ic} \quad \left(k_{Ic} = 2\sqrt{\gamma\mu/(1-\nu)}\right) \tag{8.3.5}$$

Here, γ is the true surface energy of the crystal along the cleavage plane. For metal crystals, the value of γ has an order of 1 Pa m, and hence, for aluminum, the value of k_{Ic} has an order of 10 MPa$\sqrt{mm}$ so that due to Eq. (8.3.2), the aluminum crystal is very ductile.

To characterize the brittle vs ductile behavior of crystals, it is useful to introduce the dimensionless brittleness number as follows:

$$\eta = \frac{k_1}{k_{Ic}} \tag{8.3.6}$$

A crystal is absolutely brittle if $\eta > 1$ so that the crack grows without the dislocation generation. A crystal is ductile if $\eta < 1$ so that the dislocation emission starts before the brittle crack propagation becomes possible.

Using Eqs. (8.3.2) and (8.3.6), the brittleness number can be written as follows:

$$\eta = 2.165\sqrt{\frac{\tau_0 b_e}{\gamma}} \quad (8.3.7)$$

The smaller η, the more ductile is the crystal. For crystal lattices other than the cubic ones, the coefficient in Eq. (8.3.7) can be different.

The value of η was calculated for several common crystals. For example, in diamonds, it is equal to about 1.5, and for aluminum $0.03 \div 0.04$ depending on the values of the surface energy and Schmid's constant. Typically, for most metals, the brittleness number varies in the range of 0.01–0.1.

8.4 Superplastic Materials: Ultrafine Grain Size

The discovery of superplastic materials, metglas, and graphene marked the greatest advances of material science for the last half a century. On the scale of the technological impact, each one of them is comparable with the invention of bronze and iron several thousand years ago noted down by historians as the Bronze and Iron Ages of humankind.

Superplasticity is characterized by a very large neck-free elongation to failure at elevated temperatures greater than about a half of the melting temperature T_m in Kelvins. For example, the superplastic Zn–22%Al alloy specimen can extend 23.3 times its original length at temperature 473 K and at strain rate 10^{-2} s^{-1} while the superplastic Pb–62%Sn alloy specimen can extend 76.5 times its original length at temperature 473 K and at strain rate 2.12×10^{-4} s^{-1}, see [18]. As a reminder, the melting temperature of zinc and tin which are the base metals of these alloys is equal to 693 and 505 K, correspondingly, while the melting temperature of their alloying components, aluminum and lead, is equal to 933 and 601 K. In the special literature, the base metal component is called also the host or parent metal [16–19].

The theory of superplasticity is given in Sect. 8.6 below. Usually, a material behavior is called superplastic if the elongation to failure exceeds the original length, at least, about three times. However, as shown below, the superplastic material can extend to a much greater maximum when the temperature tends to the melting point.

To make a compound liquid metal superplastic, it should be cooled down very fast from the liquid state to the solid one so that the base metal grains could be able to grow up only to the size less than some critical value, below which no stable dislocations can be born inside the solid grains and no stable microcracks can grow. The base metal grain boundary made of other special components of the alloy forms during the crystallization process as a result of pushing them out by the growing base metal grain. Thus, they create an intergranular boundary layer which prevents from the settle down of individual dislocations within the grains and from the growth of any microcracks inside the grains. And so, there should be, at least, two components in any superplastic metal alloy. Besides, the intergranular layer protects the base metal from corrosion.

Grain Size of Superplastic Materials According to fracture nanomechanics, the size of grains should be less than the value of ρ_1 characterizing the distance between the first pair of stable dislocations and a crack-tip or another stress concentrator like the vertex of a two-sided angle of the grain. And so, the grain size s should satisfy the following requirement

$$s < \alpha \frac{a\mu}{\tau_0(1-\nu)}. \tag{8.4.1}$$

here, number α varies from about 0.1 to 1 depending on the nature of the dislocation generator inside the grain and its lattice type. For the sharpest concentrator, a crack-tip, and the cubic lattice, $\alpha = 0.11$ according to Eq. (8.3.2).

Let us estimate number α in Eq. (8.4.1) using a more realistic stress singularity for the vertex of a two-sided angle between two dissimilar materials. In this case, it can be shown that the balance equation of a pair of edge dislocations near a stress concentrator is given by the following equation [7]

$$\frac{K}{r^n} = \tau_0 + \frac{\mu b_e}{4\pi r(1-\nu)} \quad (0 < n \leq 1/2) \tag{8.4.2}$$

Here, r is the distance between the vertex of the stress concentrator and the position of dislocations, n is the proper number characteristic for the angular stress singularity [5, 6], and K is the intensity of the stress concentration proportional to the external load, with the dimension being Nm^{n-2}. In the case of a crack-tip, we get $n = 0.5$ and $K = 0.327K_I(2\pi)^{-1/2}$, see [7–13].

Function $K(r)$ in Eq. (8.4.2) has one minimum at

$$r = \rho_1 = \frac{1-n}{n}\frac{\mu b_e}{4\pi\tau_0(1-\nu)}, \quad K = k_1 = \frac{\tau_0}{1-n}\left[\frac{1-n}{n}\frac{\mu b_e}{4\pi\tau_0(1-\nu)}\right]^n \tag{8.4.3}$$

These equations generalize Eqs. (8.3.1) and (8.3.2) for the crack-tip.

Hence, the first pair of stable edge dislocations emitted by this stress concentrator settles at distance $r = \rho_1$ when the external load measured by K achieves the value of k_1 in Eq. (8.4.3). No stable equilibrium dislocation position exists for $K < k_1$. According to Eq. (8.4.2), if $K > k_1$ and the first pair of stable dislocations is set down, their position r grows when K increases, until the second pair of stable dislocations is born [7–14]. Equations (8.4.3) support the estimate of coefficient α in Eq. (8.4.1) for $0.1 < n \leq 0.5$.

Using computerized calculations, the process of the dislocation generation by a crack-tip in large crystals was studied up to many thousands of individual dislocations, and the diagram of the stress intensity factor versus the crack growth was calculated, see [7–12] for more detail. In the framework of this discrete atomic model, the fracture toughness of the material was calculated as the limit of the stable growth of the crack characterized by the maximum point on this diagram, after which the unstable growth of the crack occurs.

However, a crack-tip does not grow at $k_I < k_{IC}$ and does not produce any stable dislocations, if the size of grains in a polycrystalline material is less than ρ_1. Thus, the value of ρ_1 in fracture nanomechanics characterizes *the critical ultrafine grain size of the superplastic state* of the material if $k_1 \ll k_{IC}$ which is usually the case. This value is estimated by Eq. (8.4.1).

In superplastic materials under low temperatures, the ultimate and yield strengths depend on the grain size as follows [4–6]:

$$\sigma_B\sqrt{s_0} \cong K_{IC}, \quad \sigma_Y\sqrt{s_0} \cong K_{IIC} \tag{8.4.4}$$

here, σ_B is the ultimate strength; σ_Y is the yield strength; s_0 is the characteristic grain diameter; K_{IC} is the fracture toughness; and K_{IIC} is the slip toughness of the material.

In terms of critical values of Γ_c of the invariant integral, a more general relation is valid

$$\Gamma_c \approx \frac{1-\nu}{2\mu}\left(K_{IC}^2 + K_{IIC}^2\right) \approx \frac{1-\nu}{2\mu}s_0\left(\sigma_B^2 + \sigma_Y^2\right). \tag{8.4.5}$$

For example, the decrease of the grain size from 10 μm to 10 nm makes the 31-fold increase of the yield strength of some superplastic metals [18]. The equations, analogous to Eqs. (8.4.4), are called the Petch–Hall–Straw–Cottrell equations to honor their originators. However, under high temperatures close to the melting point, the effect of surface tension dominates so that these equations turn to be invalid, see below. Also, in the nanoscale the effect of surface tension becomes essential.

It is noteworthy that a microcrack inside the grain starts on to grow as an absolutely brittle one only after the superfine stress intensity factor achieves the value of k_{IC} so that the tensile stress σ and the size of the microcrack d_c are connected as follows [13]:

$$\sigma\sqrt{d_c} = k_{IC}. \tag{8.4.6}$$

For example, in the aluminum-based grain under low tensile stress typical for the superplastic state, e.g. $\sigma = 33$ MPa, the critical size of the brittle microcrack is equal to $d_c = 100$ μm which is much greater than the grain size in the superplastic state so that the brittle failure cannot occur because $k_{IC} = 10\ \mathrm{MPa}\sqrt{\mathrm{mm}}$.

Critical Grain Size in Some Superplastic Metals According to Eqs. (8.4.1) and (8.4.3), the aluminum grain size in superplastic state should be less than about 2 μm. Evidently, this result of calculation is valid for any Al-base alloys.

Using Eq. (8.4.3) for the crack-tip concentrator and Table 6.2 in the book [13], let us provide the results of calculation of the critical size of grains in superplastic state for the following base metals (in micrometers):

Aluminum (Al): 2.3; copper (Cu): 1.8; gold (Au): 2.6; nickel (Ni): 0.9; and lead (Pb): 2;
Silver (Ag): 3.2; zinc (Zn): 5.8; ferrite (α—iron, Fe): 0.2; and white tin (Sn): 4.5.

All of these metals, except for tin and zinc, have the face-centered cubic crystal lattice (fcc), in accordance with the theoretical model served to derive Eq. (8.4.3). Tin has the tetragonal lattice, and zinc the hexagonal-close-packed lattice (hcp); therefore, the above figures for these metals, probably, need some corrections.

It is worthy of keeping in mind that these figures correspond to the minimum values of Schmid's constant observed in experiments [16, 17]. This constant which is the friction stress on the gliding planes of a pure crystal is very sensitive to any impurities so that any obstacles on these planes, beyond of atomic forces of the crystal, significantly influence its value. The interstitial atoms of alloying elements, which diameter is greater than the interatomic spacing of the lattice of the parent metal, can significantly increase Schmid's constant and, hence, decrease the critical size of grains for the superplastic state to be realized. Any distortions of the crystal lattice like original dislocations on other planes also increase this constant. On the other hand, vacancies in the crystal lattice of the base metal can decrease Schmid's constant and, hence, increase the critical size of grains. Also, according to Eq. (8.4.3), a slighter singularity of a two-sided angle can lead to an increase of this critical size as compared to the crack-tip concentrator.

The Theory of Singularities of the Elastic Field Let us provide a short resume of this theory given in [5, 6, 13]. There are point and linear singularities. The first ones are point inclusions like interstitial or foreign atoms in the lattice of a parent metal, or point holes like vacancies or small bubbles in this lattice. The second ones are the fronts of cracks and dislocations, and the vertex of the two-sided angle between dissimilar materials different from π. Also, it is necessary to distinguish the S-singularities from the N-singularities.

The S-singularities possess the self-controlled fields independent of applied loads. These are point singularities and dislocations. They obey Saint-Venant's principle and can drift in the stress field produced by applied loads without to change their own fields like some invariable particles.

The N-singularities do not possess their own independent fields—their fields are produced by applied loads and determined by them. These are crack fronts and vertices of two-sided angles. They do not obey Saint-Venant's principle. The crack front can move in the stress field, usually with changing its own field in the process of motion.

The drift of any singularity is governed by the corresponding driving force; most of these forces were discovered during the last century. The first discoveries for massive elastic solids were Irwin's law for open mode cracks, Peach–Koehler's law for edge dislocations, and Eshelby's law for point inclusions. Using invariant integrals, dozens of other laws were found out for other important field singularities including point holes, cracks and dislocations of arbitrary modes in anisotropic solids, interface cracks, cavities, cracks and bubbles between shells, plates, membranes and solids, solid–liquid contact fronts, and many others [4–6, 13, 14].

The drift of point inclusions and point holes in the field of cracks and dislocations was studied earlier in [13] under some simplifying assumptions, e.g., of zero Schmid's stress. In particular, all point inclusions and holes proved to drift into the crack front or dislocation core. It is easy to predict some effects of nonzero Schmid's stress, for example, appearance of some critical distances from the crack fronts and dislocation cores such that outside of these distances point inclusions and holes cannot drift, at all.

As to the original point singularities, like vacancies or foreign atoms in the identically stretched grains of base superplastic metals, it is evident that they do not drift in the process of flow because their driving forces equal zero for this case of the zero-gradient field of tensile stress. And so, vacancies and foreign atoms can substantially effect only upon the bulk properties of the base metal grains, for example, upon μ, τ_0, and k_{IC}. In real distorted grains, some drift of vacancies and foreign atoms toward the grain boundary occurs; however, this effect is small and ignored in the present model, see the next section. Certainly, the diffusion and migration of vacancies and foreign atoms governed by the temperature gradient play some role most essential at high temperature.

Critical Temperatures of the Superplastic State In ultrafine grains of subcritical size, the activity of any stable individual dislocations is suppressed so that the grains can deform only by the uniform flow along gliding planes [7–14, 16–18]. According to test data, the uniform neck-free flow in the superplastic state takes place for temperatures greater than about half of the melting point [13, 14, 16–19]

$$T > \frac{1}{2} T_{\mathrm{m}} \tag{8.4.7}$$

At these temperatures, the thermal fluctuations activate the random generation of many dislocations from a virtual generator inside a grain; however, all of these dislocations are unstable and disappear [7–14]. As a matter of fact, superplastic flow is creep of metals with ultrafine grains.

8.5 Superplastic Deformation and Flow of Ultrafine Grains

Let us study the extension and flow of a bar of a superplastic material at sufficiently high temperatures and low tensile loads. We assume that all grains of the material have one and same volume, and the material is incompressible. The first assumption is justified by the simplicity of the following analysis, while the latter one is close to reality for large deformations.

In polycrystalline materials, the large deformation and flow can, in principle, be caused by a relative movement of neighboring grains (the sliding mechanism). However, it forms both open and sliding interface cracks, which size in the superplastic flow would be much greater than the grain size; these cracks would grow in the process of flow so that their healing by diffusion would require an enormous concentration and activity of foreign atoms. That is why we take into account only the most probable mechanism of the superplastic flow of each grain, with no material discontinuities arising between the neighboring grains.

Grain Shape For very large deformations, as a result of stretching by tensile load, every grain acquires a shape of a prolonged prism. In order to densely pack the space, the prism cross section can be either equilateral triangle, or square, or regular hexagon. All prisms have length l and cross-sectional area A_g so that $lA_g = V_0$ where V_0 is the constant volume of any grain in this model. In the process of flow, length l increases and area A_g decreases.

The thin intergranular layer of alloying components cohering grains plays the part of surface tension for the flowing base material. Out of possible three forms of the prism cross section, square has a minimum perimeter for the prisms of the same volume and length and, hence, provides for a minimum of surface energy of prisms at any moment of flow. Therefore, according to the principle of minimum surface energy, the cross section of all prisms is the square with side s so that $A_g = s^2$ where $l \gg s$. To be densely packed in the space, both front and rear faces of the prisms should be flat. This packing order leaves no empty spacings between the prisms. All succeeding calculations are assumed for the packing of this kind.

And so, in the process of superplastic deformation, the grains of arbitrary shape become long identical prisms of square cross section, with the flat front and rear faces.

Superplastic Zero-Dilatancy Flow In the process of flow, a bar of initial length L_0 and initial cross-sectional area A_{b0} acquires length L and cross-sectional area A_b so that $LA_b = L_0 A_{b0}$. Besides, the number of grains in each cross section of the bar remains constant during this process so that we have

$$\frac{L}{l} = \frac{L_0}{s_0} \quad \left(s_0^3 = V_0\right) \tag{8.5.1}$$

Here, l is the length of each grain so that $ls^2 = V_0 = \text{const}$ for any elongation.

And so, according to Eq. (8.5.1), the grain length l assumes the function of the bar length L in terms of strain so that the strain rate $\dot{\varepsilon}(t)$ is equal to

$$\dot{\varepsilon}(t) = \frac{1}{L}\frac{\mathrm{d}L}{\mathrm{d}t} = \frac{1}{l}\frac{\mathrm{d}l}{\mathrm{d}t} \quad (t \text{ is time}) \tag{8.5.2}$$

According to Eq. (8.5.2), length l assumes as well the function of time for a given strain rate because

$$\ln\frac{l}{s_0} = \int_0^t \dot{\varepsilon}(t)\mathrm{d}t \quad (l = s_0 \text{ when } t = 0) \tag{8.5.3}$$

The volume strain rate is zero so that for the grain thickness, we get

$$\ln \frac{s}{s_0} = -\frac{1}{2} \int_0^t \dot{\varepsilon}(t) \mathrm{d}t \tag{8.5.4}$$

In this model, the deformation and flow of a single grain determines the deformation and flow of the whole specimen.

For the well-developed superplasticity, when $l \gg s$, the mass of the base metal grain is equal to $m_\mathrm{b} = \rho_\mathrm{b} l s^2$, and the mass of the alloying ingredient in the intergranular layer equals $m_\mathrm{a} = 2\rho_\mathrm{a} l s t$ per grain so that we have

$$\frac{t}{s} = \frac{\rho_\mathrm{b} m_\mathrm{a}}{2\rho_\mathrm{a} m_\mathrm{b}} \tag{8.5.5}$$

Here, t is the thickness of the intergranular layer, and ρ_a and ρ_b is the mass density of the alloying and base metals.

For example, in Pb–62%Sn eutectic alloy where tin is the base metal, and lead is the alloying ingredient, we get $t/s = 0.1$.

8.6 The Theory of Superplasticity: Elongation to Failure

The large neck-free elongation to failure is the main property describing the superplastic behavior of some materials at low stresses and high temperatures. In general terms, this phenomenon can be understood using the kinetic theory of fracture, see Chap. 2 in [13]. According to this theory, the time to failure of a stretched bar is equal to

$$t_\mathrm{F} = \tau_0 \exp \frac{U_\mathrm{F} - \sigma v_\mathrm{A}}{RT} \tag{8.6.1}$$

here, σ is the tensile stress in the bar; t_F is the time to failure; U_F is Arrhenius' chemical activation energy of the given material substance equal to about 10^2 to 10^3 KJ per gram atom or gram mole for condensed matter, which characterizes fracturing and is close to the binding energy of the substance; T is the absolute temperature in Kelvins; v_A is the bar activation volume per gram mole or gram atom, in which the active failure occurs; τ_0 is the characteristic time of an elementary thermal fluctuation, namely the propagation time of phonons on one interatomic spacing, which is about 10^{-13} to 10^{-12} s; and R is the universal gas constant equal to 8.31 J per gram mole and per Kelvin.

In this equation, Eq. (8.6.1), Boltzmann's constant is often used instead of the universal gas constant so that in such a case, U_F and v_A are taken per one atom or molecule of the substance. As a reminder, the universal gas constant equals Boltzmann's constant times the Avogadro number.

The extensive experimental investigation of the long-term strength of bars made of various common materials under stationary tensile loads demonstrated the applicability of Eq. (8.6.1) to a wide range of loading times from some microseconds to several months and absolute temperatures from close to zero to the melting point [13, 14, 16, 17]. Despite some obvious shortcomings of this equation, it provides a good common sense estimate of time to failure in an exceptionally broad range of time and temperature. However, this theory was done before the era of superplasticity so that its extrapolation for the superplastic state constitutes a hypothesis that should be independently verified.

The strain rate of the steady superplastic flow or creep of a bar is governed by the following empirical constitutive equation [13, 14, 16, 17]

$$\dot{\varepsilon} = r_0 \left(\frac{\sigma}{\sigma_0} \right)^n \exp\left(-\frac{U_c}{RT} \right) \tag{8.6.2}$$

Here, $\dot{\varepsilon}$ is the strain rate; σ is the tensile stress; n is the dimensionless creep index which for most metals usually varies in the range of 2–10, and in the superplastic state, it can be much greater; U_c is the activation energy per gram atom characterizing the creep or superplastic flow; and r_0 and σ_0 are some fitting constants of the dimension of strain rate and stress, respectively.

Suppose the strain rate and the stress do not vary in time during test which is the case in most common test procedures. In this case, from Eq. (8.5.3), it follows that

$$\ln \frac{l_F}{s_0} = \dot{\varepsilon} t_F \tag{8.6.3}$$

Here, l_F is the grain length at failure which is directly proportional to the specimen length at failure L_F. According to Eq. (8.5.1), we have

$$\frac{L_F}{L_0} = \frac{l_F}{s_0} \tag{8.6.4}$$

Replacing t_F and $\dot{\varepsilon}$ in Eq. (8.6.3) by Eqs. (8.6.1) and (8.6.2) and using Eq. (8.6.4), we come to the following equation

$$\ln \frac{L_F}{L_0} = \tau_0 r_0 \left(\frac{\sigma}{\sigma_0} \right)^n \exp \frac{\Delta U - \sigma v_A}{RT} \quad (\Delta U = U_F - U_c) \tag{8.6.5}$$

The activation energy characterizes the energy barrier which is necessary to overcome in order that a chemical or physical transformation would take place. This barrier is greater for fracturing than for creep or superplastic flow so that the value of ΔU is always positive. For the superplastic flow, quantity ΔU plays the role of the activation energy of failure/fracturing.

Equation (8.6.5) provides the specimen length at failure in terms of tensile stress. Let us write down this equation as follows:

$$y = Bx^n e^{-\lambda x} \tag{8.6.6}$$

$$\left(x = \frac{\sigma}{\sigma_0};\ y = \ln \frac{L_F}{L_0};\ B = \tau_0 r_0 \exp \frac{\Delta U}{RT};\ \lambda = \frac{\upsilon_A \sigma_0}{RT} \right)$$

Function $y = y(x)$ has the maximum value $y = y_*$ at $x = x_*$ so that

$$x_* = \frac{n}{\lambda},\quad y_* = B\left(\frac{n}{e\lambda}\right)^n = B\left(\frac{x_*}{e}\right)^n \tag{8.6.7}$$

when x grows from zero to infinity, y increases from zero to the maximum and then decreases tending to zero at infinity.

We accept as a definition that a material is in the superplastic state, if $y_* > 1$, that is, based on Eq. (8.6.7), if the following inequality is met

$$n > e\lambda B^{-1/n} \tag{8.6.8}$$

We call a material ideally superplastic if $y_* \gg 1$. The greater is the value of y_*, the more superplastic is the material. (When $L_F = eL_0$, then $y = y_* = 1$.)

According to Eqs. (8.6.6) and (8.6.7), the maximum superplastic elongation y_* in terms of temperature is given by the following function

$$y_* = \tau_0 r_0 (\delta T_*)^n \exp \frac{1}{T_*} \tag{8.6.9}$$

$$T_* = \frac{TR}{\Delta U},\quad \delta = \frac{n\Delta U}{e\sigma_0 \upsilon_A} \tag{8.6.10}$$

Function $y_* = y_*(T_*)$ in Eq. (8.6.9) has the minimum value at $T_* = 1/n$ and is infinite at zero and at infinity.

The superplastic state can take place only if $T_* > 1/n$ so that, from here and from Eq. (8.6.10), it follows that the necessary condition of superplasticity is

$$T > \frac{\Delta U}{nR} \tag{8.6.11}$$

Since $T > T_m/2$ according to test data, Eq. (8.6.11) provides a useful estimate for the activation energy of failure in the superplastic flow in terms of the melting point

$$\Delta U = \frac{1}{2} nRT_m \tag{8.6.12}$$

This fundamental relation allows us to simplify Eqs. (8.6.5) and (8.6.7) as follows:

$$\max \ln \frac{L_F}{L_0} = \tau_0 r_0 \left(\frac{nRT}{e\sigma_0 v_A} \right)^n \exp \frac{nT_m}{2T} \quad \left(\frac{\sigma}{\sigma_0} = \frac{n}{\lambda},\ T_m > T > \frac{1}{2} T_m \right) \tag{8.6.13}$$

$$\ln \frac{L_F}{L_0} = \tau_0 r_0 \left(\frac{\sigma}{\sigma_0} \right)^n \exp\left(-\frac{\sigma v_A}{RT} \right) \exp \frac{nT_m}{2T}. \tag{8.6.14}$$

According to Eq. (8.6.2), the tensile stress can be expressed in terms of the strain rate

$$\frac{\sigma}{\sigma_0} = \left(\frac{\dot{\varepsilon}}{r_0} \right)^{1/n} \exp \frac{U_c}{nRT} \tag{8.6.15}$$

From Eqs. (8.6.7) and (8.6.15), it follows that the maximum elongation is achieved when the strain rate is equal to

$$\dot{\varepsilon}_* = r_0 \left(\frac{n}{\lambda} \right)^n \exp\left(-\frac{U_c}{RT} \right) \tag{8.6.16}$$

This strain rate is called the optimum strain rate.

The comparison of this theory with test data as well as some consequences and ramifications are given in the next Sects. 8.7 and 8.8.

8.7 Characterization of Superplastic Materials: The *A*-Number

To characterize the capability of different materials in yielding the superplastic flow, let us introduce the following dimensionless number (the *A*-number)

$$A = \max \ln \frac{L_F}{L_0} \tag{8.7.1}$$

According to Eq. (8.6.13), the *A*-number is a function of temperature that can be written as follows:

$$A(T) = A(T_m) \left(\frac{T}{T_m} \right)^n \exp\left[\frac{n}{2} \left(\frac{T_m}{T} - 1 \right) \right] \quad \left(T_m \geq T > \frac{1}{2} T_m \right) \tag{8.7.2}$$

$$A(T_m) = \tau_0 r_0 e^{-n/2} \left(\frac{nRT_m}{\sigma_0 v_A} \right)^n. \tag{8.7.3}$$

Evidently, the greater the A-number, the more superplastic is the material at a given temperature. From Eq. (8.7.2), it follows that the A-number increases, i.e., the material becomes more superplastic, when temperature grows.

Maximum Superplastic Elongation of Grains at Melting Point Let us estimate the superplastic elongation of a base grain at its melting temperature when the elongation achieves maximum. We assume that the melting point of the base grain is less than that of its alloying boundary, which is usually the case. At this melting point, the solid alloying boundary of a base grain plays a role of the shell that bears all loads like the surface tension in liquids or a tensile tension in membrane shells.

Usually, the metal of the boundary layer is more strong and brittle than the base metal of the grain so that at large elongations, the boundary layer is torn by a periodic system of fractures, every one of which encircles the cross section of a long prism of the base metal grain (see below). In the process of flow, these fractures become wide gaps or cavities filled by vapor of the base metal. They cannot bear a load so that, in this limiting case of the melting temperature, it is the real surface tension of the liquid base metal that bears the tensile load.

And so, in this limiting case, the following equilibrium equation holds for the square cross section of the liquid base metal grain

$$\sigma s_{\min} = 4\gamma_{\mathrm{B}} \tag{8.7.4}$$

Here, γ_{B} is the surface tension of the base metal at melting point.

From here, using also Eqs. (8.6.4), (8.6.7), and (8.7.1), we can find

$$A(T_{\mathrm{m}}) = 2 \ln \frac{n s_0 R T_m}{4\gamma_{\mathrm{B}} v_{\mathrm{A}}} \tag{8.7.5}$$

The value of λ defined by Eq. (8.6.13) is equal to the tangent of the angle constituted by the straight lines of Eq. (8.6.2) written in terms of $\ln(\tau_{\mathrm{F}}/\tau_0)$ versus σ/σ_0 as follows:

$$\ln \frac{\tau_{\mathrm{F}}}{\tau_0} = \frac{U_{\mathrm{F}}}{RT} - \lambda \frac{\sigma}{\sigma_0} \quad (\lambda RT = v_{\mathrm{A}}\sigma_0) \tag{8.7.6}$$

These linear diagrams are well known in the kinetic theory of failure [13, 14, 16, 17].

As well, it is convenient to use the constitutive Eqs. (8.6.2) or (8.6.15) in the following shape

$$\ln \frac{\dot{\varepsilon}}{r_0} = -\frac{U}{RT} + n \ln \frac{\sigma}{\sigma_0} \tag{8.7.7}$$

It represents the superplastic behavior by straight lines on this logarithmic diagram where the creep index is the tangent of the inclination angle.

Using Eq. (8.7.5), the equation for the A-number, Eq. (8.7.2), can be written as follows:

$$A = 2\left(\frac{T}{T_{\mathrm{m}}}\right)^{n} \exp\left[\frac{n}{2}\left(\frac{T_{\mathrm{m}}}{T} - 1\right)\right] \ln \frac{n s_0 R T_{\mathrm{m}}}{4\gamma_{\mathrm{B}} v_{\mathrm{A}}}. \quad (8.7.8)$$

This is the final equation for the A-number that characterizes the superplastic property of a metal in terms of its temperature, melting point, creep index, initial grain size, surface tension of liquid metal, and activation volume.

From Eqs. (8.7.1) and (8.7.5), it follows that the upper boundary of the maximum superplastic elongation (at the melting point) is given by the following equation

$$\max \frac{L_{\mathrm{F}}}{L_0} = \left(\frac{n s_0 R T_{\mathrm{m}}}{4 v_{\mathrm{A}} \gamma_{\mathrm{B}}}\right)^2 \quad (8.7.9)$$

As a result, we have got Eqs. (8.7.1) and (8.7.8), the former providing the A-number from experiments, and the latter from the given theory.

8.8 Comparison of the Theory with Test Data

Experimental results are usually given in terms of elongation versus strain rate. In these variables, the theoretical Eqs. (8.6.13) and (8.6.14) are reduced to the following equation

$$y = Bx \exp(-\delta x^{1/n}) \quad \left(x = \frac{\dot{\varepsilon}}{r_0}\right). \quad (8.8.1)$$

here

$$y = \ln \frac{L_{\mathrm{F}}}{L_0}, \quad (8.8.2)$$

$$B = \tau_0 r_0 \exp\left(\frac{U_{\mathrm{c}}}{RT} + \frac{n T_{\mathrm{m}}}{2T}\right), \quad (8.8.3)$$

$$\delta = \frac{v_{\mathrm{A}} \sigma_0}{RT} \exp \frac{U_{\mathrm{c}}}{nRT}. \quad (8.8.4)$$

Let us compare this theory with test data for two typical superplastic alloys taken from the book by Padmanabhan and Davis [18] and from the paper [19] by Kawasaki and Langdon.

It is reasonable to choose the dimensionless constants n, δ, and B from the same test data. Also, we normalize the value of strain rate by equation $r_0 = \dot{\varepsilon}$ at the maximum of function $y = y(x)$ where the maximum elongation is achieved. Since its derivative is equal to

$$\frac{\mathrm{d}y}{\mathrm{d}x} = B\exp\left(-\delta x^{1/n}\right)\left(1 - \frac{\delta}{n}x^{1/n}\right), \tag{8.8.5}$$

we get the following two condition equations at the point $x = x_m$ of the function maximum, $y = y_m$

$$\delta = n x_m^{-1/n}, \quad B = \frac{y_m}{x_m}\mathrm{e}^n. \tag{8.8.6}$$

The only one more equation necessary to find constant n can serve the test result $y = y_*$ at the lowest value of loading $x = x_*$ so that according to Eq. (8.8.1), we have

$$B x_* = y_* \exp\left(\delta x_*^{1/n}\right). \tag{8.8.7}$$

The equation system, Eqs. (8.8.6) and (8.8.7), is reduced to the following equations:

$$2\ln 10 = -n\ln\left(1 - \frac{b}{n}\right), \tag{8.8.8}$$

$$b = \ln\frac{x_m y_*}{x_* y_m}, \quad (b > n). \tag{8.8.9}$$

After we find $n = n(b)$ from Eq. (8.8.8), the values of δ and B can be calculated by Eqs. (8.8.6). After that, the value of y determining the length of specimen at failure can be found using Eq. (8.8.1) that can be written as follows:

$$\frac{y}{y_m} = \frac{x}{x_m}\exp\left\{n\left[1 - \left(\frac{x}{x_m}\right)^{1/n}\right]\right\}. \tag{8.8.10}$$

Pb–62%Sn Eutectic Alloy [18] First, using Eq. (8.7.8), let us calculate the A-number for tin in the Pb–62%Sn alloy at $T = 252.5$, 473, and 505 K where $T_\mathrm{m} = 505$ K is the melting point of tin, the base metal.

We accept the following typical figures for white tin:

$RT_\mathrm{m} = 4.2$ KJ per gram-atom; $v_\mathrm{A} = 10^3$ cm^3 per gram-atom;
$n = 16$ (see below); $s_0 = 4.5$ μm; $\gamma_\mathrm{B} = 200$ Pa cm.

As a result, we get:

$$A(252\ \mathrm{K}) = 0.32; \quad A(T) = 4.34 \text{ at } T = 473\ \mathrm{K}; \quad A(T_\mathrm{m}) = 7.19,$$

so that $\max L_\mathrm{F}/L_0$ equals 1.38 at 252 K, 78.1 at 473 K, and 1200 at 505 K.

According to the experimental data [18], the Pb–62%Sn eutectic alloy specimen experiences the maximum length 76.5 times its original length at strain rate $2.12 \times 10^{-4}\ \mathrm{s}^{-1}$ and temperature 473 K. And so, at this point, the test results well support the theory. According to the theory, in the limit, the specimen of this superplastic alloy can extend 1200 times its original length at melting point so that its diameter becomes 35 times less than the initial value before loading.

Also, let us compare the results of calculation of superplastic elongation using Eq. (8.8.1) with other test data given in book [18] for the Pb–62%Sn eutectic alloy at $T = 473$ K:

$\dot{\varepsilon}$, s^{-1}	2.12×10^{-4}	1.06×10^{-3}	5.29×10^{-3}	2.12×10^{-2}
$\frac{\Delta L}{L_0}$ (test) (%)	7550	4600	2800	630
$\ln \frac{L_F}{L_0}$ (test)	4.34	3.85	3.36	1.99
$\ln \frac{L_F}{L_0}$ (theory)	4.34	3.99	3.08	2.08

Here, the following values of constants were used in Eq. (8.8.1):

$$n = \delta = 16; \quad r_0 = 2.12 \times 10^{-4}\ \mathrm{s}^{-1}; \quad B = \mathrm{e}^{16} \ln 76.5.$$

In this case, the maximum superplastic elongation is achieved at $\dot{\varepsilon} = r_0$.

As seen, the discrepancy between the theory and the test is about 4% in average so that we can conclude the theory is confirmed by the test results because the discrepancy is within the scatter of experimental results caused by many factors ignored either in the test or in the theory. The elongation to failure is very sensitive to these unaccounted factors, which explains wide scatter of data in tests.

Zn–22%Al Eutectoid Alloy [19] First, let us provide some figures characterizing the effect of the value of creep index on the superplastic flow at the lowest temperature $T = 0.5\, T_m$ of superplasticity. From Eq. (8.7.2), we can find:

$A(T)/A(T_m)$	0.46	0.38	0.314	0.14	0.045
n	4	5	6	10	16

And so, even at the lowest temperature, this effect is substantial.

Let us consider the test results of the extension of the Zn–22%Al eutectoid alloy specimens at 473 K for several strain rates [19]:

$\dot{\varepsilon}$, s^{-1}	10^{-4}	10^{-3}	3.3×10^{-3}	10^{-2}	3.3×10^{-2}	10^{-1}	1.0
$\frac{\Delta L}{L_0}$ (test) (%)	670	1410	1600	2230	1890	1010	520
$\ln \frac{L_F}{L_0}$ (test)	2.04	2.7	2.83	3.15	2.99	2.40	1.82

In these tests, no necking up to failure was observed. The material was originally treated by equal channel angular processing (ECAP) to produce grains of the size about 0.8 μm. After ECAP, specimens of 2 × 3 × 4 mm size were made to test.

In order to calculate the corresponding theoretical values of elongation to failure, we can use Eqs. (8.8.6) to (8.8.10) where the following figures are valid for this test:

$$x_1 = 10^{-4}\ s^{-1};\quad x_m = 10^{-2}\ s^{-1};\quad y_1 = 2.04;\quad y_m = 3.15.$$

For these figures, the root of Eq. (8.8.8) with a less than 0.5% error is given by

$$n = 23. \tag{8.8.11}$$

In accordance with the theory, when the strain rate grows, the elongation to failure increases until a certain maximum is achieved at the optimum strain rate which is equal to about $10^{-2}\ s^{-1}$, and then the elongation to failure decreases.

Using the indicated figures and Eqs. (8.8.10) and (8.8.11), let us calculate the theoretical values of y/y_m and compare them with the experimental ones:

$\frac{x}{x_m}$	10^{-2}	10^{-1}	0.33	1	3.3	10	100
$\left(\frac{y}{y_m}\right)_{\text{test}}$	0.65	0.86	0.90	1	0.95	0.76	0.58
$\left(\frac{y}{y_m}\right)_{\text{theory}}$	0.65	0.88	0.91	1	0.96	0.78	0.61

However, despite a very good compliance of this theory with the test results, we should keep in mind that Kawasaki and Langdon concluded that "grain boundary sliding is the dominant flow process during superplastic elongation" while in the present calculation of the strain rate, the boundary sliding is ignored and it is only superplastic elongation that is taken into account. And so, this compliance questions the fundamental conclusion of the original authors about the role of the pure stretching versus the boundary sliding in the mechanism of superplastic deformation. Another drastic difference is in the value of parameter n of strain rate sensitivity, which is found here to be equal $n = 23$ while Kawasaki and Langdon estimate it to be within

2–10 at different intervals of the strain rate of grains. The kinetic theory used here takes into account both stretching and sliding mechanisms leading, in sum, to the enormous values of creep index.

These cardinal differences, however, require further studies of the mechanism of superplastic deformation. The Zn–22\%Al alloy is of special interest for superplasticity because both its components, zinc and aluminum, have ultrafine grain size and are superplastic at 473 K since this test temperature is greater than half of their melting temperatures, 693 and 933 K.

8.9 Periodic System of Fractures in the Bonding Layer

An alloying metal of a thin layer that bonds two neighboring grains of a base metal usually has a greater melting temperature and is less superplastic than the base metal. As a result, at a sufficiently large extension, the bonding layer of each grain breaks with many cross-sectional fractures and turns into a periodic system of streaks of length $2L_s$ separated by the fractures.

Bonding streaks bridge the base metal grains. We assume that the streak length is much greater than its thickness t, i.e., $L_s \gg t$. Also, we assume that the x-axis is chosen along the extension direction, that is, along the stretched grain and streak, with the origin being at the center of a streak. The streak is subject to the tensile stress $\sigma(x)$ inside and the shear stress $\tau(x)$ on both interfaces of the streak bordering the base metal. These stresses bear the external load of the bar extension; they are tied by the following equilibrium equation

$$t\frac{d\sigma}{dx} = -\tau \quad (-L_s < x < +L_s) \tag{8.9.1}$$

Also, let us assume that the streak is subject to yielding under plane strain condition and use the Mises yielding criterion so that

$$\sigma^2 + 6\tau^2 = \sigma_Y^2. \tag{8.9.2}$$

Here, σ_Y is the yield strength of the alloying metal at the test temperature.

The solution of the equation system, Eqs. (8.9.1) and (8.9.2), satisfying the boundary condition $\sigma = 0$ when $x = \pm L_s$ can be written as follows:

$$\sigma = \sigma_Y \cos\frac{\pi x}{2L_s}, \quad \tau = \frac{1}{\sqrt{6}}\sigma_Y \sin\frac{\pi x}{2L_s}, \tag{8.9.3}$$

$$L_s = \pi t\sqrt{1.5}. \tag{8.9.4}$$

Hence, the streak length is about 7.7 times greater than its thickness which sufficiently well complies with the original assumption. However, we should keep in mind that the value of thickness corresponding to the developed superplastic flow is much less than it was initially.

Summary The present chapter looks into the fracturing, creep, and fatigue of polycrystalline superplastic materials at the nano- and microscales of their structure. It is shown that a rule of thumb can be used for a simple estimate of safe cycling loadings. The ultrafine grain size necessary for superplastic flow is estimated in terms of material and physical constants. The maximum elongation to failure under extension of rods is calculated using the kinetic theory and methods of the nanomechanics of fracturing. The dimensionless A-number is introduced to characterize the yielding of materials with ultrafine grains to the superplastic flow. For two popular superplastic alloys, elongations to failure are determined and compared to some test data.

Literature

1. G.P. Cherepanov, Theory of superplasticity and fatigue of polycrystalline materials based on nanomechanics of fracturing. Phys. Mesomech. **22**(1), 32–51 (2019)
2. G.P. Cherepanov, Cracks in solids. Solids Struct. **4**(4), 811–831 (1968)
3. G.P. Cherepanov, On the crack growth under cyclic loadings. J. Appl. Mech. Tech. Phys. (JAMTP) **6**, 924–940 (1968)
4. G.P. Cherepanov, H. Halmanov, On the theory of fatigue crack growth. Eng. Fract. Mech. **4**(2), 231–248 (1972)
5. G.P. Cherepanov, *Mechanics of Brittle Fracture* (Nauka, Moscow, 1974), p. 640
6. G.P. Cherepanov, *Mechanics of Brittle Fracture* (McGraw-Hill, New York, 1978), p. 950
7. G.P. Cherepanov, The start of growth of micro-cracks and dislocations. Soviet Appl. Mech. **23**(12), 1165–1183 (1988)
8. G.P. Cherepanov, The growth of micro-cracks under monotonic loading. Soviet Appl. Mech. **24**(4), 396–415 (1988)
9. G.P. Cherepanov, The closing of micro-cracks under unloading and generation of reverse dislocations. Soviet Appl. Mech. **24**(7), 635–648 (1989)
10. G.P. Cherepanov, On the foundation of fracture mechanics: fatigue and creep cracks in quantum fracture mechanics. Soviet Appl. Mech. **26**(6), 3–12 (1990)
11. G.P. Cherepanov, Invariant integrals in continuum mechanics. Soviet Appl. Mech. **26**(7), 3–16 (1990)
12. G.P. Cherepanov, A. Richter, V.E. Verijenko, S. Adali, V. Sutyrin, Dislocation generation and crack growth under monotonic loading. J. Appl. Phys. **78**(10), 6249–6265 (1995)
13. G.P. Cherepanov, *Methods of Fracture Mechanics: Solid Matter Physics* (Kluwer, Dordrecht, 1997), p. 312
14. G.P. Cherepanov, Nanofracture mechanics approach to dislocation generation and fracturing (Invited paper at the 12th U.S. National Congress of Applied Mechanics). Appl. Mech. Rev. **47**(6), Part 2, S326 (1994)
15. G.P. Cherepanov, in *Fracture Mechanics* (Moscow-Izhevsk, ICS, 2012), 872 p
16. R.G. Lerner, G.L. Trigg (eds.), *Encyclopedia of Physics*, 2nd edn. (VCH Publ, New York, 1990)
17. R.W.K. Honeycomb, *Plastic Deformations of Metals* (E. Arnold, London, 1968)
18. K.A. Padmanabhan, G.J. Davis, *Superplasticity: Mechanical and Structural Aspects, Environmental Effects, Fundamentals and Applications* (Springer, Berlin, 1980)
19. M. Kawasaki, T.G. Langdon, Grain boundary sliding in a superplastic Zinc-Aluminum alloy processed using severe plastic deformation. Mater. Trans. Jpn. Inst. Metals **49**(1), 84–89 (2008)

Chapter 9
Snow Avalanches

Abstract In this chapter, the snowpack compressed by the gravity of snow is modeled as a multilayer sandwich on a slope with several parallel planes of transverse slippage. The invariant integral of snowpack describes the critical state at the head of this structure when the avalanche starts on moving. This critical state is characterized in terms of physical and geometrical parameters of snow and slope. The solution of the avalanche motion equation was used to simulate the motion of the famous avalanche that happened on February 7, 2003, in Vallee de la Sionne in Switzerland. Also, the progressive failure of skyscrapers is studied and compared with the similar mechanics of avalanches. As applied to the WTC collapse on September 11, 2001, it is shown that its progressive failure would have taken more than 15 cek while, in fact, it took about 12 s which is characteristic for the free-fall demolition. This chapter is for those who are interested in the analysis of progressive failure leading to disasters and catastrophes.

In this chapter, the invariant $\Gamma-$integral of fracture mechanics is used to calculate the frontal pressure and resistance to the downward motion of a snow avalanche. A basic characteristic property of the snowpack termed the entrainment toughness is introduced. From an analysis of the non-entrainment frictional mechanisms of avalanches, we find the necessary condition for a fracture-entrainment regime, and from an analysis of limiting equilibrium of gravitational force and frontal resistance, the necessary condition equation for the start of avalanches. We then derive the governing equations for the dynamics of avalanches by using a point-mass approach, but taking into account entrainment. The governing equations are used to numerically simulate the Vallée de la Sionne avalanche of February 7, 2003, in Switzerland, which was carefully monitored and measured by Alpine services.

The chapter is based on the paper *A fracture-entrainment model for snow avalanches* by G. P. Cherepanov and I. E. Esparragoza published in *Journal of Glaciology*, **54**(184), 2008. The mathematical theory of avalanche-type destructions was first advanced by the present author in the paper *Mechanics of the WTC collapse* published in *Journal of Fracture*, **141**, 2006 (submitted in July 2005); see also Sect. 9.9 of this chapter and papers [1–5].

G. P. Cherepanov, *Invariant Integrals in Physics*,
https://doi.org/10.1007/978-3-030-28337-7_9

9.1 Introduction

Entrainment in snow avalanches is the key to understanding the motion of snow masses and hence to predict their speed, impact pressures, and final runout distance [6, 7]. Different avalanche entrainment mechanisms and rates have been reported by Sovilla et al. [8] who analyzed a total of eighteen avalanche events, many captured at the instrumented Vallée de la Sionne test site. In this survey of snow entrainment, three different entrainment mechanisms were identified: plowing, step entrainment, and basal erosion. The maximum entrainment rates (350 kg $m^{-2}s^{-1}$) were found to occur during frontal plowing and step entrainment.

The step entrainment mechanism was observed to occur by a fracture failure at the interface of two snow layers, while the multilayer entrainment model [8] assumes that entrainment processes are governed by snow strength, primarily the shear resistance of snow. This entrainment model is based on the earlier work of Russian snow scientists [9]. Although this model provides the correct entrainment rates, it requires shear strength values with no clear physical basis.

In this chapter, we use the invariant $\Gamma-$integral of solid and fracture mechanics for the calculation of energy balance, frontal pressure, and frontal resistance to the motion of the avalanche by the snowpack. A method to model the step entrainment process was identified by Sovilla et al. [10]. The frontal resistance is calculated, which allows us to determine the necessary condition governing the entrainment regime. For the start of avalanches, a limiting equilibrium condition equation is also found. Then, we derive the governing equations for the dynamics of avalanches in the simplest approach, taking into account entrainment, inertia, gravitation, and friction forces using the Voellmy–Salm model [11].

These governing equations are solved numerically and the Vallée de la Sionne avalanche of February 7, 2003, is simulated. Maximum possible acceleration of snow avalanches is calculated and compared with the acceleration of the avalanche-type fracturing wave that destroyed the World Trade Center in New York on September 11, 2001.

9.2 Frontal Resistance in the Fracture-Entrainment Mode

Assuming a multilayer snowpack, let us calculate the frontal resistance due to entrainment. Also, let us assume that slip fractures at the layer interfaces in the process zone govern the resistance (Fig. 9.1). The energy balance of forces in this zone is given by means of the invariant $\Gamma-$integral of fracture mechanics. The avalanche is assumed to move in the direction of the x_1- axis, with the x_2- axis being perpendicular. The x_3- axis is perpendicular to the x_1- and x_2- axes so that $Ox_1x_2x_3$ form the right triad.

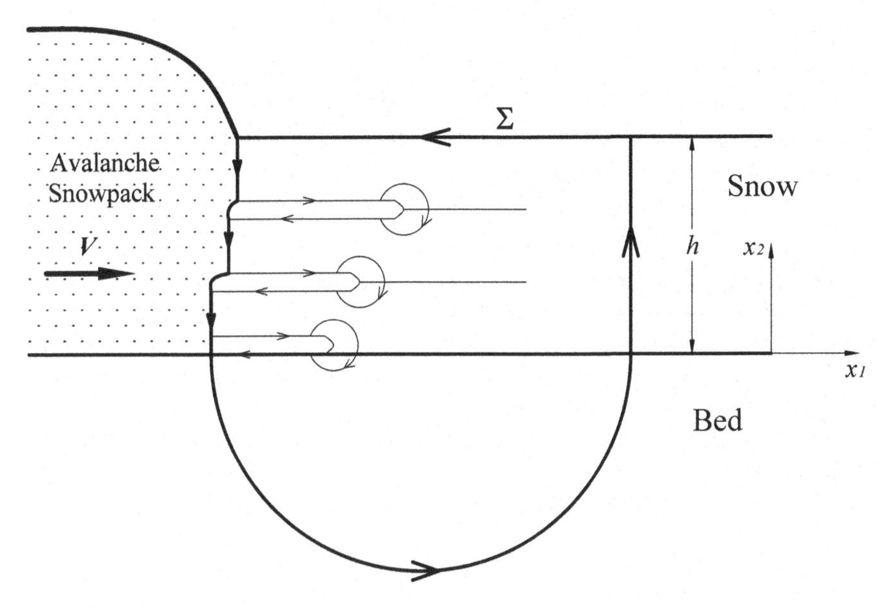

Fig. 9.1 Integration contour of the invariant Γ−integral over the process zone based on the modified SBB fracture-entrainment model. Interface slips along the boundaries of adjacent layers. Closed S contour embraces the process zone, the direction of integration being shown by arrows

Figure 9.1 represents a schematic snapshot of the process zone at a certain moment of time. The frontal resistance is a result of interactive stresses and strains in the process zone and can be calculated from the energy balance in this zone. For the purpose of such a calculation, we make the following assumptions for the process zone:

1. The bed is an elastic continuum half-space $x_2 < 0$ with one slip fracture along the interface boundary $x_2 = 0$ (Fig. 9.1).
2. Snow in front of the avalanche is a multilayer continuum with several boundaries along $x_2 =$ const being subject to slip fractures (Fig. 9.1). Inertial forces in the process zone are small in comparison to the fracture resistance.
3. All dissipative processes in the process zone are assumed to be concentrated along several slip discontinuities on interfaces at $x_2 =$ const so that the material outside of these discontinuities (fractures) is an elastic continuum.
4. The front of the snowpack is a solid line along $x_1 =$ const where the constant is different for different layers (Fig. 9.1). The shear stress on the front is zero, and the normal stress σ_{11} which is the frontal avalanche pressure of snow is equal to

$$\sigma_{11} = R_F/h. \tag{9.2.1}$$

Here, h is the thickness of resting snow layer in front of the snowpack and R_F is the frontal resistance to be found from the energy conservation law.

5. Plane strain is assumed.
   The dissipative processes on slip discontinuities including their ends are taken into account in Fig. 9.1 by the paths over the upper and lower banks of discontinuities with small circular path surrounding the ends of discontinuities. The necessity of these circular paths is a result of the fact that what happens with the material at the extreme ends under high stresses and strains in very small volumes is unknown.

We can only characterize the dissipative process inside a small circular zone by a specific dissipation energy spent to advance the discontinuity of unit length. Thus, the snow cover in the process zone is subject to comparatively small deformations up to failure, with both elastic and inelastic components produced by the front pressure of the avalanche.

The law of energy conservation in a solid continuum inside a closed contour can be written in the form of the invariant Γ–integral [12–14] taken over the closed contour S embracing the process zone with discontinuities (Fig. 9.1).

$$\oint_S \left(Wn_1 - \sigma_{ij}n_j u_{i,1}\right)\mathrm{d}S = 0, \quad (i, j = 1, 2, 3). \tag{9.2.2}$$

Here, W is the volume density of deformation work, n_j are the components of the outer unit normal to contour S, σ_{ij} are the stress tensor components, and u_i are the displacement vector components.

The integral over the upper surface of snow layer $x_2 = h$ is equal to zero because $n_1 = 0, n_2 = 1$ and $\sigma_{ij}n_j = 0$ for $i, j = 1, 2, 3$ since there is no loading on the free surface.

The integration path in the bed can be taken in the form of a circle of large radius R where $R \gg h$. The stress–strain field in the bed far from the process zone tends to that of concentrated force $(R_\mathrm{F}, 0)$. The Γ–integral over this circle represents the Γ–residue of the concentrated force, and it is equal to zero in this case.

The Γ–integral over the front of the snowpack where $x_1 = \text{const}$, $n_1 = 1$, $n_2 = 0$, $\sigma_{12} = 0$, and $\sigma_{11} = R_\mathrm{F}/h$ is equal to

$$\int_0^h \left(-W + \sigma_{11}u_{1,1}\right)\mathrm{d}x_2 = h\left(W - \sigma_{11}u_{1,1}\right) = \left(-1 + 2\nu^2 - 2\nu^3\right)\frac{R_\mathrm{F}^2}{2Eh} \tag{9.2.3}$$

because from Hooke's Law for the plane strain, it follows that

$$\begin{aligned} u_{1,1} &= \frac{1-\nu^2}{E}\sigma_{11}, \ \sigma_{33} = \nu\sigma_{11}, \ u_{1,2} = u_{2,1} = 0, \\ u_{2,2} &= -\frac{\nu(1+\nu)}{E}\sigma_{11}, \quad W = \left(\frac{1}{2} + \nu^3\right)\frac{1}{E}\sigma_{11}^2, \end{aligned} \tag{9.2.4}$$

since $\sigma_{22} = 0$ in the common thin plate approximation as applied to the snow layer.

Now, let us calculate the integrals over the slip discontinuities along $x_2 = \text{const}$ where:

(i) $n_1 = 0$ and $n_2 = \pm 1$ ("plus" for the upper bank and "minus" for the lower bank);
(ii) $\sigma_{22} = 0$ in the thin plate approximation; and
(iii) $\sigma_{12} = \tau_{i,i+1}$.

Here, $\tau_{i,i+1}$ is the limiting shear stress on the slip discontinuity between the ith and the $(i+1)$th layer. The integral over the upper and lower banks of the ith discontinuity is equal to $\Delta_i \tau_{i,i+1}$ where $\Delta_i = 2\int_0^{L_i} u_{1,1}\,\mathrm{d}x_1$ is the summary displacement jump between the upper and lower banks of the discontinuity accumulated at the front of the snowpack, the so-called transverse shear crack distortion (here, L_i is the length of the ith discontinuity).

The integral over the small circular path surrounding the singular end of the ith discontinuity is equal to $\Gamma_{\mathrm{C}i}$ where $\Gamma_{\mathrm{C}i}$ is the dissipation energy spent to increase the ith discontinuity by a unit length.

Combining these particular calculations in Eq. (9.2.2) provides

$$\left(1 - 2\nu^2 + 2\nu^3\right)\frac{R_{\mathrm{F}}^2}{2Eh} = \sum_{i=1}^{N} \Delta_i \tau_{i,i+1} + \sum_{i=1}^{N} \Gamma_{\mathrm{C}i} \tag{9.2.5}$$

Here, N is the number of discontinuities (in our particular case, we have $N = 3$).

Equation (9.2.5) allows us to formulate the frontal resistance R_{F} in terms of h and structural material constants as follows:

$$R_{\mathrm{F}} = K_{\mathrm{E}}\sqrt{2h}, \tag{9.2.6}$$

where

$$K_{\mathrm{E}} = \sqrt{\frac{E}{1 - 2\nu^2 + 2\nu^3}\left(\sum_{i=1}^{N} \Gamma_{\mathrm{C}i} + \sum_{i=1}^{N} \Delta_i \tau_{i,i+1}\right)}. \tag{9.2.7}$$

In the case of identical limiting shear stresses when $\tau_{i,i+1} = \tau_s$ for any i, we get

$$K_{\mathrm{E}} = \sqrt{\frac{E}{1 - 2\nu^2 + 2\nu^3}\left(\sum_{i=1}^{N} \Gamma_{\mathrm{C}i} + \Delta\tau_s\right)}. \tag{9.2.8}$$

Here, $\Delta = \sum_{i=1}^{N} \Delta_i$ is the summary shear displacement in the process zone near the front of the snowpack.

The parameter K_{E} termed the *entrainment toughness* characterizes the resistance capabilities of the material in the process zone in front of the avalanche. The determination of this value from actual avalanche data is required to predict avalanche motion with entrainment. According to Eq. (9.2.7), the fracture work lost by an avalanche on a unit length of its path can be expressed in terms of the entrainment toughness as follows:

$$\sum_{i=1}^{N} \Gamma_{\mathrm{Ci}} + \sum_{i=1}^{N} \Delta_i \tau_{i,i+1} = \frac{1 - 2\nu^2 + 2\nu^3}{E} K_{\mathrm{E}}^2. \tag{9.2.9}$$

Therefore, this formula describes the arresting capabilities of the process zone in front of the avalanche.

9.3 The Fracture-Entrainment Threshold

The fracture-entrainment mechanism described in the preceding section provides only the frontal resistance force of the snowpack, R_{F}, due to the frontal fracturing. Another possible entrainment mechanism is by shearing the interfaces between snow layers. To determine the resisting force of this friction process, first let us study the limiting equilibrium of the ith snow layer with a through slip plane CD inclined by angle β_i to the x_1- axis (Fig. 9.2). The tangential (shear), τ_n, and normal, σ_n, components of stress on this interfacial slip plane obey Coulomb's Law

$$\tau_n = \tau_i + |\sigma_n| \tan \varphi_i. \tag{9.3.1}$$

Here, τ_i is the adhesion constant, and φ_i is the angle of internal friction in the ith layer.

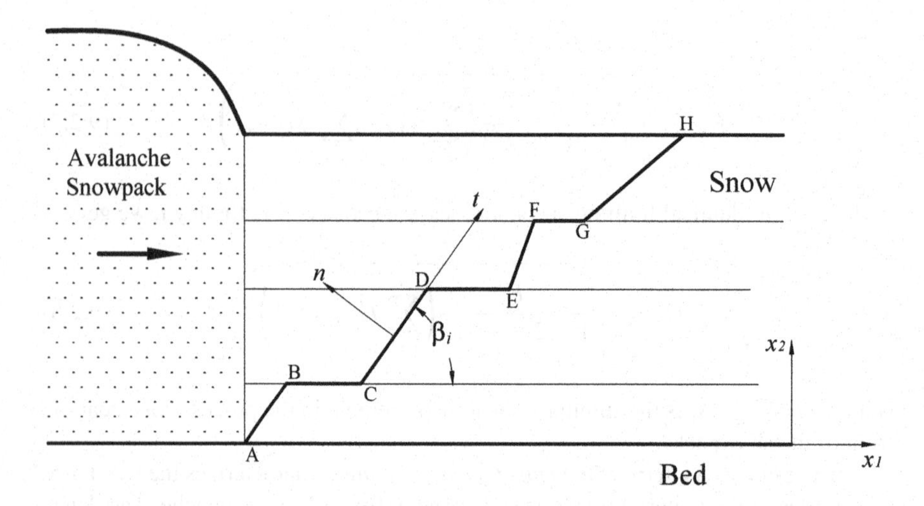

Fig. 9.2 Non-entrainment mode of frontal resistance in the SBB multilayer model. Through slips AB, CD, EF, and GH and interface slips BC, DE, and FG along the boundaries of layers

Values of τ_i and φ_i for snow avalanches derived from chute experiments can be found in [15]. Taking into account that $\sigma_{22} = 0$ in the thin plate approximation, we have the following two equations of equilibrium:

$$|\sigma_n|\cos\beta_i = |\tau_n|\sin\beta_i, \quad R_i = h_i|\sigma_n| + h_i|\tau_n|\frac{\cos\beta_i}{\sin\beta_i}. \tag{9.3.2}$$

Here, h_i is the thickness of the ith layer, and R_i is the component of the friction resistance caused by the ith layer.

Solving the equation system, Eqs. (9.3.1) and (9.3.2), we obtain:

$$|\tau_n| = \frac{\tau_i}{1-\tan\varphi_i\tan\beta_i}, \quad |\sigma_n| = \frac{\tau_i\tan\beta_i}{1-\tan\varphi_i\tan\beta_i}; \tag{9.3.3}$$

$$R_i = \frac{2h_i\tau_i}{\sin(2\beta_i)(1-\tan\beta_i\tan\varphi_i)}. \tag{9.3.4}$$

Let us analyze R_i as a function of β_i. This function tends to plus infinity when $\beta_i \to 0+$ and $\beta_i \to \left(\frac{\pi}{2} - \varphi_i\right)-$. Hence, it has, at least, one minimum in $\beta_i \in \left[0, \frac{\pi}{2} - \varphi_i\right]-$. Equating the derivative to zero, we find the following equation for the minimum point:

$$\tan\beta_i = \frac{1-\sin\varphi_i}{\cos\varphi_i}. \tag{9.3.5}$$

From here and Eq. (9.3.4), it follows that:

$$\beta_i = \frac{\pi}{4} - \frac{\varphi_i}{2}, \tag{9.3.6}$$

$$R_i = 2h_i\tau_i\cot\left(\frac{\pi}{4} - \frac{\varphi_i}{2}\right). \tag{9.3.7}$$

Summing up the resistance of all N layers, we arrive at the friction resistance, R_{F}^{+}, of the process zone

$$R_{\mathrm{F}}^{+} = \sum_{i=1}^{N} 2h_i\tau_i\cot\left(\frac{\pi}{4} - \frac{\varphi_i}{2}\right) + \sum_{i=1}^{N} d_{i,i+1}\tau_{i,i+1}. \tag{9.3.8}$$

Here, $\tau_{i,i+1}$ is the limiting shear stress on the boundary between the ith and $(i+1)$th layers, and $d_{i,i+1}$ is the length of the interface slip between the ith and $(i+1)$th layers.

All terms in the second sum are positive; they increase the value of friction resistance. Hence, $d_{i,i+1} = 0$ for any i, and the absolute minimum with respect to all β_i

where $i = 1, 2, \ldots, N$ provides the frontal resistance R_F of the process zone in the non-entrainment mode as follows:

$$R_F = 2\sum_{i=1}^{N} h_i \tau_i \cot\left(\frac{\pi}{4} - \frac{\varphi_i}{2}\right). \tag{9.3.9}$$

If this value is less than that given by Eq. (9.2.6) for the fracture-entrainment mode, then shearing entrainment is the more likely entrainment mode.

Therefore, fracture entrainment occurs if and only if,

$$K_E\sqrt{2h} < 2\sum_{i=1}^{N} h_i \tau_i \cot\left(\frac{\pi}{4} - \frac{\varphi_i}{2}\right). \tag{9.3.10}$$

Equation (9.3.10) provides an important estimate for the upper bound of the entrainment toughness K_E characterizing the frontal pressure and frontal resistance to the avalanche snowpack.

9.4 Estimate of Arresting Capabilities of Entrainment

Let us use Eqs. (9.2.9) and (9.3.10) in order to estimate the arresting capability and the gravity work in the entrainment processes.

For a simple estimate, we assume that $i = 1,\ h = 1$ m. Because of great diversity of snow properties, the cohesion constant τ_i can vary from 1 kPa to about 100 kPa and the friction angle φ_i from 10° to 40°. From Eq. (9.3.10), we find that the entrainment toughness K_E can vary from 0.01 to 10 MPa $m^{1/2}$.

Young's modulus E of snow can vary from 0.01 GPa to about 1.0 GPa depending on the snow type and density. Using this range of values of E for snow and the above estimate of the entrainment toughness K_E, we can find that the specific dissipation energy of entrainment front per unit area according to Eq. (9.2.9) can vary in a very large range, from 0.01 Jm^{-2} to about 10 MJm^{-2}. Similar to other fracture processes, the entrainment toughness is the main property to measure the arresting capability of a snow slope to an avalanche.

Let us compare this value with the work done by gravity per unit area which is about MgH/A where MgH is the potential energy of gravitation of the avalanche mass M and A is the total area covered by the moving snow mass. Using the data from [8], the following estimates are acceptable for avalanches: M varies from 0.01 to 0.1 Megaton and A varies from 0.1 to about 1 km^2. If we assume that $H = 1000$ m, then the specific work of avalanches will vary from 0.1 MJm^{-2} to about 10 MJm^{-2}.

As seen, under common snow conditions the dissipation energy of entrainment front is much less than the work done by gravity. Only in the case of a well-consolidated snow cover, that it is comparable with the gravity work of an avalanche. A similar result was obtained in [16] for the entrainment of woody debris by avalanches. This remarkable property is common for all brittle fractures: some little, low-energy cracks open door to the large-scale destruction phenomena, like keys do. This is why fracture mechanics is so important.

9.5 A Model of Frictional Resistance with Entrainment

Let us formulate the model conditions of frictional resistance with entrainment.

A mountain of height H is covered by a layer of snow.

Designate: t is time; x is the vertical axis directed downward so that $x = 0$ is the top of the mountain and $x = H$ is the bottom of the mountain; y is the horizontal axis beginning at the top of the mountain under study where $y = 0$; and $\{x = x(s), y = y(s)\}$ is the parametric equation of the curvilinear bed of the mountain on which the snow is lying, where s is the length of the curvilinear path along the bed and $h(s)$ is the thickness of the snow layer (Fig. 9.3).

We assume that the bed is rigid and the snow layer is thin so that $|h(s)| \ll H$. Suppose $M(t)$ is the snowpack mass moving downhill under the gravity force along the curvilinear bed $x = x(s)$, $y = y(s)$. We assume that mass M has the shape of a parallelepiped with dimensions $a \times b \times c$ where c is the snow mass thickness normal to the bed surface, a is the frontal dimension normal to the motion direction, and b is the depth of the snow mass along its motion path.

Therefore, we have

$$M = \rho abc. \tag{9.5.1}$$

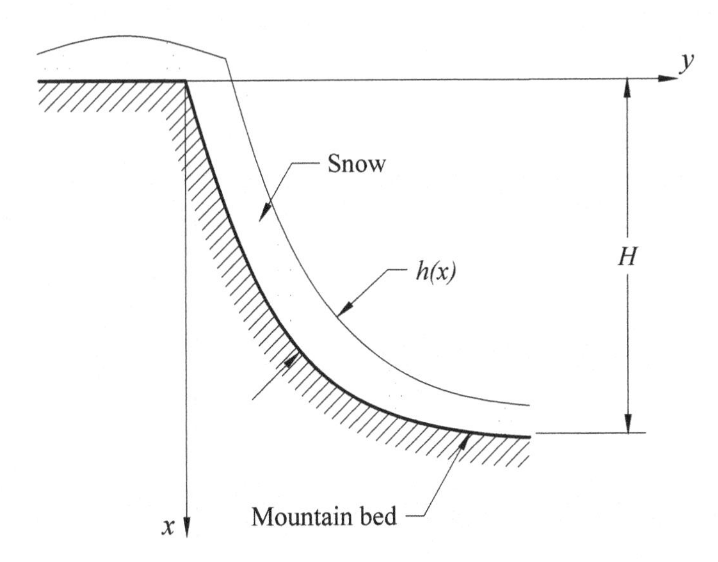

Fig. 9.3 Coordinate system of problem under study

Here, ρ is the snow density in moving mass M.

Due to the entrainment, when mass $M(t)$ moves down, it absorbs $\rho_0 h_* a ds$ amount of a new snow for time dt, such that

$$\frac{\mathrm{d}M}{\mathrm{d}t} = \rho_0 h_* a \frac{\mathrm{d}s}{\mathrm{d}t}. \tag{9.5.2}$$

Here, ρ_0 is the density of the intact snow on the mountain and $s(t)$ is the length of the avalanche path.

Some snow can also be deposited from the avalanche. The value of $h_* = h_*(s)$ in Eq. (9.5.2) therefore represents an effective difference between the snow influx and snow deposition, usually at the tail of the avalanche [17]. When the avalanche is well developed, the deposition equals the influx and $h_* = 0$.

We assume that

(1) ρ_0 is a constant or a known function of s so that $\rho_0 = \rho_0(s)$.
(2) $h(s)$ is a constant or a known function of s.

Applying Newton's Law, the equation of motion for mass M, taking into account the entrainment, is:

$$\frac{\mathrm{d}}{\mathrm{d}t}\left(M\frac{\mathrm{d}s}{\mathrm{d}t}\right) = Mg\cos\alpha - R. \tag{9.5.3}$$

Here, $s(t)$ is the location of mass M on the bed; α is the angle between the $x-$ axis and the motion direction so that dy = (ds)cosα and dx = (ds)sinα; $g = 9.81$ ms^{-2}; and R is the resistance force equal to the sum of the friction force plus the frontal resistance R_F caused by the entrainment of the intact snow layer (Fig. 9.4).

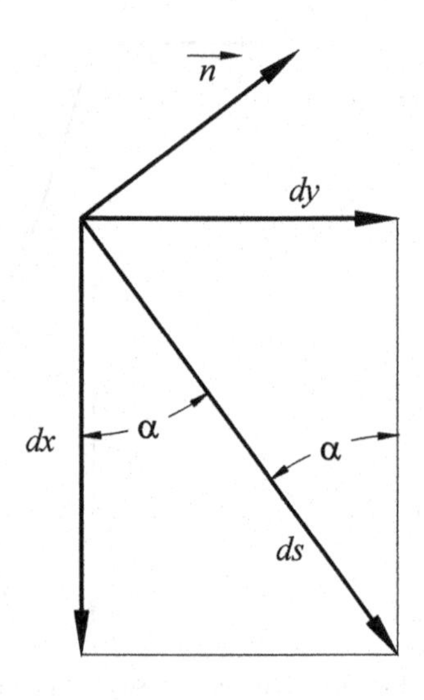

Fig. 9.4 An element of the motion path

For the friction force, we use the Voellmy–Salm model [11, 18, 19], which decomposes R into dry Coulomb friction and "turbulent" velocity-dependent resistance.

Therefore, we have

$$R = f\left(Mg\sin\alpha + \mathrm{Mkv}^2\right) + \rho\mathrm{gacv}^2/\xi + aR_\mathrm{F} \quad (v = \mathrm{d}s/\mathrm{d}t). \tag{9.5.4}$$

Here, f is the Coulomb coefficient of dry friction on the snow–bed interface, ξ is the coefficient of "turbulent" friction of the snow flow, $k = \mathrm{d}\alpha/\mathrm{d}s = 1/r$ is the curvature of the mountain bed, and r is the radius of curvature.

The term Mkv^2 describes the centrifugal force directed along the normal to the bed, which, depending on the curvature, either increases or decreases the normal force of interaction between the snow mass and the bed. The "turbulent" friction term physically represents different velocity-dependent drag forces, for example, air resistance at the front of the avalanche. According to Eq. (9.2.6), the frontal resistance R_F is equal to $K_\mathrm{E}\sqrt{2h(s)}$ where K_E is the entrainment toughness and $h = h(s)$ is the incumbent snow thickness in the process zone.

The resistance force R substantially depends on the speed $\mathrm{d}s/\mathrm{d}t$ of the moving mass M (Fig. 9.5). The entrainment toughness K_E and therefore the frontal resistance R_F are the greater, the denser and older the snow is. We assume that K_E is a constant depending on geographic and seasonal snow conditions. Also, we assume that a

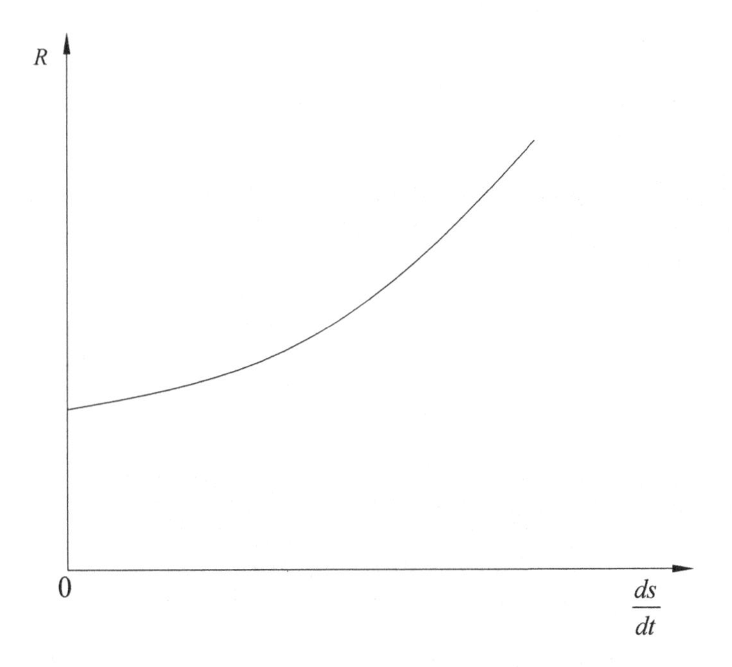

Fig. 9.5 Resistance force versus speed of moving mass

is a constant determined by the specific geometry of the track topography. Frontal resistance may be especially important during the initial stage of avalanche when the avalanche speed is small.

9.6 Governing Equations

Under these assumptions, from Eqs. (9.2.6) and (9.5.2)–(9.5.4) we get the closed equation system for $s = s(t)$ and $M = M(t)$ as follows:

$$\frac{\mathrm{d}M}{\mathrm{d}t} = F(s)\frac{\mathrm{d}s}{\mathrm{d}t}, \quad (F = \rho_0 a h_*); \tag{9.6.1}$$

$$\frac{\mathrm{d}}{\mathrm{d}t}\left(M\frac{\mathrm{d}s}{\mathrm{d}t}\right) = M\left\{g\cos\alpha - f\left[g\sin\alpha + k\left(\frac{\mathrm{d}s}{\mathrm{d}t}\right)^2\right]\right\} - \rho g a c \frac{1}{\xi}\left(\frac{\mathrm{d}s}{\mathrm{d}t}\right)^2 - aK_{\mathrm{E}}\sqrt{2h(s)}. \quad (s = 0,\ \ \mathrm{d}s/\mathrm{d}t = 0 \ \text{ when } \ t = 0) \tag{9.6.2}$$

Here, $\cos\alpha = \mathrm{d}x/\mathrm{d}s$ and $\sin\alpha = \mathrm{d}y/\mathrm{d}s$ are some known functions of s found from the equation of the mountain bed, and $k = \mathrm{d}\alpha/\mathrm{d}s$.

Integrating Eq. (9.6.1) yields

$$M = M(s) = \int_0^s F(s)\,\mathrm{d}s + M_0 = \int_0^s \rho_0 a h_*(s)\,\mathrm{d}s + M_0. \tag{9.6.3}$$

Here, M_0 is the initial mass of the release zone. The avalanche mass M is therefore a function of s that is determined by Eq. (9.6.3).

Let us designate:

$$G\left(s, \frac{\mathrm{d}s}{\mathrm{d}t}\right) = M\left\{g\cos\alpha - f\left[g\sin\alpha + k\left(\frac{\mathrm{d}s}{\mathrm{d}t}\right)^2\right]\right\} - \rho g a c \frac{1}{\xi}\left(\frac{\mathrm{d}s}{\mathrm{d}t}\right)^2 - aK_{\mathrm{E}}\sqrt{2h(s)}. \tag{9.6.4}$$

Using the functions determined by Eqs. (9.6.3) and (9.6.4), Eq. (9.6.2) can be written as

$$\frac{\mathrm{d}}{\mathrm{d}t}\left[M(s)\frac{\mathrm{d}s}{\mathrm{d}t}\right] = G\left(s, \frac{\mathrm{d}s}{\mathrm{d}t}\right). \tag{9.6.5}$$

It is the governing equation determining the motion of the snowpack with time.

We will apply the following procedure for the numerical treatment of this differential equation. In a first series of simulations, we solve Eq. (9.6.5) assuming that the turbulent friction is neglected and curvature k is zero so that $G = G(s)$. In this case, we have

$$v(s) = \frac{ds}{dt} = \frac{1}{M(s)}\sqrt{2\int_0^s M(s)G(s)\,ds + v_0^2 M_0^2}, \tag{9.6.6}$$

$$t = \int_0^s \frac{M(s)ds}{\sqrt{2\int_0^s M(s)G(s)\,ds + v_0^2 M_0^2}}. \tag{9.6.7}$$

When the curvature and/or the turbulent friction are taken into account so that $G = G\left(s, \frac{ds}{dt}\right)$, Eq. (9.6.5) can be transformed into the following form:

$$\frac{dF}{ds} = 2M(s)G\left(s, \frac{\sqrt{F(s)}}{M(s)}\right), \quad F(s) = [M(s)v(s)]^2. \tag{9.6.8}$$

Initial conditions are $s = 0,\ F = 0,\ v = ds/dt = 0$ when $t = 0$.

Equation (9.6.8) can be integrated numerically and the avalanche speed $v = v(s)$ can be found as a certain function of s so that

$$t = \int_0^s [1/v(s)]\,ds. \tag{9.6.9}$$

This equation provides coordinate s of mass M as an implicit function of time.

In the well developed, steady-state regime when $h_* = 0$ and $d^2s/dt^2 = 0$, the speed of the avalanche is determined by equation $R(v) = Mg\cos\alpha$ where $R(v)$ is given by Eq. (9.5.4).

9.7 Numerical Simulation

In this section, we present a numerical simulation of the Vallée de la Sionne avalanche event of February 7, 2003 [19]. For this purpose, Eq. (9.6.5) is solved taking into account both dry and turbulent friction, bed curvature, and the frontal entrainment ignored by other models such as the Voellmy–Salm model.

The data for this event are reported in [8]. We summarize the data required for the numerical simulation using the following information:

a. The release mass is 11.15×10^6 kg, and the deposit mass is 17.16×10^6 kg.
b. Density 200 kgm^{-3}.

c. Coulomb friction is taken $f = 0.26$ [15].
d. The slope angle and avalanche width are approximated as a function of the path length from the data given in [8].
e. In accordance with the entrainment data in [8], the entrainment depth h_* that is the effective difference between snow influx and snow deposition, was assumed to be constant at the beginning of the avalanche up to 800 m of the path length, then decrease linearly to zero between 800 and 1000 m of the path length, remain zero between 1000 and 1600 m of the path length, and due to flank entrainment regains the original constant after 1600 m of the path length.
f. The ratio $\rho gac/\xi$ in the turbulent resistance term was assumed to remain constant. The value of $\xi = 800\,\mathrm{ms}^{-2}$ was used here and taken from [20].
g. The frontal resistance R_F was assumed to remain constant $R_F = 0.1$ Mpam.
h. The curvature of the mountain bed was determined from the slope angle curve given in [8]. This value was approximated and used over sections of the path length.

The result of the simulation using Eq. (9.6.5) based on the above assumptions is shown in Fig. 9.6. As seen, this simulation model is well confirmed by the practical measurement data reported by Sovilla et al. [8].

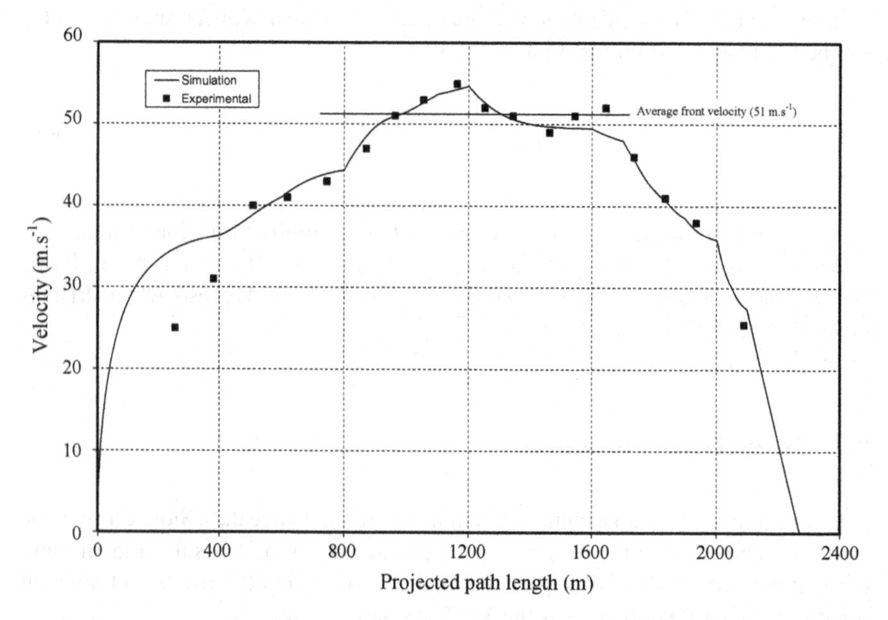

Fig. 9.6 Velocity simulation of the Vallée de la Sionne avalanche event of February 7, 2003, using the data found in Sovilla et al. [8]. The analytical results are compared with the observed data

9.8 Maximum Acceleration of Avalanches—The WTC Collapse

It is useful to get a simple estimate of the avalanche acceleration. Let us assume that $\alpha = \text{const}$ so that the mountain bed is represented by a straight linear slope (Fig. 9.7), and let us ignore the frontal and "turbulent" resistance. Also, we assume that $M_0 = 0,\ v_0 = 0$ that is initial mass and speed of avalanche are zero. Besides, we assume that $a\rho_0 h_* = F = \text{const}$.

In this case, we can get from Eqs. (9.6.1) and (9.6.2) that

$$M = Fs, \quad s = \frac{g}{6}(\cos\alpha - f\sin\alpha)t^2. \tag{9.8.1}$$

And so, the acceleration a_A of the avalanche cannot exceed

$$a_\text{A} = \frac{g}{3}(\cos\alpha - f\sin\alpha). \tag{9.8.2}$$

It will be a bit less if the frontal and "turbulent" resistance forces are taken into account.

For example, when $\alpha = 45^0$ and $f = 0.26$ the maximum possible acceleration of the snow avalanche is equal to $a_\text{A} = 0.17\,g$, which is about five times less than the acceleration of the collapse of the World Trade Center towers in New York on September 11, 2001 [1], that occurred in the regime close to free fall.

Mechanics of the WTC collapse [1, 3–5]. The progressive failure of skyscrapers and towers occurs in the fastest avalanche-type mode similar to snow avalanches. However, it is much slower than free fall that happens after a tower is destructed by explosives, for example, in a demolition procedure.

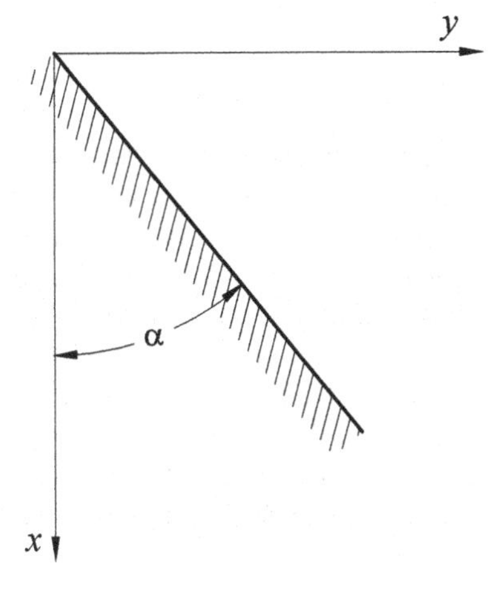

Fig. 9.7 Simplest slope scheme

For the purpose of comparison, let us consider a different regime of fracturing, namely the progressive failure of a tower of height H. Let x be the vertical coordinate directed down so that the top of the tower corresponds to $x = 0$ and the ground floor to $x = H$. Designate by $m = m(x)$ the mass of the building per unit of height.

Suppose $M(t)$ is the mass of the upper structure that moves down under the gravity force, and $H - x(t)$ is the height of the underlying structure intact at the moment t (t is time). Mass $M(t)$ increases with time because it absorbs the underlying structure so that

$$\frac{\mathrm{d}M}{\mathrm{d}t} = m\frac{\mathrm{d}x}{\mathrm{d}t}. \tag{9.8.3}$$

The motion equation of mass M is

$$\frac{\mathrm{d}}{\mathrm{d}t}\left(M\frac{\mathrm{d}x}{\mathrm{d}t}\right) = Mg - R. \tag{9.8.4}$$

Here, R is the force of the resistance of the underlying, intact structure to the motion of the upper mass M falling down. The system of Eqs. (9.8.3) and (9.8.4) describes the regime of progressive failure [1].

Let us estimate the minimum possible time of the progressive failure, which corresponds to the destruction of no resistance when $R = 0$. We get

$$\frac{\mathrm{d}M}{\mathrm{d}t} = m\frac{\mathrm{d}x}{\mathrm{d}t}, \quad \frac{\mathrm{d}}{\mathrm{d}t}\left(M\frac{\mathrm{d}x}{\mathrm{d}t}\right) = Mg. \tag{9.8.5}$$

This equation system describes the fastest regime of progressive failure.

Let us show that it is still much slower than the free fall caused by the demolition process using explosives. The time t_{D} of the free fall of the demolition of a building of height H is equal to

$$t_{\mathrm{D}} = \sqrt{2H/g}. \tag{9.8.6}$$

The World Trade Center was about 417 m high. Hence, it would take about 9.3 s for its free fall, if it would be caused by the explosives destructing the whole building at the very beginning. In practice, the demolition time is about 10–20% higher because explosives are usually placed mostly in the foundation, and not uniformly in the whole building as it is assumed by Eq. (9.8.6).

In fact, the time of the WTC collapse on September 11, 2001, was about 12 s, which practically coincides with the time of common demolitions by explosives. Meanwhile, the progressive failure caused by a local fracturing would take much more time. We prove that using, for a comparison, the fastest regime of progressive failure, Eq. (9.8.5), starting from the top of the WTC building so that:

$$x = 0, \quad \frac{\mathrm{d}x}{\mathrm{d}t} = 0, \quad M = 0 \quad \text{when} \quad t = 0. \tag{9.8.7}$$

Let us indicate some exact solutions of Eqs. (9.8.5) and (9.8.7); for more detail on some solutions and analysis, see [1, 3–5].

(1) Suppose the mass of the building is distributed uniformly so that m is a constant. In this case, we get

$$x = \frac{1}{6}gt^2, \quad M = \frac{1}{6}mgt^2, \quad T_F = \sqrt{3}\, t_D. \tag{9.8.8}$$

In this mode, the acceleration of the moving mass M is equal to $g/3$, one third of the gravitational acceleration of free fall. The collapse would take time T_F which is $\sqrt{3}$ times the free-fall time.

(2) Suppose the mass of the building is distributed linearly like a pyramid so that $m = ax$ where a is a constant. In this case, we get

$$x = \frac{1}{10}gt^2, \quad M = \frac{1}{200}ag^2t^4, \quad T_F = \sqrt{5}\, t_D. \tag{9.8.9}$$

In this mode, the acceleration of the moving mass is equal to $g/5$ and the collapse would take $\sqrt{5}$ times the free-fall time.

These results prove that the collapse in the regime of progressive failure is much slower than the free fall even if we ignore the resistance of the underlying intact structure. The comprehensive data of NIST evidence the free-fall regime of the WTC collapses on September 11, 2001, which means that the WTC towers were disintegrated at the very beginning of the collapse. Progressive failure describes avalanche-type phenomena similar to the much slower snow avalanches.

Summary. Based on this entrainment model, the energy balance in the process zone was studied using the invariant integral of fracture mechanics, and the frontal resistance and critical snow pressure were calculated. We introduced the entrainment toughness K_E as a basic property characterizing the frontal resistance of the process zone and the frontal pressure in the avalanche. Some estimates of the entrainment toughness were derived from the physical properties of snow and from avalanche data [8]. The entrainment model presented here can be implemented in simulation tools. The governing equations of the dynamics of avalanches were derived in the simplest approximation taking into account entrainment, inertia, gravitation, and Voellmy–Salm friction.

Literature

1. G.P. Cherepanov, Mechanics of the WTC collapse. J. Fract. **141**, 287–290 (2006)
2. G.P. Cherepanov, I.E. Esparragoza, A fracture-entrainment model for snow avalanches. J. Glaciol. **54**(184), 182–189 (2008)
3. G.P. Cherepanov, Fracture waves revisited. J. Fract. **159**, 81–85 (2009)
4. G.P. Cherepanov, I.E. Esparragoza, A hybrid model of WTC collapse. J. Appl. Mech. Engng. **12**(3), 575–585 (2008)

5. G.P. Cherepanov, I.E. Esparragoza, Progressive collapse of towers: the resistance effect. J. Fract. **143**, 203–207 (2007)
6. P. Gauer, D. Issler, Possible erosion mechanism in snow avalanches. Ann. Glaciol. **38**(2004), 384–392 (2004)
7. D. Issler, Modelling of snow entrainment and deposition in power-snow avalanches. Ann. Glaciol. **26**, 253–258 (1998)
8. B. Sovilla, P. Burlando, P. Bartelt, Field experiments and numerical modeling of mass entrainment in snow avalanches. J. Geophys. Res. **111**(FO3007), 1–16 (2006)
9. M.E. Eglit, K.S. Demidov, Mathematical modeling of snow entrainment in avalanche motion. Cold Reg. Sci. Technol. **43**, 10–23 (2005)
10. B. Sovilla, S. Margreth, P. Bartelt, On snow entrainment in avalanche dynamics calculations. Cold Reg. Sci. Technol. **47**(1-2), 69–79 (2007)
11. P. Bartelt, B. Salm, U. Gruber, Calculating dense-snow avalanche runout using a Voellmy-fluid model with active/passive longitudinal straining. J. Glaciol. **45**(150), 242–254 (1999)
12. G.P. Cherepanov, *Mechanics of Brittle Fracture* (Mc-Graw-Hill, New York, 1979), p. 940
13. G.P. Cherepanov, *Fracture Mechanics of Composites* (Nauka, Moscow, 1983), p. 300
14. G. P. Cherepanov, *Fracture Mechanics* (Moscow-Izhevsk, ICS, 2012), pp. 1–840
15. K. Platzer, P. Bartelt, M.A. Kern, Measurements of dense snow avalanche basal shear to normal stress ratios. Geophys. Res. Lett. **34**(7), L07501 (2007)
16. P. Bartelt, V. Stockli, The influence of tree and branch fracture, overturning and debris on snow avalanche flow. Ann. Glaciol. **32**, 209–216 (2001)
17. P. Bartelt, O. Buser, K. Platzer, Starving avalanches: frictional mechanisms at the tails of finite-sized mass movements. Geophys. Res. Lett. **34**, L20407 (2007)
18. B. Salm, A. Burkard, H. Gubler, Berechnung von Fliesslawinen, eine Anleitung für Praktiker mit Beispielen, *Mitteilungen des Eidgenössischen Institutes für Schnee und Lawinenforschung*, No. 47 (Davos, Switzerland, 1990)
19. A. Voellmy, Über die Zerstörungskraft von Lawinen, *Schweizerische Bauzeitung*. Jahrg. **73**, Heft **12** (159–162), **15**(212–217), **17** (246–249), **19** (280–285) (1955)
20. F. Tiefenbacher, M.A. Kern, Experimental devices to determine snow avalanche basal friction and velocity profiles. Cold Reg. Sci. Technol. **38**, 17–30 (2004)

Chapter 10
Relativistic Electron Beams

Abstract In this chapter, the invariant integral of electrodynamics is used to derive the interaction force of electric charges moving at superluminal velocities (superluminal electrons discovered by Cherenkov found many applications). As a result, Coulomb's Law for static electric charges is generalized; now, it depends also on the speed of charges. According to the new law, superluminal electrons attract one another. The beams of superluminal electrons form superdense clusters can easily penetrate and destroy any solids. Therefore, relativistic electron beams will be the main weapon of future star wars in cosmos. This chapter may be of special interest for military scientists.

Just a matter of scientific curiosity today, relativistic electron beams will be the main weapon of star wars in the future. A special interest causes the superluminal electron beams discovered and studied by Russian scientist P. A. Cherenkov still in 1934. As distinct from slow electrons obeying the classic Coulomb's Law, superluminal electrons attract one another, with getting packed into dense clusters that can easily cut any solid, see *G. P. Cherepanov and A. A. Borzykh, Theory of the electron fracture mode in solids, J. Appl. Physics,* ***74****(12), 1993.*

The new law says that the interaction force F of two electric charges q_1 and q_2 moving along the same straight line at speed V is equal to

$$F = \frac{q_1 q_2}{\varepsilon R^2} \frac{\dfrac{V^2}{a^2} - 1}{1 - \dfrac{V^2}{c^2}}. \tag{10.01}$$

Here R is the distance between the charges in the proper coordinate frame, c is the speed of light in vacuum, a is the speed of light in a medium, and ε is the dielectric constant that shows how much the interaction force in the medium is less than in vacuum ($a \leq c$, $\varepsilon \geq 1$ and $\varepsilon = 1$ in vacuum). The force of attraction is assumed to be positive and the force of repulsion negative. E. M. Lifschitz called this equation the Coulomb–Borzykh–Cherepanov law; therefore, we call it the CBC law, in short.

Equation (10.01) is valid for both charges when $V < a < c$, that is, when the speed is subluminal. However, for superluminal charges when $a < V < c$, that is,

G. P. Cherepanov, *Invariant Integrals in Physics*,
https://doi.org/10.1007/978-3-030-28337-7_10

when the speed of charges exceeds the speed of light in the medium, the force F acts only upon the rear charge; in this case, this force becomes attractive for charges of one and same sign. Particularly, the CBC law allows one to understand and predict unusual fracturing capabilities of relativistic electron beams that can create cuts like brittle fractures even in soft biological tissues.

For slowly moving charges, when $V \ll a$, Eq. (10.01) provides the classic Coulomb's Law

$$F = -\frac{q_1 q_2}{\varepsilon R^2}. \tag{10.02}$$

Below, we derive the CBC law and some of its implications using invariant integrals of electrodynamics. For more detail on electron mode fracturing, the reader refers to the above-mentioned paper in J. Applied Physics.

10.1 Introduction

In the mid-1960s, high-power pulse electron-beam accelerators having a voltage of some millions of volts were invented and later used to fracture various materials. In 1980, the CBC law was discovered that was applied to collective relativistic interactions in beams of superluminal electrons [1, 2]. A new, electron fracture mode was introduced when caused by the formation of dense clusters of superluminal electron beams that act as blades or wedges producing crack-like cuts in any materials.

The electron fracture mode is characterized by the following unusual features [2]:

(i) Initial macrocracks do not affect the threshold of fracture;
(ii) Fracture of all, even liquid materials occurs in an absolutely brittle way; and
(iii) The splitting cracks propagate with supersonic velocities.

All these and other peculiarities of electron fracturing mode were explained and described using invariant integrals and the new, CBC law of electron interaction, see [2, 3] for more detail.

In Sect. 10.2 below, we provide the invariant Γ-integrals for electromagnetic fields in dielectrics, and in Sect. 10.3, the field equations of a moving charge.

Then, in Sect. 10.4, we derive the new, CBC law of interaction of moving relativistic charges, generalizing Coulomb's Law, and apply it to the chain of superluminal electrons in Sect. 10.5.

10.2 Invariant Integrals of Electromagnetic Field

The state of electromagnetic field in dielectrics is characterized by field vectors **E** and **D**, and **H** and **B** which satisfy the following Maxwell's equations (in SI units):

$$\gamma_{ijk} E_{j,k} + \frac{\partial B_i}{\partial t} = 0, \quad \gamma_{ijk} H_{j,k} - \frac{\partial D_i}{\partial t} = J_i, \tag{10.2.1}$$

$$D_{i,i} = \delta, \quad B_{i,i} = 0, \quad J_{i,i} + \frac{\partial \delta}{\partial t} = 0. \tag{10.2.2}$$

Here, J_i is the current density vector component, δ is the charge density, t is time, and $\gamma_{123} = \gamma_{231} = \gamma_{312} = 1$, $\gamma_{132} = \gamma_{321} = \gamma_{213} = -1$, all other γ_{ijk} being equal to 0 where $i, j, k = 1, 2, 3$.

In a dielectric medium where $\delta = 0$ and $J_i = 0$, and the electromagnetic field is stationary or steady in a moving coordinate frame, Maxwell's equations in Eqs. (10.2.1) and (10.2.2) acquire the following shape

$$E_{j,k} = E_{k,j}, \quad H_{j,k} = H_{k,j}, \quad D_{i,i} = 0, \quad B_{i,i} = 0. \tag{10.2.3}$$

The invariant Γ-integrals of this electromagnetic field used in the moving coordinate frame with respect to which the physical field is steady can be written as follows [1–3]:

$$\Gamma_i = \oiint\limits_S \left[(D_j E_i + B_j H_i) n_j - \frac{1}{2} (E_j D_j + H_j B_j) n_i \right] \mathrm{d}S \quad \text{as } i,j = 1, 2, 3. \tag{10.2.4}$$

In the case of superluminal velocities of a charge, the perturbation domain represents a moving cone, with the charge being at the apex of the cone.

In Sect. 1.6, Maxwell's equations were derived from the invariant Γ-integrals for arbitrary nonlinear dielectric and diamagnetic media. In the case of linear constitutive equations for isotropic homogeneous matter when

$$D_i = \varepsilon_0 \varepsilon E_i \quad \text{and} \quad B_i = \mu_0 \mu H_i \quad \text{(in the SI units)}, \tag{10.2.5}$$

the proof is quite simple.

Suppose there are no charges and currents in any volume V inside surface S. In this case, $\Gamma_i = 0$ and the application of the divergence theorem to the integral in Eq. (10.2.4) provides

$$\int\limits_V \left[D_{j,j} E_i + B_{j,j} H_i + \varepsilon_0 \varepsilon E_j (E_{i,j} - E_{j,i}) + \mu_0 \mu H_j (H_{i,j} - H_{j,i}) \right] \mathrm{d}V = 0. \tag{10.2.6}$$

From here, it follows that Maxwell's equations, Eq. (10.2.3), are valid in V since E_i and H_i are arbitrary.

10.3 Field Equations of a Moving Charge

For the study of collective interactions, let us use the known solution of the Maxwell equations for the proper field of a charge moving at constant velocity V which is greater than the phase speed of light in medium a (but less than the speed of light in vacuum c). According to the Lorentz transformations, the asymptotic steady-state field of a superluminal negative charge e moving along the x_3-axis in the proper space–time frame of coordinates, is [1, 4]:

$$E_i = \frac{eM^2 x_i}{2\pi\epsilon'}\left(x_3^2 - M^2 r^2\right)^{-3/2} \quad \text{where } i = 1, 2; \tag{10.3.1}$$

$$E_3 = -\frac{eM^2 x_3}{2\pi\epsilon'}\left(x_3^2 - M^2 r^2\right)^{-3/2}; \tag{10.3.2}$$

$$H_1 = Ax_2, \quad H_2 = -Ax_1, \quad H_3 = 0. \tag{10.3.3}$$

Here,

$$A = \frac{\mu' e V M^2\left(1 - \dfrac{a^2}{c^2}\right)}{2\pi\left(1 - \dfrac{V^2}{c^2}\right)}\left(x_3^2 - M^2 r^2\right)^{-3/2}. \tag{10.3.4}$$

$$M^2 = \frac{\left(V^2/a^2\right) - 1}{1 - \left(V^2/c^2\right)} = \frac{\mu\epsilon V^2 - c^2}{c^2 - V^2} > 0, \quad (V > a) \tag{10.3.5}$$

$$a = \frac{c}{\sqrt{\mu\varepsilon}}, \quad \mu' = \mu\mu_0, \quad \epsilon' = \epsilon\epsilon_0, \quad r^2 = x_1^2 + x_2^2 \tag{10.3.6}$$

$$\left(c = 1/\sqrt{\varepsilon_0\mu_0} = 3 \times 10^8 \text{ m/s} \quad \text{in the SI units}\right)$$

Here, E_1, E_2, E_3, H_1, H_2 and H_3 are the electromagnetic field components in the $x_1x_2x_3$-system of coordinates, and M is the relativistic Mach number ($\varepsilon' = \varepsilon\varepsilon_0$ and $\mu' = \mu\mu_0$ are the absolute dielectric and magnetic permeabilities of the medium).

The field described by Eqs. (10.3.1)–(10.3.6) is defined inside the Mach cone, that is, in the region $x_3^2 > M^2 r^2$, $x_3 < 0$. Outside this cone, the field of the charge e vanishes.

Solution of Maxwell's equations. Linear Maxwell's equations, Eqs. (10.2.1), (10.2.2), and (10.2.5), have the analytical solution for any number of point charges arbitrarily moving in the infinite space of the medium. Due to the principle of superposition, it is sufficient to find this solution for a single point charge. To this aim, the scalar φ and vector $\boldsymbol{P}$ potentials are introduced as follows:

$$\boldsymbol{B} = \text{rot}\,\mathbf{P}, \quad \mathbf{E} = -\text{grad}\varphi - \partial\mathbf{P}/\partial t \quad \text{(in the SI units)} \tag{10.3.7}$$

Using these potentials, Maxwell's equations, Eqs. (10.2.1), (10.2.2) and (10.2.5), are reduced to the following system of wave equations

$$\Delta\varphi - \frac{1}{a^2}\frac{\partial^2\varphi}{\partial t^2} = -\frac{\delta}{\varepsilon'}, \tag{10.3.8}$$

$$\Delta\mathbf{P} - \frac{\mu\varepsilon}{c^2}\frac{\partial^2\mathbf{P}}{\partial t^2} = -\mu'\mathbf{j}, \quad \left(a^2 = \frac{c^2}{\mu\varepsilon}\right) \tag{10.3.9}$$

if the Lorentz' condition equation is satisfied

$$\mathrm{div}\mathbf{P} + \frac{\mu\varepsilon}{c^2}\frac{\partial\varphi}{\partial t} = 0. \tag{10.3.10}$$

Here, Δ is Laplace's operator.

For arbitrarily distributed charges and currents in the volume V_1 of the infinite space, the solution of wave equations, Eqs. (10.3.8) and (10.3.9), is written as follows:

$$\varphi(x, y, z\ t) = \frac{1}{4\pi\varepsilon'}\int\limits_{V_1}\frac{1}{r}\delta\left(x_1, y_1, z_1, t - \frac{r}{a}\right)\mathrm{d}V_1, \tag{10.3.11}$$

$$\boldsymbol{P}(x, y, z, t) = \frac{\mu'}{4\pi}\int\limits_{V_1}\frac{1}{r}\mathbf{j}\left(x_1, y_1, z_1,\ t - \frac{r}{a}\right)\mathrm{d}V_1. \tag{10.3.12}$$

Here:

$$r^2 = (x - x_1)^2 + (y - y_1)^2 + (z - z_1)^2, \quad \mathbf{j} = \mathbf{j}(j_x, j_y, j_z) \tag{10.3.13}$$

Let us consider a particular case of this problem when a point charge e (electron) moves along the z-axis at the constant speed V which is greater than the speed of light in this medium so that

$$a = \frac{c}{\sqrt{\varepsilon\mu}} < V < c. \tag{10.3.14}$$

In this case, the integrand functions in Eqs. (10.3.11) and (10.3.12) can be written in terms of the Dirac delta function as follows:

$$\delta = e\delta(x_1)\delta(y_1)\delta(z_1 - Vt), \tag{10.3.15}$$

$$j_x = j_y = 0, \quad j_z = eV\delta(x_1)\delta(y_1)\delta(z_1 - Vt) \tag{10.3.16}$$

Substituting these values into Eqs. (10.3.8) and (10.3.9) provides:

$$P_x = P_y = 0, \quad P_z = \varepsilon \frac{V}{c} \varphi. \tag{10.3.17}$$

Equation (10.3.10) becomes

$$\frac{\partial P_z}{\partial z} + \frac{\mu\varepsilon}{c^2} \frac{\partial \varphi}{\partial t} = 0. \tag{10.3.18}$$

Using Eqs. (10.2.5), (10.3.7), (10.3.17), and (10.3.18), we can determine all components of this electromagnetic field in terms of potential φ which, based on Eqs. (10.3.11) and (10.3.15), can be written as follows:

$$\varphi = \frac{e}{2\pi\varepsilon'} \left(x_3^2 - M^2 r^2\right)^{-1/2} \quad \text{as} \quad x_3 < Mr, \tag{10.3.19}$$

$$\varphi = 0 \quad \text{as } x_3 < Mr, \tag{10.3.20}$$

where

$$x_3 = z - Vt, \quad r^2 = x_1^2 + x_2^2. \tag{10.3.21}$$

Here, $Ox_1x_2x_3$ is the proper coordinate frame of the moving electron taking into account the relativistic contraction of the space. In this frame, the electromagnetic field is steady state, and this electron is at the coordinate origin.

10.4 Generalized Coulomb's Law (the CBC Law)

In a dielectric material, let us consider an individual electron with negative charge e which moves at a constant superluminal velocity V in a homodeneous, external electromagnetic field

$$E_{10} = E_{20} = 0, \ E_{30} \neq 0, \quad H_{10} = H_{20} = H_{30} = 0. \tag{10.4.1}$$

The proper singular field $\mathbf{E}_s$ and $\mathbf{H}_s$ of this electron is determined by Eqs. (10.3.1)–(10.3.6).

According to the principle of superposition based on the linearity of Maxwell's equations, the complete electromagnetic field near the individual electron is

$$\mathbf{E} = \mathbf{E}_0 + \mathbf{E}_s, \quad \mathbf{H} = \mathbf{H}_0 \tag{10.4.2}$$

Here, the components of $\mathbf{E}_0$ and $\mathbf{H}_0$ are given by Eq. (10.4.1).

Let us calculate the force F_3 acting upon the moving charge e from the external field of Eq. (10.4.1). Let us represent this charge as a limit $e = \lim(q\Delta)$ when

$q \to \infty,\ \Delta \to 0$ where q is the linear density of charge distributed along the segment $0 > x_3 > -\Delta$ of length Δ along the x_3-axis. This field is evidently axisymmetrical.

Let us consider the surface $S = S_t + S_c$ where S_t is the butt end of the cylinder $x_1^2 + x_2^2 < \delta^2$ at $x_3 = -\Delta$, and S_c is the surface of the cylinder $x_1^2 + x_2^2 = \delta^2$ when $0 > x_3 > -\Delta$ and $\delta \ll \Delta$ so that $\lim(\delta/\Delta) = 0$ when $\delta \to 0$, $\Delta \to 0$. The force acting upon the charge is equal to the Γ-integral Γ_3 in Eq. (10.2.4) over the closed integration surface S. This force is directed along the x_3-axis.

Based on Eqs. (10.3.1)–(10.3.6), the field of **H** is ignorably small inside S, and the axisymmetrical field of **E** and **D** in the dielectric inside S is determined by the following potential

$$\varphi = \frac{q}{2\pi\varepsilon'} \ln r \quad \text{where} \quad r^2 = x_1^2 + x_2^2. \tag{10.4.3}$$

As a reminder, the field is stationary or steady state in the proper coordinate system, and it does not depend on x_3 inside S since q does not depend on x_3.

Taking account of the axial symmetry and the rule of Γ-integration, the Γ-integral Γ_3 in Eq. (10.2.4) is reduced to the following one

$$\Gamma_3 = \int\limits_{S_c} E_3 D_r \mathrm{d}S = \int\limits_{S_c} E_{30} D_r^s \mathrm{d}S = 2\pi\delta \int\limits_{-\Delta}^{0} E_{30} D_r^s \mathrm{d}x_3. \tag{10.4.4}$$

Here, it was taken into account that the integral over the butt end S_t of the cylinder tends to zero when $\delta/\Delta \to 0$, and that the normal components are equal to $n_3 = 0$, $n_r = 1$ on S_c while the external field D_r^0 does not contribute to Γ_3.

According to Eq. (10.4.3), the field in the dielectric inside S is:

$$D_r^s = \varepsilon' E_r^s = \varepsilon' \frac{\partial\varphi}{\partial r} = \frac{q}{2\pi r}. \tag{10.4.5}$$

Substituting D_r^s in Eq. (10.4.4) by Eq. (10.4.5) provides

$$\Gamma_3 = eE_{30} \tag{10.4.6}$$

because $q\Delta \to e$ when $q \to \infty$, $\Delta \to 0$.

Equation (10.4.6) coincides with the corresponding equation in the case of an electron or a charge at rest. Up to a constant coefficient, it follows also from the dimensional analysis. Evidently, it is valid for any electric charge.

Let two charges q_1 and q_2 move along one and same straight line, with having the superluminal speed V so that $c > V > a$. The rear charge q_2 is in the Mach cone of the frontal charge q_1. Hence, the rear charge is acted upon by the force $q_1 E_{30}$ along their common axis where E_{30} is the field created by the frontal charge q_1 at the place of the rear charge q_2. The latter value of E_{30} is given by Eq. (10.3.2) at $r = 0$, $x_3 = -R$ so that the force acting upon the rear charge is equal to

$$\Gamma_3 = \frac{q_1 q_2 M^2}{\varepsilon R^2} = \frac{q_1 q_2}{\varepsilon R^2} \frac{\frac{V^2}{a^2} - 1}{1 - \frac{V^2}{c^2}}. \quad (\varepsilon = 2\pi\varepsilon') \tag{10.4.7}$$

Here, R is the distance between the moving charges.

And so, the charge moving in the wake of a frontal like charge at the superluminal speed is acted upon by the force of attraction determined by Eq. (10.4.7). This equation is Coulomb's Law generalized for relativistic velocities (the CBC law).

10.5 Self-compression of a Chain of Superluminal Electrons

Let us consider the behavior of a one-dimensional, semi-infinite chain of superluminal electrons equidistantly separated by the interval b at the initial instant. In this case, forces exist directed only along the axis of the chain. We denote the force acting on the mth electron and due to the nth electron by $f_{mn} (n < m)$ so that the resultant force acting on the mth electron is

$$F_m = \sum_{n=0}^{m-1} f_{mn}. \tag{10.5.1}$$

According to Eqs. (10.4.7) and (10.5.1), at the initial instant, we obtain

$$F_m(b) = \frac{e^2 M^2}{2\pi\epsilon' b^2} \sum_{n=0}^{m-1} (n+1)^{-2}. \tag{10.5.2}$$

As is known,

$$1 < \sum_{n=0}^{m-1} (n+1)^{-2} < \pi^2/6. \tag{10.5.3}$$

Hence,

$$F_1(b) < F_m(b) < \frac{\pi^2}{6} F_1(b). \tag{10.5.4}$$

From Eq. (10.5.4), it follows that for any m, the forces $F_m(b)$ differ little from $F_1(b)$. Therefore, we can obtain a simple estimate of the deformation of a chain system by considering the motion of s single electron e_1 in the field produced by e_0.

Taking Eq. (10.4.7) into account, the relativistic equation of motion of electron e_1 in the moving coordinate system has the form [1, 2]

$$\frac{d^2 z}{dt^2} = \frac{e^2\left(\frac{V^2}{a^2} - 1\right)}{2\pi\epsilon' m_0\left(1 - \frac{V^2}{c^2}\right)z^2}\left[1 - \frac{1}{c^2}\left(\frac{dz}{dt}\right)^2\right]^{3/2}. \tag{10.5.5}$$

Let us limit ourselves to the case of small relative particle velocities when $dz/dt \ll c$ in Eq. (10.5.5). Let us solve Eq. (10.5.5) in this case using the following initial conditions

$$z = -b \quad \text{and} \quad \frac{dz}{dt} = 0 \quad \text{at} \quad t = 0. \tag{10.5.6}$$

The resulted solution is given by the following equation

$$tK^{1/2} = b^{1/2}(-z)^{1/2}(z + b)^{1/2} + b^{3/2}\arcsin\left(1 + \frac{z}{b}\right)^{1/2} \tag{10.5.7}$$

where

$$K = \frac{e^2\left(\frac{V^2}{c^2} - \epsilon^{-1}\right)}{\pi\epsilon_0 m_0\left(1 - \frac{V^2}{c^2}\right)} \quad \left(a^2 = \frac{c^2}{\mu\epsilon}, \mu = 1 \text{ for dielectrics}\right) \tag{10.5.8}$$

Let us estimate the time τ that is necessary for e_1 to approach e_0 for a very short distance. In this way, a dense system of two electrons is formed; quantum effects are essential in this dense system, just as in a solid.

Substituting $z = 0$ into the solution, Eq. (10.5.7), we obtain the time τ for joining two electrons

$$\tau = \frac{\pi b^{3/2}}{2K^{1/2}}. \tag{10.5.9}$$

We note that quantities b, t and τ are considered in the coordinate system fixed at the first electron. Using the Lorentz transformation and the laboratory frame,

$$b' = b\left(1 - \frac{V^2}{c^2}\right)^{1/2}, \quad t' = t\left(1 - \frac{V^2}{c^2}\right)^{-1/2}, \tag{10.5.10}$$

we obtain from Eq. (10.5.9)

$$(\tau')^2 = \frac{\pi^3\epsilon_0 m_0 (b')^3}{4e^2\left(1 - \frac{V^2}{c^2}\right)^{3/2}\left(\frac{V^2}{c^2} - \epsilon^{-1}\right)}. \tag{10.5.11}$$

The dependence of τ' on V^2/c^2 for a dielectric material is also shown in [1, 2]. As we can see, the distance between the two electrons contracts most substantially in a narrow region of energies (velocities) of the particles, where τ' is small. For $V^2/c^2 = (3+2\epsilon)/(5\epsilon)$, the time of condensation τ' takes the minimum value τ'_{min}

$$\tau'_{\mathrm{min}} = \frac{c_m (b')^{3/2}}{\left(1-\epsilon^{-1}\right)^{5/4}} \tag{10.5.12}$$

where

$$c_m = \frac{\left(\pi 5^{5/4}\right)^{3/2} \epsilon_0^{1/2} m_0^{1/2}}{12^{3/4} e}. \tag{10.5.13}$$

For an electron, we have

$$c_m = 6.437 \times 10^{-2}\,\mathrm{m}^{-3/2}\mathrm{s}$$

Hence, a chain of superluminal electrons condenses in a solid-like state during a time span τ' defined by Eq. (10.5.11). For instance, for $b' = 1\,\mu\mathrm{m}$ (the order of the distance between electrons in pulsed electron beams), we obtain $\tau'_{\mathrm{min}} = 10^{-10}$ s. This means that electron plasma clusters form at a depth on the order of 1 mm beneath the surface of an irradiated material.

For the estimate of fracturing properties of superluminal electron beams, the reader is referred to the detailed paper [1].

Literature

1. G.P. Cherepanov, A.A. Borzykh, Theory of the electron fracture mode in solids. J. Appl. Phys. **74**(12), 7134–7154 (1993). In this paper, the reader can find references to about 80 papers on electron fracturing, the subject beyond that of the current book
2. A.A. Borzykh, G.P. Cherepanov, Collective relativistic interactions in electron beams, J. Exp. Theor. Phys. (Soviet Physics JETP) **78**(1) (1980)
3. G.P. Cherepanov, *Methods of Fracture Mechanics: Solid Matter Physics* (Kluwer, Dordrecht, 1997), pp. 1–314

Chapter 11
Cosmology

Abstract Cosmology is the philosophy and astronomy of the universe. In this chapter, cosmology is built on the invariant integral describing the law of energy conservation for the masses which are capable of experiencing not only Newtonian attraction, but also repulsion which appeared to be proportional to the distance between two masses. The present approach denies the general relativity. The evolution equation of the universe was solved using the data collected by WMAP and Planck missions. The solution provided the asymptotic description of the Big Bang, the history of the early decelerating universe, the expansion at a constant rate in the middle age, and the current stage of the accelerated expansion. The calculated age of the universe appeared to be close to that determined by the FLRW and ΛCDM models based on the general relativity. The critical masses of neutron stars and Black Holes, orbital velocities of stars in galaxies, and other important values of the universe were calculated; the results are very close to that observed by astrophysicists. The new model of the revolving universe provided a simple explanation of the Dark Energy, the most mystic concept of the former cosmology. The cosmological constant is defined in terms of the angular velocity of the universe which is shown to be a rotating, expanding, prolate spheroid of constant eccentricity. This chapter is for everybody because everybody has some interest in cosmology and metaphysics of nature.

Contrary to the common approach of the general relativity, this chapter uses the invariant integral of cosmology to model and study the universe at the large scale of about 100 MPc in the Euclidian space. The flatness of the universe proven by numerous probes of the WMAP and Planck satellite missions necessitates this approach. From the invariant integral of cosmology, the interaction force of two point masses in the cosmic gravitational field is derived. This force is proven to be a sum of two terms, one being the Newtonian gravity and the other the repulsion force caused by the cosmological constant. Both terms make up the right-hand part of the evolution equation of the dynamic universe.

Qualitatively in agreement with the FLRW and ΛCDM models, and with the WMAP and Planck mission data, the exact solution of this equation has provided the history of the early decelerating universe and the asymptotic description of the Big

G. P. Cherepanov, *Invariant Integrals in Physics*,
https://doi.org/10.1007/978-3-030-28337-7_11

Bang, the expansion at an almost constant rate in the middle age, and the current stage of the accelerated expansion of the universe. The age of the universe is found to be equal to 12.3 billion years. It is shown that neutron stars become stable Black Holes when their masses are greater than 6.7 $M_{\odot}$. Then, it is assumed that the universe not only expands but also revolves, and the evolution equations of the revolving and expanding universe are advanced, with the cosmological constant being defined in terms of the angular velocity of the universe. A singular solution of these evolution equations has described the history of the revolving and expanding universe.

Orbital velocities of stars in the Milky Way and similar galaxies are calculated to be about 250 km/s independent of the distance of stars from the galaxy center. Using the equation of the fractal dimension of the universe as a power-law fractal, the thickness of a disk-shaped universe is found. The graviton of minimum frequency is hypothesized to be the smallest elementary particle and the building block of everything. The shape of spheroidal universe and the super-photon hypothesis of the universe origin are analyzed.

The chapter is based on this author's papers *The Large-scale Universe: the Past, the Present and the Future* published in *Journal of Phys. Mesomechanics*, **20**(1), 2017, and *A Neoclassic Approach to Cosmology Based on the Invariant Integral* published in *Horizons in World Physics,* vol. 288, Nova, New York, 2017.

11.1 Introduction

Cosmology is a speculative science/philosophy about the universe/cosmos based on astrophysical observations and human imagination. It feeds the curiosity of human beings in quest of their place in the world and fate in the future. Sometimes, they hoped to find the answers in what could be seen in the sky. The chain of events there appeared more persistent than anything on the Earth, "eternal," which made them use the heaven for worshiping as well as for measuring the time and coordinates on the Earth.

Still the ancient Egyptians could see with the naked eye in the sky what *Aristotle* (388–323BC), an ancient Greek polymath and a disciple of Plato, set forth later in his treatise *On the Heaven*. He viewed the Earth as the center of the Cosmos, with Moon, Mercury, Venus, Sun, Mars, Jupiter, Saturn, and stars being at an ever-increasing distance from it—everything enclosed by a prime solid sky. In his masterpiece *The Sand Reckoner*, *Archimedes* (287–212BC), another Greek genius, calculated the total number of indivisible grains in the universe to be about 10^{64} in modern designation. Amazingly, this figure coincides with the current estimate of the mass of the universe about 10^{56}g, if we assume the mass of one grain to be equal about 10^{-9}g! Archimedes possessed some knowledge of calculus two thousand years before *Leibnitz* (1646–1716), a German polymath, and *Newton* (1642–1727), a British genius.

It was almost two thousand years ago that *Ptolemy* (100–170AD), an ancient Greek–Roman astronomer, created the astronomical tables and geographical maps used during the next 1500 years throughout the world. With his famous treatise *Elements*, *Euclid* (circa 320–260BC), an ancient Greek mathematician and one of "the giants on which shoulders Newton stood," created the geometry unshakeable for more than two thousand years until *Riemann* (1826–1866), a German genius, and *Lobachevski* (1792–1856), a Russian mathematician, advanced the non-Euclidian geometry.

The ancient Greek community, less in number than today's Luxembourg, for a short time brought up Pythagoras, Socrates, Plato, Aristotle, Archimedes, Euclid, Ptolemy, and a dozen of other brilliant brains that eclipsed the achievements of all great minds of the last four centuries bred in the pool which is 10^4 larger than the ancient Greece. The ancient Greek theory about the geocentric Aristotelian/Ptolemaic cosmology reigned in the world during two thousand years and was only destroyed by the self-sacrificing deeds of *Copernicus* (1473–1543), a German astronomer, *Bruno* (1568–1600), an Italian hero, and *Galileo* (1564–1642), an Italian great. Meanwhile, still Archimedes recognized the heliocentric system called now the Copernican one. Today, we see that bitter controversy only as the difference of opinion concerning a convenient origin of the frame of reference so that both Aristotle and Copernicus are right. (Everywhere in this chapter, the national identification of a person corresponds to the state of his birth.)

In 1907, *Minkowski* (1864–1909), a Russian mathematician and Einstein's teacher, coined the term *space–time* and suggested the following Minkowski metric of the flat space–time [1]

$$\mathrm{d}s^2 = -c^2\mathrm{d}t^2 + \mathrm{d}x^2 + \mathrm{d}y^2 + \mathrm{d}z^2 \tag{11.1.1}$$

Here, c is the speed of light in vacuum; t is time; and x, y, and z are the Cartesian coordinates. In mathematical terms, this metric characterizes the special relativity theory earlier advanced by *Lorentz* (1853–1928) and *Einstein* (1879–1955), German physicists, and by *Poincare* (1854–1912), a French mathematician.

Declaring c to be an absolute constant, Einstein advanced the following dualisms of mass–energy and space–time as the physical properties of everything

$$E = Mc^2, \quad L = cT. \tag{11.1.2}$$

Here, E, M, L, and T are the energy, mass, length, and time. These dualisms have since revolutionized the science/philosophy of humankind.

In 1916, Einstein generalized the Minkowski metric and advanced the general relativity theory of the curved space–time in order to describe the Newtonian gravity as a geometrical property of the curved space–time, with a greater curvature being caused by a denser matter. He derived the following Einstein equations [2]

$$G_{\alpha\beta} = \frac{8\pi G}{c^4} T_{\alpha\beta} \tag{11.1.3}$$

$$\left(G_{\alpha\beta} = R_{\alpha\beta} - \frac{1}{2}Rg_{\alpha\beta}; \quad R_{\alpha\beta} = R^{\mu}_{\alpha\mu\beta}; \quad \alpha, \beta, \mu = 1, 2, 3, 4\right)$$

Here, G is the gravitational constant; $T_{\alpha\beta}$ is the energy–momentum tensor of matter; x^1, x^2, x^3, and x^4 are the redesignated t, x, y, and z; $g_{\alpha\beta}$ is the covariant metric tensor of the space–time $ds^2 = g_{\alpha\beta}dx^{\alpha}dx^{\beta}$; $G_{\alpha\beta}$ is the Einstein tensor; $R_{\alpha\beta}$ is the Ricci curvature tensor; $R = R^{\alpha}_{\alpha}$ is the Ricci scalar; and $R^{\mu}_{\alpha\mu\beta}$ is the Riemann curvature tensor of the space–time.

In 1922, *Friedmann* (1880–1926), a Russian mathematician, applied the general relativity to the homogeneous and isotropic universe and derived the following Friedmann equations [3]

$$\left(\frac{\dot{a}}{a}\right)^2 = \frac{8\pi}{3}G\rho - k\frac{c^2}{a^2}, \quad \frac{\ddot{a}}{a} = -\frac{4\pi}{3}G\left(\rho + 3\frac{p}{c^2}\right) \tag{11.1.4}$$

Here, $a = a(t)$ is the scale factor of space–time; t is time; ρ is the mass density of the universe; p is the pressure defined by a cosmological model chosen; and k, the normalized curvature index of the universe ($k = 0$ for the flat universe, $k = 1$ for the sphere, and $k = -1$ for the hyperboloid).

Although Einstein was a referee of the Friedmann paper, he failed to appreciate its value for cosmology. Equation (11.1.4) was later modified and called the *FLRW* equations after Friedmann and also after *Lemaitre* (1894–1966), a Belgian astronomer, *Robertson* (1903–1961), an American physicist, and *Walker* (1909–2001), a British mathematician. The *FLRW* equations have been also used in the ΛCDM model called the *standard cosmological model* which is mostly recognized today.

From the second Friedmann equation, it follows that for positive p the universe contracts, which corresponds to the original view of Einstein. However, later *Hubble* (1889–1953), an American astronomer, measured the value of the following Hubble parameter [4]

$$H = \dot{a}/a. \tag{11.1.5}$$

He proved that this parameter was a positive constant (*the Hubble law*) so that the universe appeared to be expanding at a constant rate. The further studies supported this law, but corrected the value of the constant obtained by Hubble.

During the next 50 years, Einstein and other theoreticians tried to modify the general relativity and the Friedmann equations in order to match the theory to the numerous astronomical discoveries of many new galaxies, quasars, pulsars, Black Holes, Dark Matter, supernovae, and so forth. Then, as a result of the long-term *WMAP satellite mission*, the curvature of the universe was proven to be equal to zero with only 0.4% margin of error. Moreover, in the past when the universe was much denser, the curvature of the universe was much closer to zero contrary to the general relativity, according to which the denser matter made the curvature of the space–time greater.

And so, the universe is, and was at any time, flat so that the general relativity is wrong because it is built on the wrong assumption of a non-flat universe. In the literature, this controversy is politely called the *flatness problem* of the general relativity. In an attempt to save the general relativity, even a new fantastic inflation theory was made up.

In the course of the *Supernova Cosmology Project* and the *WMAP and Planck satellite missions*, it was also established that in the cosmos, beyond the gravity, some tensile forces of an unclear nature provide the observed acceleration of the expansion of the universe. These forces were called the Dark Energy. This discovery as well contradicts to the general relativity. Many other particular facts, for example, the almost constant orbital velocity of stars in spiral galaxies, have been unexplainable by the general relativity.

In the present chapter, we will, in main, revive the Aristotelian/Ptolemaic view of the cosmos and study the large-scale universe using the invariant integral of cosmology introduced by this author. Our intention is, in particular, to explain and make clear the basic astronomical phenomena recently discovered but not understood in the framework of the general relativity such as the Dark Energy, the Dark Matter, the Black Hole, and others. For short, we will call this new approach the NEOC (the Neoclassic Cosmology).

11.2 Basic Assumptions: The Large-Scale Universe

At first, we need to evaluate the reasonable time horizon determined by natural biological limits of the human race lifetime. Homo Sapience appeared forty thousand years ago and its civilization culminated in the epoch of the ancient Egypt and Greece, when the major part of the life's important knowledge we use today was created. Since then, in the struggle for survival on this small planet, the Homo Sapience has been degenerating at an ever-increasing rate. This struggle makes them create more and more sophisticated weapons sufficient to annihilate themselves, altogether. Culture of humanism, life support, and unselective self-reproduction as well contribute genetically to this process of degeneration.

Today's world population is ten thousand times larger in number than the ancient Greece population. However, for the last three to four centuries who can be on a par with Pythagoras, Socrates, Plato, Aristotle, Archimedes, Euclid, and Ptolemy? Nobody can. Obviously, the Homo Sapience time horizon for the future being is less than about ten thousand years. Therefore, any cosmological theory beyond this time limit of the future man existence has neither sense nor meaning. Certainly, a theory looking forward beyond this time may entertain, but not more than that, because it cannot ever be proven or refuted by human beings.

However, to make sure we assume that the time horizon of the human race is about ten million years. For ten million years, light will have covered the distance which is much less than 100 MPc (the latter is less than about $1/300$ of the diameter of the universe). Based on the astrophysical data, at the scale of 100 MPc the current universe can be considered homogeneous and isotropic. In the NEOC approach, at any time the distance of about $1/300$ of the universe diameter, or its time equivalent, will be called the epoch. It is the large scale we accept in this chapter to treat the universe as homogeneous and isotropic at any time. This scale will play the role of an elementary cell resembling similar notions of elementary cells in continuum physics.

Based on the *WMAP and Planck missions* data, within less than 0.4% margin of error, the universe is proven to be flat in the previous and current epochs so that its FLRW metric in the spherical coordinates is reduced to the following Minkowski metric in the Cartesian coordinates

$$\begin{aligned} \mathrm{d}s^2 &= -c^2\mathrm{d}t^2 + a(t)^2\left(\mathrm{d}r^2 + r^2\mathrm{d}\theta^2 + r^2\sin^2\theta\mathrm{d}\varphi^2\right) \\ &= -c^2\mathrm{d}t^2 + \mathrm{d}x^2 + \mathrm{d}y^2 + \mathrm{d}z^2 \end{aligned} \tag{11.2.1}$$

At the scale of $\Delta x \sim \Delta y \sim \Delta z \sim 100\,\mathrm{MPc}$ and $\Delta t \sim 10^7$ years, the Minkowski metric is reduced to the Euclidian metric

$$\mathrm{d}s^2 = \mathrm{d}x^2 + \mathrm{d}y^2 + \mathrm{d}z^2 \tag{11.2.2}$$

because the spatial terms in Eq. (11.2.1) a hundred times greater than the temporal term. For one hundred thousand years, light has run less than $1/10$ of an elementary cell (100 MPc) of the large-scale structure of the universe.

In other words, at this large scale the universe on the average will be exactly the same even during the time, which is thousand times greater than the reasonable lifetime of the human race. Particularly, any cosmological theories of the future beyond this lifetime cannot ever be supported or refuted, and hence, all of them are speculative. Within our epoch, that is, during the Homo Sapience lifetime the large-scale structure of the universe is always one and the same—the last man on the Earth will see the same universe the first man saw.

And so, at the large scale in the previous, current, and future epochs, the universe can be considered flat, homogeneous, and isotropic. Within our epoch, we, mainly, remain in the framework of the static Aristotelian and Ptolemaic universe described by the Euclidian space. The latter space is accepted in this chapter for any epoch.

11.3 Invariant Integral of Cosmology

Let the physical matter interact by the field potential $\varphi(x_1, x_2, x_3)$ where x_1, x_2, and x_3 are the Cartesian coordinates of the Euclidian space with the origin of the reference frame at the point of observation. We can write the law of the energy conservation in this system by means of the following invariant $\Gamma-$ integral [5]:

$$\Gamma_j = (4\pi G)^{-1} \oint_S \left(\frac{1}{2}\varphi_{,i}\varphi_{,i}n_j - \varphi_{,i}n_i\varphi_{,j} + 4\pi G\Lambda\varphi n_j\right)\mathrm{d}S$$

$$(i, j = 1, 2, 3) \qquad (11.3.1)$$

Here, the surface integral is taken over an arbitrary closed surface S in the Euclidean space, dS is an element of S, n_i is the ith component of the outer unit vector normal to S, Λ is the cosmological constant, and G is the gravitational constant.

Vector $\mathbf{\Gamma}(\Gamma_1, \Gamma_2, \Gamma_3)$ represents the force acting on the matter inside S, which is equal to the work spent to move the matter inside S per init length. Particularly, $\mathbf{\Gamma} = 0$ if there are no matter inside S. The dimension of φ is "energy/mass."

The first term in Eq. (11.3.1) represents the flow of the field gravitational energy through S, the second term the work of the field gravitational intensity on S, and the third term the flow of the Dark Energy through S.

When $\Lambda = 0$, Eq. (11.3.1) provides the theory/model of the Newtonian gravitation of the matter. When $\Lambda \neq 0$, Eq. (11.3.1) delivers the model of the cosmic gravitational matter, in which the term $\Lambda\varphi n_j$ represents the flow of the Dark Energy.

The field of potential $\varphi(x_1, x_2, x_3)$ introduced by Eq. (11.3.1) is called the cosmic gravitational field, or the *CG* field. It accounts for both the gravitational matter (the ordinary or baryonic matter plus the Dark Matter) and the anti-gravitational matter (the Dark Energy of the "vacuum" uniformly distributed in the space). The gravitational matter can collapse and create point-like sources of the field while the anti-gravitational matter can never collapse due to the self-repulsion.

11.4 Interaction Forces in the Cosmic Gravitational Field

Using the divergence theorem, let us convert Eq. (11.3.1) to the following shape:

$$\Gamma_j = (4\pi G)^{-1} \iiint \left(\varphi_{,ii} + 4\pi G\Lambda\right)\varphi_{,j}\mathrm{d}v \quad (i, j = 1, 2, 3) \qquad (11.4.1)$$

Here, the integral is taken over the volume inside S so that dv is an elementary volume.

Let us consider first the homogeneous and isotropic matter, every part of which volume is self-balanced, so that in this matter the integral in Eq. (11.4.1) is equal to zero over an arbitrary volume. From Eq. (11.4.1), it follows that the following differential equation is valid at every point inside the volume of such a matter

$$\varphi_{,ii} = -4\pi G\Lambda \qquad (11.4.2)$$

This is Poisson's equation for the Newtonian potential accounting for a negative mass uniformly distributed inside the volume under consideration so that Λ is the density of this anti-gravitational matter called the Dark Energy. And so, in this model the Dark Energy is the negative mass uniformly distributed in the space.

Let us find the law of interaction of gravitational masses in this field of the Dark Energy. Suppose a point mass M to be at point $(0, 0, 0)$. From Eq. (11.4.2), it follows that this mass and the Dark Energy create the following field:

$$\varphi = -\frac{GM}{r} - \frac{2\pi}{3} G\Lambda r^2 \ \left(\text{where} \ \ r^2 = x_i x_i\right) \tag{11.4.3}$$

The first term in Eq. (11.4.3) corresponds to the gravitational potential of mass M and the second term to the potential of the Dark Energy.

Suppose mass m to be at an arbitrary point inside S. Let us shrink the closed surface S onto this point and calculate the singular integral in Eq. (11.3.1) by means of the Γ-integration rule. As a result, we get the following equation of the force acting upon mass m

$$\Gamma_i = -m\varphi_{,i} \tag{11.4.4}$$

Here, $\varphi_{,i}$ is the intensity of the external field at the point of the mass location.

Using Eq. (11.4.3), we find the field intensity at point $(R, 0, 0)$ created by mass M at point $(0, 0, 0)$ and by the Dark Energy

$$\varphi_{/1} = \frac{GM}{R^2} - \frac{4\pi}{3} G\Lambda R, \quad \varphi_{,2} = 0, \quad \varphi_{,3} = 0. \tag{11.4.5}$$

From Eqs. (11.4.4) and (11.4.5), it follows that mass m at point $(R, 0, 0)$ is acted upon by the following field force [5]:

$$\Gamma_1 = -\frac{GmM}{R^2} + \frac{4\pi}{3} mG\Lambda R, \quad \Gamma_2 = \Gamma_3 = 0. \tag{11.4.6}$$

Similarly, we can derive that mass M at point $(0, 0, 0)$ is acted upon by the following field force:

$$\Gamma_1 = \frac{GmM}{R^2} - \frac{4\pi}{3} MG\Lambda R, \quad \Gamma_2 = \Gamma_3 = 0. \tag{11.4.7}$$

Equations (11.4.6) and (11.4.7) provide a new law of interaction of two point masses. The first term in this law accounts for the attraction caused by Newton's gravity while the second term accounts for the repulsion of the Dark Energy. The gravity decreases when the distance between the masses increases and tends to zero at a very large distance; the force of repulsion grows limitlessly when this distance increases.

For any mass m, there exist a critical distance $R = R_m$, such that for $R < R_m$ the Newtonian gravity dominates, and for $R > R_m$, the repulsion force of the Dark Energy dominates over the Newtonian gravity. According to Eq. (11.4.6) where M is assumed to be the mass of the universe, we have

$$R_m = \left(\frac{3M}{4\pi \Lambda} \right)^{1/3} \tag{11.4.8}$$

The astrophysical measurements proved that the radius of the universe at the current epoch is greater than R_m because the repulsion of the Dark Energy currently dominates at a very large scale. It is the cause of the accelerated expansion of the universe, the most significant discovery of the recent past.

11.5 Gravitational Matter and the Dark Energy

The mass of the gravitational matter of the homogeneous isotropic universe inside a sphere of radius R is equal to

$$M = \frac{4\pi}{3} \rho R^3 \tag{11.5.1}$$

where ρ is the average density of the gravitational matter (the baryonic matter plus the Dark Matter). Let us consider any mass m on the edge of this sphere. The field force acting upon this mass is, evidently, given by Eq. (11.4.6).

From Eqs. (11.4.6) and (11.5.1), it follows that any mass m moves away with some acceleration if $\Lambda > \rho$, that is, if the density of the Dark Energy is greater than the average density of the gravitational matter. It is the fundamental property of the expanding universe at the current epoch. Acceleration of the expansion is determined by the value of difference $\Lambda - \rho$. Though this expansion increases the distance between objects that are under shared gravitational influence, it does not increase the size of these objects, e.g., galaxies.

From Eqs. (11.4.6) and (11.4.7), it follows also that, if $\Lambda < \rho$, the universe experiences a deceleration, and if $\Lambda = \rho +$, the universe expands at a constant velocity which corresponds to the principle of Galileo.

At the scale less than 100 MPc, the gravitational matter is inhomogeneous, and thus, under the prevailing force of gravitation, it can collapse and form Black Holes, galaxies, quasars, pulsars, stars, planets, and various particles. At any scale, the competition of the gravity versus repulsion force is characterized by the dimensionless number $\mathrm{Ch} = M/\Lambda L^3$ where L and M are the specific linear size and gravitational mass of the system under consideration. When $\mathrm{Ch} \gg 1$, we can ignore the Dark Energy, and when $\mathrm{Ch} \ll 1$, we can neglect the Newtonian gravity.

Let us provide some figures for the current epoch assuming the cosmological constant to be equal to $\Lambda = 10^{-26}$ kg/m^3:

The solar system: $M = 2 \times 10^{30}$ kg, $L \sim 10^{11}$ m, $\mathrm{Ch} \sim 10^{23}$;
The Milky Way (our galaxy): $M = 1.4 \times 10^{42}$ kg, $L \sim 10^{21}$ m, $Ch \sim 10^5$;
A supercluster of million galaxies:

$$M = 10^{48}\,\mathrm{kg}, \quad L \sim 3 \times 10^{24} m \sim 100\,\mathrm{Mpc}, \quad \mathrm{Ch} \sim 1;$$

The universe: $M = 10^{53}$ kg, $L = (10)^{1/3} \cdot 10^{26}$ m, Ch = 1.

And so, in the scale of 100 MPc and greater the effect of the Dark Energy is, at least, of the same order as that of the Newtonian gravity. In the smaller scales, the effect of the Dark Energy can be ignored. In the space beyond the universe, the Dark Energy dominates because number Ch may be much less than 1.

According to the recent most accurate measurements of the *Planck satellite mission*, the ordinary matter plus the Dark Matter make up 31.7% of the mass of the universe, and the other 68.3% accounts for the Dark Energy. The ordinary matter accounts only for 4.9% of the total mass. At the current epoch, the repulsion forces can dominate at the large scale while the gravity dominates at the smaller scales [6].

The universe represents a community of the gravitational matter and the anti-gravitational Dark Energy; for the scales less than 100 MPc at the current epoch, the universe is bound by the prevailing forces of gravitation. The Dark Energy is uniformly distributed in the universe. For example, in the Earth there is about 0.01 g of the Dark Energy.

11.6 The Dynamic Universe: The Evolution History

Let us study the radial motion of an arbitrary mass in the cosmic gravitational field of the universe. The outward motion of the masses forms the expansion of the universe. During one epoch, this dynamic expansion is very small as compared to 100 MPc. Hubble was first to measure its rate using the *Doppler effect* that describes the decrease of frequency of a receding source of waves (*redshift*) and the increase of frequency of an approaching source (*blueshift*) in the spectrum of hydrogen, helium, and other chemical elements of the source.

By means of Eqs. (11.4.6) and (11.5.1), the radial motion equation of any probe mass at the edge of the universe of radius R is written as follows:

$$\frac{\mathrm{d}^2 R}{\mathrm{d}t^2} = \frac{4\pi}{3} G R(\Lambda - \rho). \tag{11.6.1}$$

This serves as the evolution equation of the expanding universe.

Let us accept the following assumptions of our model:

1. The universe is flat, homogenous, and isotropic at any time of its history.
2. The cosmological constant Λ is one and the same at any time.
3. The gravitational mass M of the universe is constant at any time.

Integrating Eqs. (11.5.1) and (11.6.1) under these assumptions, we come to the following basic equation

$$H^2 = \left(\frac{1}{R}\frac{\mathrm{d}R}{\mathrm{d}t}\right)^2 = \frac{8\pi}{3} G\left(\rho + \frac{1}{2}\Lambda\right) - K\frac{c^2}{R^2} \quad \left(\rho = \frac{3M}{4\pi R^3}\right) \tag{11.6.2}$$

Here, H is the Hubble parameter, and K is a constant.

At the current epoch, we have

$$t = t_0, \quad R = R_0, \quad \dot{R}/R = H_0. \tag{11.6.3}$$

Using Eqs. (11.6.2) and (11.6.3), we get K

$$Kc^2 = R_0^2\left(\frac{4\pi}{3}G\Lambda + \frac{2MG}{R_0^3} - H_0^2\right). \tag{11.6.4}$$

It is amazing that Eq. (11.6.2) almost coincides with the basic equation of the modern *FLRN* and ΛCDM models derived from the general relativity based on the wrong assumption of the curved universe (the difference is in the meaning of K).

The integration of Eq. (11.6.2) provides the following evolution of the universe at any epoch from the birth when $t = 0$ to the infinite future when $t \to \infty$

$$t = \int_0^R \left(\frac{4\pi}{3}G\Lambda R^2 + \frac{2MG}{R} - Kc^2\right)^{-1/2} \mathrm{d}R \tag{11.6.5}$$

The analysis of basic equations of evolution, Eqs. (11.6.1) and (11.6.2), shows that in the life of the universe there were three different stages characterizing the early universe, the middle-age universe, and the old-age universe.

The earliest universe had a small size $R \ll R_m$. Its rate of expansion, infinite at the birth, was decreasing with time growing. At the earliest age, $\rho \gg \Lambda$ so that the density of the gravitational matter was much greater than the density of the Dark Energy.

At the middle age, the universe was growing at the close-to-constant rate. At this age, $\rho \sim \Lambda$ and the rate of expansion was minimal at $R = R_m$.

At the old age, $\Lambda > \rho$ and the rate and acceleration of expansion are increasing when time grows. We live in the beginning of this stage which is characterized by the following exponential expansion when $t \to \infty$

$$\frac{\mathrm{d}R}{\mathrm{d}t} = \left(\frac{4\pi}{3}G\Lambda\right)^{1/2} R, \quad R = R_0 \exp\left[\sqrt{\frac{4\pi}{3}G\Lambda}(t - t_0)\right] \tag{11.6.6}$$

Here, $t = t_0$ and $R = R_0$ at the current epoch.

All these predictions of the present NEOC model have been supported by the well-known astronomical discoveries for the last 30 years. Let us notice some of them:

a. According to Eq. (11.6.5), $R \to 0$ and $\dot{R} \to \infty$ when $t \to 0$, that is the universe was born from nothing and grew at an infinite rate (*the Big Bang*).
b. The early universe was decelerating, and at the current epoch, it is accelerating. This phenomenon was discovered by a large team of outstanding astronomers

who for about 20 years have collected a tremendous amount of the supporting evidence; three of them were awarded the Nobel Prize (*A. G. Riess, S. Perlmutter, and B. P. Schmidt,* American astrophysicists).

At the time when $t \to \infty$, the universe will disintegrate into a great number of independent communities running one from another by the Dark Energy but keeping their own constituents owing to the gravity.

11.7 The Age of the Universe

Let us estimate the current age of the universe using Eq. (11.6.5). At the present epoch, $R = R_0$ so that we have

$$T_{\mathrm{U}} = \int_0^{R_0} \left(\frac{4\pi}{3} G \Lambda R^2 + \frac{2MG}{R} - Kc^2 \right)^{-1/2} \mathrm{d}R \qquad (11.7.1)$$

Here, T_{U} is the current age of the universe.
Let us transform Eq. (11.7.1) to the following shape:

$$T_{\mathrm{U}} = T \int_0^1 \left[x^2 + 2\alpha \left(\frac{1}{x} - 1 \right) + \beta \right]^{-1/2} \mathrm{d}x \qquad (11.7.2)$$

Here,

$$T = \sqrt{\frac{3}{4\pi G \Lambda}}, \quad \alpha = \frac{\rho_0}{\Lambda}, \quad \beta = \frac{3H_0^2}{4\pi G \Lambda} - 1. \qquad (11.7.3)$$

Parameters α and β are dimensionless. Let us call them the Poincare number and the Hubble number to honor these great contributors into cosmology. The Poincare number is directly proportional to the Ch number calculated for the current epoch at the scale of the universe. And so, the current age of the universe is determined by the Hubble and Poincare numbers and by one constant of the time dimension depending only on $G\Lambda$.

To determine the age of the universe, let us use the results of observations collected by the *Planck satellite mission* for many years. According to these data, at the current epoch the universe contains the Dark Energy 68.3/4.9 times more than the ordinary (baryonic) matter, and the average density of the baryonic matter is equal to $\rho_0 = 4.5 \times 10^{-28}$ kg/m^3. From here, it follows that the density of the Dark Energy is equal to

$$\Lambda = 0.63 \times 10^{-26}\ \mathrm{kg/m}^3. \qquad (11.7.4)$$

Using Eqs. (11.7.3) and (11.7.4) and other data obtained by the *Planck and WMAP satellite missions*, we get the Poincare number and other parameters

$$T = 0.756 \times 10^{18}\,\mathrm{s} = 2.4 \times 10^{10}\,\text{years},$$

$$\alpha = \frac{31.7}{68.3} = 0.464, \quad H_0 = 68.3\frac{\mathrm{km}}{\mathrm{s}}\,\text{per MPc}. \tag{11.7.5}$$

From here and from Eqs. (11.7.3) and (11.7.4), we find the Hubble number

$$\beta = 1.8. \tag{11.7.6}$$

Using Eqs. (11.7.2), (11.7.5) and (11.7.6), we calculate the current age of the universe

$$T_{\mathrm{U}} = 12.26 \times 10^9\,\text{years} \sim 12.3\,\text{billion years}. \tag{11.7.7}$$

According to the ΛCDM model, the age of the universe is about 13.8 billion years. It is striking that the ΛCDM theory based on the wrong assumption of the general relativity about the non-flat universe could achieve the result that is so close to the correct one!

11.8 The Neutron Universe, the Dark Matter, and Black Holes

Let us study the early universe. From Eq. (11.6.2), it follows that

$$\frac{\mathrm{d}R}{\mathrm{d}t} = \sqrt{\frac{2MG}{R}} \quad \text{when} \quad t \to 0 \tag{11.8.1}$$

so that

$$R = (2MG)^{1/3}\left(\frac{3}{2}t\right)^{2/3} \quad \text{when} \quad t \to 0. \tag{11.8.2}$$

Using Eq. (11.8.2), we can find the density of the early universe

$$\rho = \frac{1}{6\pi G t^2} \quad \text{when} \quad t \to 0. \tag{11.8.3}$$

And so, the density of the early universe was the function of only G and t. Evidently, in the early universe $\rho \gg \Lambda$ so that in this model it consisted mostly of the gravitational matter.

Neutrons and protons make up nuclei of atoms of all elements. They have mass about 1.67×10^{-27} kg and radius about 10^{-15} m. Hence, their density ρ_B is equal to

$$\rho_B = 4 \times 10^{17}\,\text{kg/m}^3. \tag{11.8.4}$$

From Eq. (11.8.3), it follows that in about 45 μs after the universe was born, it had the density which was about the density of atom nuclei.

It is reasonable to assume that at sometimes around $t \sim 45\,\mu$s the universe represented a dense gluon "soup" of *down* quarks and of a double amount of *up* quarks, and it was covered by a dense "atmosphere" of photons and neutrinos trapped by the gravitation forces of the universe. It was a gigantic Black Hole we call the neutron universe. Other elementary particles like electrons, positrons, muons, and others, both bosons and fermions, were also trapped by the gravity, although being either unstable or insignificant in amounts. Because the universe is electro-neutral in average, proton charges were annihilated by electrons so that the ratio of the number of *down* quarks and *up* quarks was typical for neutrons.

If the gravitational mass of the universe was about 10^{54} kg at any time, then according to Eqs. (11.8.2) and (11.8.4), the radius of the neutron universe at that time was equal to

$$R_N = 3.8 \times 10^7\,\text{km} \tag{11.8.5}$$

It is about the distance between the Sun and Mercury, a very small part of the solar system.

All gravitating particles near a big-mass object, including photons and neutrinos, are trapped by the gravity of the big object, if

$$2GM \geq c^2 R \tag{11.8.6}$$

Here, M and R are the mass and radius of the big object like a neutron star or the neutron universe.

Evidently, the gravitating object meeting the condition equation, Eq. (11.8.6), is a Black Hole. In terms of the density ρ_B of the Black Hole, Eq. (11.8.6) can be written as

$$8\pi G \rho_B R^2 \geq 3c^2 \tag{11.8.7}$$

For the neutron universe, the condition equation, Eq. (11.8.6), is certainly satisfied so that all photons and neutrinos were trapped in the "atmosphere" of the earliest universe; it was a gigantic Black Hole.

The gravitational stresses σ_{ik} inside a Black Hole or a dead neutron star are distributed as follows:

$$\sigma_{ik} = \left[p + \frac{2\pi}{3}\rho_B^2 G\left(R^2 - r^2\right)\right]\delta_{ik} \tag{11.8.8}$$

Here, r is the distance from the center of a spherical Black Hole or dead neutron star, p is the pressure of the neutrino–photon atmosphere ($p \sim 0$ for dead neutron stars), and δ_{ik} is the Kronecker delta function.

Equation (11.8.8) is valid both for heavy ideal fluids and for heavy elastic solids which Poisson's ratio is equal to 0.5 due to the big pressure.

According to Eq. (11.8.7), the neutron stars that are less in size than R_* are losing energy and fast dying

$$R_* = \frac{c}{2}\sqrt{\frac{3}{2\pi G\rho_{\mathrm{B}}}} = 20\,\mathrm{km} \tag{11.8.9}$$

Hence, the neutron stars of radius $R > 20\,\mathrm{km}$ turn out to be stable Black Holes keeping all their energy.

From Eq. (11.8.6), it follows that the critical mass of stable Black Holes is equal to

$$M_* = 1.34 \times 10^{31}\,\mathrm{kg} = 6.7\,M_{\odot} \tag{11.8.10}$$

Here, $M_{\odot}$ is the mass of the Sun.

And so, the mass $6.7\,M_{\odot}$ is the maximum value of the mass of neutron stars and the minimum value of the mass of stable Black Holes. It is useful to remember another important critical value, the Chandrasekhar limit $1.4\,M_{\odot}$ which separates the stars of lesser mass, turning into dead white dwarfs in the long run, from the neutron stars of higher mass energy that originally rotate extremely fast and create hot rotating clouds of the gravitational matter but fast lose energy and die [7].

As a reminder, common stars usually go through a long evolution of more than 10 billion years synthesizing hydrogen from quarks, burning hydrogen into helium and then, in Red Giant Stars, helium into carbon, oxygen and heavier elements. Finally, they explode as supernovae which emit a tremendous amount of photons, neutrinos, and electromagnetic radiation in pulsars. After that, they get cool, rotate slower, and die fast [8].

Dead neutron stars which do not revolve cannot be easily detected. However, despite its size may be about that of a big meteorite, a neutron star will disturb the solar system if moves within its reach. If this happens, the humankind will inevitably perish. The smallest dead neutron stars are most probable "candidates" to meet with the solar system.

The Black Holes which mass is greater than $6.7\,M_{\odot}$ are much more powerful. They carry a tremendous energy, and they keep trapped any photons, neutrinos, and other particles and adsorb any external emission received. It is hard to detect them. They can be discovered only by the nebulae of clouds of the revolving gravitational matter in their gravitational field. Gigantic Black Holes can have the mass of many million $M_{\odot}$ and form galaxies like the Milky Way. Some hope for the direct monitoring of the Black Holes provides the so far undetected Hawking effect that follows from the quantum mechanics [9].

At the large cosmological scale, dead neutron stars and Black Holes form the Dark Matter that reveals itself only by the gravitational effect. It is mostly from this dangerous matter that our gravitational universe is made. The ordinary matter which can be, in principle, observed makes up only 4.9% of the content of the universe.

11.9 The Planck Epoch and the Big Bang

The neutron stars and Black Holes are the densest cosmological objects of nature. And so, we can only guess about the state of the universe at times less than 45 μs after it was born. More than a hundred years ago, *Max Planck* (1858–1947), a German physicist and the reluctant originator of quantum mechanics, offered a mesmerizing guess about some absolute units of time and space.

The point is that the only dimensionless combination of time t, the gravitational constant G, the speed of light c, and the Planck constant h is the following one:

$$\frac{t^2 c^5}{hG} \tag{11.9.1}$$

From here, the specific time t_P and the specific length l_P that can be called the Planck time and length are equal to [10]

$$t_P = \sqrt{\frac{hG}{c^5}} = 1.3 \times 10^{-43}\,\text{s}, \quad l_P = ct_P = 4 \times 10^{-35}\,\text{m} \tag{11.9.2}$$

These values are of great interest because they are made from the fundamental constants characterizing the main physical phenomena of nature, namely the gravitation, the relativity, and the quantum property of the micro-world. If these constants are absolute, then the Planck time and length may be the quanta of the space–time in the future Unified Theory.

According to Eq. (11.8.2), we have $R = 3 \times 10^{-14}$ m when $t = t_P$ (the Planck epoch). And so, at that time of the Big Bang the pro-universe was a little bit greater in size than the nucleus of the helium atom at our epoch.

As seen, qualitatively the NEOC approach supports all substantial points of the Big Bang theory, the *FLRW* model, and ΛCDM model which are based on the general relativity of a mystically curved universe, while the current theory uses only well-measured data within the framework of the classical mechanics of the common flat space. However, both fail to explain the nature of the Big Bang.

11.10 Revolution of the Universe and the Dark Energy

In planetary systems of stars, as well as in galaxies, clusters, and super-clusters, the gravitating masses revolve around a center of gravity of the corresponding system. As "a birth defect," this revolution is caused by an asymmetry of an original system arisen, for example, when a cloud of dust and gas collapsed. Due to the law of the angular momentum conservation, the denser collapsed system has the greater angular velocity of its rotation. And so, the greatest angular velocities are characteristic for neutron stars and Black Holes, the densest objects of nature.

The axis of rotation always goes through the center of gravity of the original cloud, and the gravitation force acting upon each body is directed toward this center. The revolution creates the centrifugal force acting upon each revolving body. However, this force can balance only the component of the gravity force which is perpendicular to the axis of rotation. The component of the gravity force directed along the axis of rotation is unbalanced; this component moves all revolving bodies onto one and same plane which is perpendicular to the axis of rotation and goes through the center of gravity.

And so, as a result, all revolving systems of gravitating bodies become flat or close to flat depending on the age of the system. The solar system is practically flat, and all planets lie in one and same plane. The fractal dimension of the Milky Way is about 2.2 so that this galaxy is an area-like fractal close to a pancake by its general shape. The decrease of the angular momentum of a revolving system can occur only due to an emission of energy from the system.

This general property of gravitating systems makes us suggest that the universe as well revolves in the space around a certain axis going through the center of gravity of the universe. This *hypothesis* is especially alluring because it allows us to give a very simple explanation of the Dark Energy, the most obscure and even mystic subject of cosmology at the present time.

Indeed, in the revolving universe every mass experiences the centrifugal force which is equal to the product of the mass, the square of its angular velocity, and its distance to the axis of rotation. Let us assume that this *centrifugal force of the revolving universe is the repulsion force of the Dark Energy acting upon any gravitating mass.*

From here, using Eqs. (11.4.6) and (11.4.7), we get

$$\omega^2 = \frac{4\pi}{3} G\Lambda \tag{11.10.1}$$

Here, ω is the angular velocity of the universe. It should be emphasized that it is an average quantity for all gravitating objects of the universe; the values of ω for particular objects can differ very much.

Substituting Λ in Eq. (11.10.1) by Eq. (11.7.4), we have

$$\omega = 1.33 \times 10^{-18}\ \text{s}^{-1} \tag{11.10.2}$$

This small angular velocity of the universe cannot be ever detected by human beings. At such an angular velocity, for 12.3 billion years the universe would turn only by angle 30°.

The same and even more difficult problem is to detect the center of gravity of the universe and the axis of its rotation. It should be mentioned that the notions of the gravity center and rotation axis of the universe contradict to the Copernican principle accepted in the general relativity. These are most essential objections against the present neoclassical theory. However, they do not matter, since in the NEOC approach all observers within the 100 MPc distance around the Earth and about one million years apart are equivalent because they would observe one and same picture of the large-scale universe.

And so, it is, probably, impossible to prove or disprove the current simple approach to the Dark Energy using direct measurements of the angular velocity or angular displacement of the universe. Still, this NEOC outcome has some evident merits because the man used to prefer understandable things to mystic ones—this is our life experience.

The hypothesis of the revolving universe leads us as well to the conclusion that the fractal dimension of the universe should be less than 3, and in the long run, it should approach to 2 similar to clusters, galaxies, planetary systems, and any revolving systems of gravitating masses.

11.11 Modified Evolution Equations of the Universe

The revolution can drastically change the dynamics of the evolution of the expanding universe, particularly, because it makes the cosmological constant vary with time. Let us derive the basic equations of the revolving and expanding universe. Because the symmetric expansion is unstable, any asymmetry in the process of the expansion of the universe from the Big Bang could produce a moment of force and revolution.

The moment of force causing the revolution is equal to the product of an eccentric force and its distance from the axis of rotation. From the dimensional analysis, it follows that the perturbation force is directly proportional to the gravity of eccentric masses, and that their distance from the axis is proportional to the radius of the universe. Based on these assumptions, the equation of the rotational dynamics of the universe can be written as follows:

$$\frac{\mathrm{d}}{\mathrm{d}t}\left(c_i M R^2 \omega\right) = \frac{G M m_\mathrm{e}}{R^2} \cdot c_\mathrm{e} R \qquad (11.11.1)$$

Here, M, R, and ω are the mass, maximum radius, and average angular velocity of the universe; m_e is some eccentric mass; and c_i and c_e are some dimensionless coefficients depending on the shape of the universe and on the position of eccentric masses.

In particular, the value of c_i is equal to 0.4 for solid spheres, 0.2 for thin circular plates, and $0.2 + 0.2(b/R)^2$ for oblate solid spheroids of the maximum thickness $2b(0 \leq b \leq R)$. The latter two are some possible shapes the universe can get due to the revolution.

The left-hand part of Eq. (11.11.1) represents the rate of the angular momentum, and the right-hand part the moment of eccentric forces. Let us rewrite Eq. (11.11.1) as follows:

$$r\frac{\mathrm{d}}{\mathrm{d}\tau}\left(ar^2\right) = P_{\mathrm{E}} \quad \text{where} \quad a = a(\tau) \tag{11.11.2}$$

Here, r, τ, a, and P_{E} are the following dimensionless parameters

$$r = \frac{R}{R_*}, \quad \tau = \frac{t}{T_*}, \quad a = \omega T_*, \quad P_{\mathrm{E}} = \frac{G m_{\mathrm{e}} c_{\mathrm{e}} T_*^2}{c_i R_*^3}, \tag{11.11.3}$$

where R_* and T_* are the radius and age of the expanding and revolving universe at some specific epoch.

Parameter P_{E} called the eccentricity parameter characterizes the dynamics of the revolving universe. Its value is determined by some unknown disturbances of the symmetric distribution of masses in the early universe at the time of the Big Bang, with this "birth defect" asymmetry being remained forever. In the current NEOC model of the expanding and revolving universe, the value of P_{E} is an empirical constant like the gravitational constant.

Based on Eqs. (11.5.1), (11.10.1), and (11.11.3), the main equation, Eq. (11.6.1), of the universe expansion takes the following shape:

$$\frac{\mathrm{d}^2 r}{\mathrm{d}\tau^2} = ra^2 - \frac{G_*}{r^2} \tag{11.11.4}$$

Here, G_* is the dimensionless gravitational number of the universe at some specific epoch. It is equal to

$$G_* = \frac{GMT_*^2}{R_*^3} = \frac{4\pi}{3}\rho_* G T_*^2 \tag{11.11.5}$$

Here, ρ_* is the average density of the universe at the specific epoch.

The equation system, Eqs. (11.11.2) and (11.11.4), determines the evolution of the revolving and expanding universe in this NEOC model. The solution of this system for $0 < \tau < \infty$ should meet the following boundary conditions:

$$r = 0, \quad a \to \infty \quad \text{when} \quad \tau = 0; \tag{11.11.6}$$

$$r = 1, \quad \text{when} \quad \tau = 1. \tag{11.11.7}$$

It is easy to find the following singular solution to Eqs. (11.11.2) and (11.11.4) satisfying the boundary conditions, Eqs. (11.11.6) and (11.11.7)

$$r = \tau^{2/3}, \quad a = 3P_{\mathrm{E}}\tau^{-1}, \quad P_{\mathrm{E}} = \frac{1}{3}\sqrt{G_* - \frac{2}{9}} \tag{11.11.8}$$

From here, it follows that the Hubble parameter corresponding to this singular solution is equal to

$$H = \frac{2}{3t}. \tag{11.11.9}$$

According to Eqs. (11.11.5), (11.11.8) and (11.11.9), this singular solution is valid when $G_* > 2/9$ so that the density of the universe at the specific epoch is greater than a certain critical value ρ_{c}

$$\rho_* > \rho_{\mathrm{c}} = \frac{3H_*^2}{8\pi G} = \frac{1}{6\pi G T_*^2}. \tag{11.11.10}$$

It is interesting that this critical density of the revolving and expanding universe coincides with the density of the early universe in the model of the expanding, but not revolving universe at this specific epoch; see Eq. (11.8.3).

11.12 Orbital Velocities of Stars in Spiral Galaxies

The average orbital velocity V of a planet of mass m in the solar system is determined by the balance equation of the inertia force mV^2/R and the force GmM/R^2 of the attraction to the Sun of mass M where R is the average distance between the planet and the Sun. And so, $V = (GM/R)^{1/2}$ so that the orbital velocity decreases and tends to zero when this distance increases. Since $\mathrm{Ch} \gg 1$ for the Milky Way and all other galaxies, the orbital velocity of stars in galaxies seems to be described by the same law. However, based on the astrophysical data, this velocity practically does not depend on the distance between a star and the galaxy center, and it is usually equal to about 220–260 km/s.

This paradox evoked a number of theories. The modified Newton's dynamics (MOND) theory rejects Newton's Law of inertia and replaces it by another law, according to which the inertia force equals $\xi\left(mV^2/R\right)^2$ where ξ is a mass coefficient. From here and the balance equation, it follows that the orbital velocity does not depend on the distance. According to another theory, this paradox is caused by the effect of the Dark Energy; however, it is not true because $\mathrm{Ch} \gg 1$ in this case.

Meanwhile, the flat, spiral structure of the Milky Way and other galaxies allows us to provide a simple explanation and prediction of the paradox within the framework of the classical mechanics and Newton's Law of gravity. Indeed, it is reasonable to assume that the gravitational matter of the Milky Way is uniformly distributed along some logarithmic spirals with the common pole at the center of the disk of the Milky Way. These spirals are well-documented.

The length of an arc of a logarithmic spiral is equal to $(R_2 - R_1)/\cos\beta$ where β is the spiral constant equal to the angle between the radius vector of a point on the spiral and the tangent to the spiral at the point, and R_2 and R_1 is the distance between the pole and the ends of the arc. When $R_2 \gg R_1$, the length of the arc is directly proportional to the distance $R = R_2$ between this point and the center of the galactic disk which is the pole of the logarithmic spirals.

And so, mass M of the galactic disk inside radius R is directly proportional to R for any number of spirals

$$M = kR \tag{11.12.1}$$

Here, k is a galactic constant.

For the Milky Way, $M = 1.4 \times 10^{42}$ kg when $R \sim 1.5 \times 10^{21}$ m so that

$$k \approx 10^{21}\ \text{kg/m} \tag{11.12.2}$$

From Eq. (11.12.1), it follows that the gravitational force attracting a star of mass m to the center of the galaxy is equal to kmG/R. This force is balanced by the centrifugal force of inertia of the star which is equal to mV^2/R. From here, we get the orbital velocity of stars in spiral galaxies [5]

$$V = \sqrt{kG} \tag{11.12.3}$$

According to Eqs. (11.12.2) and (11.12.3), the average orbital velocity of stars in the Milky Way is equal to about $250\frac{\text{km}}{\text{s}}$. This result of calculation is confirmed by astrophysical data.

From Eq. (11.12.1), it follows that the distribution of gravitating masses in spiral galaxies obeys the following law [5]:

$$\rho = \frac{k}{2\pi R} \tag{11.12.4}$$

Here, ρ is the gravitating mass per unit area of the galactic disk, and R is the distance from the galactic center.

The NEOC law expressed by Eq. (11.12.3) is, evidently, valid also for any galaxies which gravitational matter, including the Dark Matter, is distributed in the galactic disk similar to Eq. (11.12.4). And so, the present NEOC approach to this problem discards the viewpoint of *Rubin (1928–2015),* an American astronomer, who explained this anomaly by the effect of the Dark Matter [11].

11.13 The Fractal Universe

As it was first shown by *Mandelbrot* (1924–2010), a French scientist, all geometrical objects of nature including the universe represent some fractals which length, area, and volume depend on the scale of measurement. For example, the length of the coastline of the continents on the Earth substantially depends on whether we measure it using maps in the scale of 1 mile, or 100 miles (per inch of a map), or in the scale of 1 ft by a walker.

The results of the measurements of the coastline length can be approximated as follows [12]:

$$\frac{L}{L_0} = \left(\frac{\Delta}{\Delta_0}\right)^{1-d_f}, \quad \text{or,} \quad \log\frac{L}{L_0} = (1-d_f)\log\frac{\Delta}{\Delta_0}. \tag{11.13.1}$$

Here, L is the length in the scale of Δ; L_0 is the length in the scale of Δ_0; and d_f is the fractal dimension. Fractals which are characterized by the power-law function of Eq. (11.13.1) are called the power-law fractals.

For the line-like fractals like a coast line, the fractal dimension varies in the range

$$1 \le d_f \le 1+\theta \quad \text{where} \quad 0 \le \theta < 1. \tag{11.13.2}$$

The greater is the fractal dimension, the longer is the length L. For example, the fractal dimension of the Australian coast equals $d_f = 1.14$ while that of the Norway coast equals $d_f = 1.49$. If $\theta = 0$, then $d_f = 1$; this is the common metric dimension of length that does not depend on the scale of measurement. Fractal dimension is a measure of the geometrical complexity of a system.

Let us apply this approach to the universe which is surely the most complex fractal object. Suppose we can measure mass M of the universe using some astrophysical devices of various precision allowing us to study the universe within some device-dependent radius R which is the scale of measurement in this case. Evidently, the result of measurement depends on this scale.

As it was shown in the previous section, any rotating system of gravitating masses acquires a shape close to a disk which metric dimension equals 2. Hence, in the long run the fractal dimension of the universe should be close to 2. Based on this assumption, let us interpolate the measurements of the fractal universe by the following power-law function

$$\frac{M}{M_0} = \left(\frac{R}{R_0}\right)^{2-d_f} \quad \text{where} \quad d_f = 2+\delta \quad (\delta \ll 1) \tag{11.13.3}$$

Here, M is the mass of the universe in the scale of R; M_0 is the mass of the universe in the scale of R_0; and d_f is the fractal dimension of the universe.

And so, according to Eq. (11.13.3), the universe has the shape of a thin circular disk of radius R and thickness h, which fractal dimension is close to 2. Although some astrophysical observations support this assumption, the results are still inconclusive and contradictory because the methods of measurement differ very much. Therefore, Eq. (11.13.3) can only serve as a *hypothesis.*

Since $M = \pi \rho h R^2$ in metric dimensions, from here and from Eq. (11.13.3), we can derive

$$h = \frac{\zeta}{\pi \rho} R^{-d_f} \quad \text{where} \quad \zeta = \frac{M_0}{R_0^{2-d_f}}. \tag{11.13.4}$$

This equation allows one to estimate the thickness of the universe using the empirical constant ζ and the fractal dimension of the universe.

11.14 Gravitons and the Unified Theory

Gravitons and gravitational waves are still subject to even more speculations because of their very low intensity. We mention here only one hypothesis, according to which gravitons are the gravitational waves of the maximum possible length λ which has the value of the order of the radius of the universe [5]

$$\lambda = 10^{26}\,\text{m}, \quad \nu = c/\lambda = 3 \times 10^{-18}\,\text{s}^{-1}, \quad m_g = h\nu/c^2 = 10^{-68}\,\text{kg} \tag{11.14.1}$$

Here, ν is the wave frequency, and m_g is the mass energy of gravitons.

According to Eq. (11.14.1), this hypothetical graviton is the smallest elementary particle of nature. *No physical particle can be less.* Its frequency corresponds to the period of the wave of the order of ten billion years which is about the age of the universe. From this hypothesis, it follows that our universe consists of about 10^{120} gravitons.

Each "elementary" particle represents a cluster of a huge number of gravitons. In particular, we get the following figures:

Neutrino mass equals $0.32\,\text{eV}/c^2 = 0.58 \times 10^{-36}\,\text{kg} = 0.26 \times 10^{32}\,m_g$;
Up quark mass equals $2.3\,\text{MeV/c}^2 = 4.1 \times 10^{-30}\,\text{kg} = 7 \times 10^{37}\,m_g$;
Down quark mass equals $4.8\,\text{MeV}/c^2 = 8.6 \times 10^{-30}\,\text{kg} = 2 \times 10^{38}\,m_g$;
Electron and positron mass equals $9.1 \times 10^{-31}\,\text{kg} = 2 \times 10^{38}\,m_g$; and
Gamma photon effective mass equals $1.8 \times 10^{-30}\,\text{kg}$ for $1MeV$ photons.

And so, these gravitons as the smallest elementary particles can pretend to the role of building blocks in the future Unified Theory.

However, this role imposes some strict constrains on the topology of the developing universe because, according to Eq. (11.14.1), the value of the maximum radius of the universe becomes an absolute constant like the speed of light or the gravitational constant. It means that in the earliest stage, the universe represented a rotating whip-like set of gravitating masses of the fractal dimension close to 1 which length was always equal to one and same constant. In the course of time, the shape of the universe gradually changed its geometry from an almost 1D whip to the current almost 2D disk of the maximum radius that is equal to the same constant. To a certain extent, this hypothesis is similar to the Aristotelian concept of "solid sky."

The Planck dimensional analysis that led him to the Planck length and time can be supplemented by the Planck mass, force, energy, stress, momentum, and so on. For example, in terms of the absolute constants, we get:

$$\begin{aligned} &m_\mathrm{p} = \sqrt{\mathrm{ch}/G}, \quad F_\mathrm{p} = c^4/G, \quad E_\mathrm{p} = c^2 m_\mathrm{p}, \\ &\sigma_\mathrm{p} = c^7/hG^2, \quad I_\mathrm{p} = c\sqrt{\mathrm{ch}/G} \end{aligned} \tag{11.14.2}$$

Here, m_p, F_p, E_p, σ_p and I_p are the Planck mass, force, energy, stress, and momentum, respectively. And any physical quantity which depends on time, length and mass can be represented by a similar equation.

From Eq. (11.14.2), we find:

$$\begin{aligned} &m_\mathrm{p} = 5.4 \times 10^{-16}\,\mathrm{kg}, \; F_\mathrm{p} = 1.2 \times 10^{44}\,\mathrm{N}, \\ &E_\mathrm{p} = 48.6\,\mathrm{J}, \qquad\qquad I_\mathrm{p} = 16.3\,\mathrm{N\,s} \end{aligned} \tag{11.14.3}$$

And so, the Planck energy and momentum have some earthly values as distinct from time, length, mass, and force.

If we assume, following the Planck way of reasoning, that the cosmological constant has to be determined by the absolute constants, then we come to the following result

$$\Lambda_\mathrm{p} = \frac{c^5}{hG^2} = 0.82 \times 10^{96}\,\mathrm{kg/m^3}. \tag{11.14.4}$$

This value of the Planck cosmological constant 10^{120} times greater than its value according to all astrophysical data! From this paradox, it may follow that the cosmological constant can also be an absolute constant independent of other absolute constants, or there exist some new undiscovered independent variables beyond mass, length, and time. The universe and nature present us many unresolved puzzles that are far beyond the human imagination.

11.15 Spheroidal Universe

According to Eqs. (11.10.1) and (11.11.2), the angular velocity and the cosmological constant of the universe varied with time. It is reasonable to assume that due to the rotation the universe should, at least at the initial stage of a very dense state, have the shape of an oblate or prolate spheroid, with the axis of rotation being the main axis of the spheroid. For the expanding and revolving universe, the evolution equation,

Eq. (11.11.4), must be modified in order to take into account the rotation and the spheroidal shape of the universe.

The initial stage of the universe history. Suppose that at the initial stage of its history, just after the Big Bang, the homogeneous universe revolving around axis x_3 is inside the following spheroid

$$\frac{x_1^2 + x_2^2}{R^2} + \frac{x_3^2}{\lambda^2 R^2} = 1 \tag{11.15.1}$$

Here, λ is the eccentricity parameter of the spheroid expressed as follows:

$$\text{for the oblate spheroid} \quad \lambda = \sqrt{1 - e^2} \quad (0 \le e \le 1, \quad \lambda \le 1), \tag{11.15.2}$$

$$\text{for the prolate spheroid} \quad \lambda = 1/\sqrt{1 - \kappa^2} \quad (0 \le \kappa \le 1, \quad \lambda \ge 1). \tag{11.15.3}$$

The mass of this spheroid is equal to

$$M = \frac{4\pi}{3} \lambda \rho R^3 \tag{11.15.4}$$

The components X_1, X_2 and X_3 of the gravitation force upon a unit point mass on the surface of this spheroid are equal to [13]:

For the oblate spheroid:

$$X_i = -2\pi \rho G x_i \sqrt{1 - e^2} \frac{1}{e^3} \left(-\sqrt{1 - e^2} + \frac{\text{arcsine}}{e} \right) \quad (i = 1, 2) \tag{11.15.5}$$

$$X_3 = -4\pi \rho G x_3 \frac{1}{e^3} \left[e - \sqrt{1 - e^2} \arctan \left(\frac{e}{\sqrt{1 - e^2}} \right) \right] \tag{11.15.6}$$

For the prolate spheroid:

$$X_i = -2\pi \rho G x_i \frac{1}{\kappa^3} \left[\kappa - \frac{1}{2} (1 - \kappa^2) \ln \frac{1 + \kappa}{1 - \kappa} \right] \quad (i = 1, 2) \tag{11.15.7}$$

$$X_3 = -2\pi \rho G x_3 \frac{1 - \kappa^2}{\kappa^3} \left(-2\kappa + \ln \frac{1 + \kappa}{1 - \kappa} \right) \tag{11.15.8}$$

For the sphere, Eqs. (11.15.5)–(11.15.8) provide

$$X_i = \frac{4\pi}{3} \rho G x_i \quad (i = 1, 2, 3) \tag{11.15.9}$$

From Eqs. (11.15.5) to (11.15.8), we can also derive the following useful asymptotes.

For the thin disk of radius R, when $e \to 1$:

$$X_1 = X_2 = 0, \quad X_3 = -3\frac{GM}{R^2} \quad (x_1 = x_2, x_3 \to 0) \tag{11.15.10}$$

$$X_1 = -\frac{3\pi}{4}\frac{GM}{R^2}, \quad X_2 = X_3 = 0 \quad (x_1 = R, x_2 = x_3 = 0) \tag{11.15.11}$$

For the equivalent cylinder of radius R and length $L \gg R$, when $\kappa \to 1$:

$$X_1 = X_2 = 0, \quad X_3 = \frac{32}{9}\frac{GM}{L^2} ln(1-\kappa) \tag{11.15.12}$$

$$\begin{aligned} &\left(x_1 = x_2 = 0, \quad x_3 = L = \frac{4R}{3\sqrt{1-\kappa^2}}\right); \\ &X_1 = -2\pi\rho\, GR, \quad X_2 = X_3 = 0 \\ &\left(x_1 = R = \frac{3}{4}L\sqrt{1-\kappa^2}, \quad x_2 = x_3 = 0\right). \end{aligned} \tag{11.15.13}$$

According to Eqs. (11.15.5) and (11.15.7), the gravitational force upon a unit probe mass at the edge of the universe when $x_3 = 0$, $x_1^2 + x_2^2 = R^2$ is equal to

$$F = -\eta\frac{GM}{R^2}, \tag{11.15.14}$$

where

$$\eta = \frac{3}{2e^3}\left(-e\sqrt{1-e^2} + \text{arcsine}\right) \quad \text{for the oblate spheroid,} \tag{11.15.15}$$

$$\eta = \frac{3}{2}\frac{\sqrt{1-\kappa^2}}{\kappa^3}\left[\kappa - \frac{1}{2}(1-\kappa^2)\ln\frac{1+\kappa}{1-\kappa}\right] \quad \text{for the prolate spheroid.} \tag{11.15.16}$$

Coefficient η describes the effect of eccentricity of the universe. For the oblate spheroid, it monotonously grows from $\eta = 1$ when $e = 0$ (sphere) to $\eta = 3\pi/4$ when $e = 1$ (thin flattened spheroid). For the prolate spheroid, it monotonously decreases from $\eta = 1$ when $\kappa = 0$ (sphere) to $\eta = 3R/L$ for the spheroid length $L \gg R$ when $\kappa \to 1$. For a mass on the edge of the oblate universe, this effect monotonously grows when eccentricity e increases so that the gravity of the thin disk is $3\pi/4 \approx 2.35$ times greater than the gravity of the sphere of the same mass and radius. The gravitation force upon the unit probe mass at $x_1 = R, x_2 = x_3 = 0$ on the surface of a very long prolate spheroid of mass M is equal to $3GM/(RL)$ where L is the length of the spheroid.

Let us place a probe mass at point $x_1 = x_2 = 0, x_3 = \lambda R$ on the axis of rotation at the edge of the universe. Because the centrifugal force is zero at this point, due to Eqs. (11.15.1), (11.15.6), (11.15.8), and (11.15.9), the equilibrium equation of the probe mass takes the following shape

$$\frac{\mathrm{d}^2}{\mathrm{d}t^2}(\lambda R) = -\zeta \frac{MG}{R^2}, \tag{11.15.17}$$

where

$$\zeta = \frac{3}{e^3}\left[e - \sqrt{1-e^2}\arctan\left(\frac{e}{\sqrt{1-e^2}}\right)\right] \quad \text{for the oblate spheroid,} \tag{11.15.18}$$

$$\zeta = \frac{3}{2}\frac{1-\kappa^2}{\kappa^3}\left(-2\kappa + \ln\frac{1+\kappa}{1-\kappa}\right) \quad \text{for the prolate spheroid.} \tag{11.15.19}$$

For the oblate spheroid, coefficient ζ monotonously grows from $\zeta = 1$ when $e = 0$ (sphere) to $\zeta = 3$ when $e \to 1$ (thin spheroidal disk).

For the prolate spheroid, coefficient ζ monotonously decreases from $\zeta = 1$ when $\kappa = 0$ (sphere) to $3\lambda^{-2}\ln\lambda \to 0$ when $\lambda \to \infty$ $(\kappa \to 1)$ for a very long spheroid.

For the probe mass at $x_3 = 0, x_1^2 + x_2^2 = R^2$, due to Eq. (11.15.14), the equilibrium equation has the following shape:

$$\frac{\mathrm{d}^2 R}{\mathrm{d}t^2} = a^2 R - \eta \frac{MG}{R^2} \tag{11.15.20}$$

Here, η is defined by Eqs. (11.15.15) and (11.15.16).

The equation system, Eqs. (11.11.2), (11.11.4) and (11.15.15) to (11.15.20), describes the evolution of the expanding and revolving universe in time, with $R(t)$, $a(t)$ and $e(t)$ or $\kappa(t)$ characterizing the change of the angular velocity, size, and shape of the universe in terms of time at the initial stage of its development.

In the current model, the initial stage means that $0 < t < T_*$ and $0 < R < R_*$ where $t = 0$, $R = 0$ corresponds to the Big Bang, and $t = T_*$, $R = R_*$ corresponds to a specific epoch that characterizes the transition of the growth of the universe from the initial stage to the next stage. The values of T_* and R_* should be found using the conditions of the smooth transition.

The final stage is characterized by one evolution equation, Eq. (11.15.20), where $a = \omega = \text{const}$ is determined by Eq. (11.10.1) and coefficient η has to be found from the smooth transition conditions that can be written as follows:

$$\text{when} \quad t = T_* : \quad a = \omega, \quad R = R_*, \quad [e] = 0, \quad [\dot{R}] = 0 \tag{11.15.21}$$

The last two equations require the continuity of the expansion rate and shape of the universe at the transition epoch.

Let us study the initial stage controlled by the system of the following three equations:

$$\frac{d^2 R}{dt^2} = a^2 R - \eta \frac{MG}{R^2}, \quad \frac{d^2}{dt^2}(\lambda R) = -\zeta \frac{MG}{R^2}, \quad R \frac{d}{dt}\left(a R^2\right) = \frac{R_*^3}{T_*^2} P_E \quad (11.15.22)$$

Here, coefficient λ is a function of e or κ given by Eqs. (11.15.2) or (11.15.3). The solution of this equation system is given by the following theorem:

Theorem 1 *The exact solution of the equation system,* Eq. (11.15.22), *satisfying the initial boundary value conditions $R \to 0, a \to \infty$ when $t \to 0$, is given by the following functions*

$$R = \mu(p)(MG)^{1/3} t^{2/3}, \quad a = \delta(p) t^{-1}, \quad \lambda = \lambda(p) \quad (t \le T_*) \quad (11.15.23)$$

where functions $\mu(p)$, $\delta(p)$ and $\lambda(p)$ *are to be found, and*

$$p = \frac{MGT_*^2}{P_E R_*^3}. \quad (11.15.24)$$

Proof According to Eq. (11.15.22) and to the initial boundary value conditions, the problem-solving functions R, a and λ can depend only on $t(s)$, $MG\left(m^3/s^2\right)$, and $P_E R_*^3/T_*^2\left(m^3/s^2\right)$. From here, based on the simple analysis of dimensions, it follows that the solution of the equation system, Eq. (11.15.22), satisfying the initial boundary value conditions can be written in the only shape of Eqs. (11.15.23) and (11.15.24). The theorem is proven.

According to Eq. (11.15.23), the universe occasionally shaped as an ellipsoid at the Big Bang retains its eccentricity, at least, at an initial stage of its development.

Substituting R, a and λ in Eq. (11.15.22) by the corresponding functions in Eq. (11.15.23) provides the following three algebraic equations:

$$-\frac{2}{9}\mu(p) = \mu(p)\delta^2(p) - \eta\mu^{-2}(p) = -\frac{\zeta}{\lambda}\mu^{-2}(p), \quad (11.15.25)$$

$$p\delta(p)\mu^3(p) = 3. \quad (11.15.26)$$

Also, the first two smooth condition equations in Eq. (11.15.21) are met, if

$$R_* = \mu(p)(MG)^{1/3} T_*^{2/3}, \quad \delta(p) = \omega T_*. \quad (11.15.27)$$

Then, other two smooth condition equations in Eq. (11.15.21) are satisfied automatically.

Five algebraic equations, Eqs. (11.15.25)–(11.15.27), serve to determine five unknown functions p, μ, δ, R_* and T_*.

Let us transform three equations, Eqs. (11.15.25) and (11.15.26), to the following ones

$$\delta(p) = \frac{2\lambda}{3p\zeta}, \quad \mu^3(p) = \frac{9\zeta}{2\lambda}, \quad p^2 = \frac{2\lambda^2}{\zeta[\eta\lambda - \zeta]}. \quad (11.15.28)$$

Using Eq. (11.15.27) and the first equation in Eq. (11.15.28), let us find R_* and T_* in terms of p and λ

$$R_*^3 = \frac{2MG\lambda}{\omega^2 p^2 \zeta}, \quad T_* = \frac{2\lambda}{3\omega p \zeta}. \tag{11.15.29}$$

gravitational force upon a unit

Substituting R_* and T_* in Eq. (11.15.24) by Eq. (11.15.29) provides

$$p = \frac{2\lambda}{9P_{\mathrm{E}}\zeta}. \tag{11.15.30}$$

Now, using Eq. (11.15.30) and the last equation in Eq. (11.15.28), we find the characteristic equation which determines the eccentricity of the universe in terms of the original "birth defect" coefficient P_E

$$40.5P_{\mathrm{E}}^2\zeta = \eta\lambda - \zeta. \tag{11.15.31}$$

Let us analyze Eqs. (11.15.2) to (11.15.3), (11.15.15) to (11.15.16), and (11.15.18) to (11.15.19) in terms of e or κ. Evidently, for the oblate spheroid we have $\lambda < 1$ and $\zeta > \eta$ so that the characteristic equation, Eq. (11.15.31), *cannot be met.* For the prolate spheroid, we have $\lambda > 1$ and $\eta > \zeta$ so that Eq. (11.15.31) can be satisfied, and we come to the following theorem.

Theorem 2 *The universe has the shape of the prolate spheroid which eccentricity* κ *is uniquely determined in terms of the "birth defect" coefficient* P_{E} *by the following characteristic equation:*

$$P_E^2 = \frac{2}{81}\left[\lambda(\kappa)\frac{\eta(\kappa)}{\zeta(\kappa)} - 1\right] \quad (0 \le \kappa \le 1) \tag{11.15.32}$$

Here, functions $\lambda(\kappa)$, $\eta(\kappa)$ and $\zeta(\kappa)$ are given by Eqs. (11.15.3), (11.15.16) and (11.15.19).

Function $C(\kappa)$ given by the following equation

$$C(\kappa) = \lambda(\kappa)\frac{\eta(\kappa)}{\zeta(\kappa)} - 1 \tag{11.15.33}$$

monotonously grows when κ increases, from $C = 0$ when $\kappa = 0$ (sphere) to infinity when $\kappa \to 1$ (cylinder) so that the "birth defect" coefficient equals zero for the spherical, "defectless" universe. And so, one and only one value of eccentricity κ corresponds to any value of parameter P_E. Theorem 2 is proven. At the initial stage of its development, the universe in this model had the shape of a prolate spheroid with eccentricity $\kappa(P_{\mathrm{E}})$ defined by Eqs. (11.15.32) and (11.15.33).

11.16 The Super-Photon Hypothesis

The Big Bang theory and the NEOC cosmology cannot answer the question of the origin of the universe because these theories are phenomenological. Most advantageous and alluring is the idea that the current universe was born from a certain primordial elementary particle. Let us analyze this idea.

From the position of the generally recognized Standard Model of the modern physics, everything is built from elementary particles, the "zoo" of which consists of 61 "species" including quarks, neutrinos, photons, electrons, gluons, muons, and others, plus their antiparticles, the Higgs boson, and the hypothetical graviton. Every elementary particle is characterized by its mass, charge, spin, and by some other properties of the second order. In principle, every particle can be split into some other elementary particles in a fission process being controlled by the conservation laws of energy, momentum, and angular momentum.

Of most interest are stable elementary particles which lifetime is infinite. These are photon, gluon, neutrino, electron, and some quarks. Mass and charge are very elusive properties, being easily converted into energy. Neutrino has been recently proved to have mass $0.32 \pm 0.08\,\mathrm{eV/c^2}$ which is the least among all particles with some mass. Therefore, photons and gluons that the only ones that have zero masses can be the first in the row to pretend to the role of the primordial elementary particle. However, gluons "glue" massive quarks and hence can be omitted from this consideration.

And so, it is reasonable to suggest that *everything came from a primordial photon* we will call the *super-photon.* The super-photon had zero mass, zero charge, and the spin equal to $+1$ or -1, if the current universe rotates. If the current universe does not rotate, a pair of the super-photons of different spin could be the original universe.

Based on the energy conservation law, the following equation is valid

$$E^2 = m^2c^4 + p^2c^2 \tag{11.16.1}$$

Here, E, m and p are the total energy, mass, and momentum of any elementary particle, and c is the speed of light in vacuum.

For the zero mass super-photon, we have

$$p = \frac{E}{c}, \quad E = h\nu, \quad \lambda = \frac{c}{\nu} \tag{11.16.2}$$

Here, λ and ν are the wavelength and frequency of the super-photon, and h is Planck's constant equal to 6.6×10^{-34} J s.

Let us estimate the total energy of the universe at the Big Bang including only the ordinary matter and the Dark Matter. We assume that the total energy is equal to the kinetic energy plus the rest energy what is strictly right only for velocities much less than the speed of light. The rest energy of the universe is equal to Mc^2 where M is the gravitational mass of the universe (the ordinary and Dark Matter). The kinetic energy at the Big Bang in the NEOC cosmology depends only on M, MG, and c. Hence, it is equal to ζMc^2 where ζ is a number which can be estimated using the assumption that at the Big Bang, the velocity $v(r)$ varied linearly with radius and achieved the speed of light at the edge of the universe. (As a reminder, in the NEOC model the velocity of matter inside the universe was out of consideration.) Then, we get

$$\int_0^R 0.5\rho v^2(r)\mathrm{d}V = 2\pi c^2 \int_0^R (r/R)^2 r^2 \mathrm{d}r = 0.3Mc^2 \quad (11.16.3)$$

Thus, the total energy of the universe at the Big Bang was equal to 1.3 Mc^2 in this approximation assuming that heat energy of the universe is much less than its mass energy. This energy should be the same at any time in the history of the universe, if there have been no losses of energy as we assume.

According to the current knowledge, the mass of the ordinary matter of the universe is equal to about 10^{53} kg so that based on the last data of the Planck mission the total mass of the gravitational matter of the universe including the Dark Matter is 31.7/4.9 times greater, that is equal to $M = 2 \times 10^{54}$ kg. From here and Eq. (11.16.3), it follows that the primordial super-photon had the following energy, frequency, and wavelength

$$E = h\nu = 1.3Mc^2 = 2.3 \times 10^{71} J,$$
$$\nu = \frac{1.3Mc^2}{h} = 3.5 \times 10^{104} \frac{1}{s},$$
$$\lambda = \frac{c}{\nu} = 1.4 \times 10^{-96} \text{ m}.$$

The following possible chain of physical transformations of the super-photon leading to the current universe might go through myriad scenarios which all could well comply with modern physical knowledge.

Literature

1. H. Minkowski, Raum und Zeit. Physicalische Zeitschrift **10**, 75–88 (1907)
2. A. Einstein, Die Grundlage der allgemeinen Relativitatstheory. Ann. Phys. **49**, 769–822 (1918)
3. A.A. Friedman, Uber die Krummung des Raumes. Z. Angew. Phys. **10**(1), 377–386 (1922)
4. E.P. Hubble, A relation between distance and radial velocity among extra-galactic nebulae. Proceed. Natl. Acad. Sci. **15**(3), 168–173 (1929)
5. G.P. Cherepanov, Some new applications of the invariant integral in mechanics. J. Appl. Math. Mech. (JAMM) **76**(5), 519–536 (2012)
6. C.L. Bennet, Wilkinson microwave anisotropy probe (WMAP), in *Questions of Modern Cosmology: Galileo's Legacy*, ed. by M. D'Onofrio, C. Burigana (Springer, Berlin, 2009), 525 pp
7. M. Camenzind, *Compact Objects in Astrophysics: White Dwarfs, Neutron Stars and Black Holes* (Springer, Berlin, 2007), 588pp. ISBN 978-3-540-25770-7
8. S. Weinberg, *Cosmology* (Oxford University Press, 2008), 616 pp

9. S.W. Hawking, Black holes in general relativity. Commun. Math. Phys. **25**(2), 152–166 (1972)
10. Max Planck, Uber irreversible Strahlungsvergange. Sitzungsberichte der Koniglich Preusischen Akademie der Wissenschaften zu Berlin **5**, 440–480 (1899)
11. V.C. Rubin, *Bright Galaxies, Dark Matter* (Woodbury, New York, 1997)
12. B. Mandelbrot, *The Fractal Geometry of Nature* (Freeman, New York, 1982)
13. F.R. Moulton, *An Introduction to Celestial Mechanics* (Macmillan, New York, 1902), 285 pp
14. G.P. Cherepanov, *Methods of Fracture Mechanics: Solid Matter Physics* (Kluwer, Dordrecht, 1997), 312pp. ISBN 0-7923-4408-1
15. G.P. Cherepanov, *Fracture Mechanics* (ICS Publishers, Moscow-Izhevsk, 2012), 850 pp (in Russian). ISBN 978-5-4344-0036-7
16. G.P. Cherepanov, The invariant integral: some news, in *Proceedings of the ICF-13*, vol 6, *Sir Alan Howard Cottrell Symposium* (Bejing, China, 2013), pp. 4953–4962. ISBN 978-1-6299-3369-6
17. G.P. Cherepanov, The invariant Integral of physical mesomechanics as the foundation of mathematical physics: some applications to cosmology, electrodynamics, mechanics and geophysics. Phys. Mesomech. **18**(1), 3–21 (2015)
18. G.P. Cherepanov, The mechanics and physics of fracturing: application to thermal aspects of crack propagation and to fracking. Philos. Trans. A Proc. Roy. Soc. **A373**, 0119 (2014). https://doi.org/10.1098/rsta.2014.0119
19. G.P. Cherepanov, A neoclassic approach to cosmology based on the Invariant Integral, Chapter 1 in*: Horizons of World Physics*, vol 288 (Nova Science Publ., New York, 2016), pp 3–35. ISBN: 978-1-63485-905-9
20. G.P. Cherepanov, The large-scale Universe: the past, the present and the future. Phys. Mesomech. **19**(4), 3–20 (2016)

Biographical Sketch and Some Publications of the Author Relevant to Invariant Integrals in Physics

G.P. Cherepanov, Some new applications of the invariant integrals of mechanics. J. Appl. Math. Mech., **76**(5), 519–536 (2012); Crack propagation in continuous media. J. Appl. Math. Mech. **31**(3), 503–512 (1967); Invariant Γ-integrals and some of their applications. J. Appl. Math. Mech., **41**(3), 475–493 (1977); Invariant Γ-integrals. J. Engng Fract. Mech. **14**(1), 3–25 (1981); Invariant integrals in continuum mechanics. Soviet Appl. Mech. **26**(7), 3–16 (1990); The invariant integral of physical mesomechanics as the foundation of mathematical physics: some applications to cosmology, electrodynamics, mechanics and geophysics. Phys. Mesomech. **18**(3), 203–212 (2015); Invariant integrals: from Euler and Eshelby to most recent application, in *Advances in Mathematics Research* (Nova Publ., 2018), pp. 1–21.

G.P. Cherepanov, A.A. Borzyh, Theory of the electron fracture mode in solids. J. Appl. Phys. **74**(12), 7134–7154 (1993).

G.P. Cherepanov, Rolling: an introduction to tribology, in *Advances in Tribolgy Research* (Nova Publ., 2018), pp. 1–32.

G.P. Cherepanov, On a non-linear problem in the theory of analytical functions. Dokl. USSR Acad. Sci. (Math.), **147**(3), 566–568 (1962).

G.P. Cherepanov, Stresses in an inhomogeneous plate with cracks. Notices USSR Acad. Sci. (Mech.), **1**, 131–138 (1962) (in Russian).

G.P. Cherepanov, On a method of solution of elastic-plastic problems. J. Appl. Math. Mech. **27**(3), 428–435 (1963).

G.P. Cherepanov, The Riemann-Hilbert problem for cuts along a straight line or circumference, Dokl. USSR Acad. Sci. (Math.), **156**(2), 275–277 (1964).

G.P. Cherepanov, Boundary value problems with analytical coefficients. Dokl. USSR Acad. Sci. (Math.), **161**(2), 312–314 (1965).

G.P. Cherepanov, Some problems with unknown boundaries in the theory of elasticity and plasticity, in *Applications of the Theory of Functions in Continuum Mechanics,* vol 1 (Nauka Publ., Moscow, 1965), pp 135–150 (in Russian).

G.P. Cherepanov, On the problem of non-uniqueness in the theory of plasticity. Dokl. USSR Acad. Sci. **218**(4), 1124–1126 (1974).

G.P. Cherepanov, *Mechanics of Brittle Fracture* (McGraw Hill, New York, 1978), 950pp.

G. P. Cherepanov, *Invariant Integrals in Physics*,
https://doi.org/10.1007/978-3-030-28337-7

G.P. Cherepanov, *Fracture Mechanics* (ICR Publ., Moscow, 2012), 872pp (in Russian).
G.P. Cherepanov, *Methods of Fracture Mechanics: Solid Matter Physics*, (Kluwer, Dordrecht, 1997), 300pp.
G.P. Cherepanov, Theory of rolling: solution of the Coulomb problem. J. of Appl. Mech. and Tech. Physics, **55**(1), 182–189 (2014).
G.P. Cherepanov, Contact problem of the theory of elasticity with stick-and-slip zones, with application to the theory of rolling and tribology. J. Appl. Math. Mech., **79**(1), 112–143 (2015).
L.A. Galin, G.P. Cherepanov, Contact elastic-plastic problem for plates. Dokl. USSR Acad. Sci. (Mech.) **177**(1), 56–58 (1967).

Biographical Sketch

Affiliation: The New York Academy of Sciences, New York, USA.
A Honorary Life Member elected on Dec. 8, 1976 together with Linus Pauling and George Polia.
Education: BS from the Moscow Institute of Physics and Technology (1960), Ph.D. from the Moscow State University (1962), Dr. Sci. in Physics and Mathematics from the USSR Academy of Sciences (1964), Full Professor of Mathematics certified by the USSR Highest Commission (1970).
Research and Professional Experience: More than 300 refereed papers in two dozens of the American and Russian journals on physics, mechanics and applied mathematics from 1960 to 2018. Many monographs and textbooks on mechanics and physics.
Professional Appointments: Senior Scientist, the Institute of Mechanics, the USSR Academy of Sciences; Full Professor of Mathematics, the Moscow Mining Academy; Head of Laboratory, the Pacific Ocean Studies Institute, the USSR Academy of Sciences; Distinguished Research Associate, Harvard University; and Professor of Mechanical Engineering, Florida International University. Retired since 2000.
Honors: Lenin Komsomol Prize (1972), Fulbright Prize (2000).
Some Publications for last 3 years:
G.P. Cherepanov, Phys. Mesomech. **18**(2), 174–178 (2015); **18**(3), 203–212 (2015); **18**(4), 391–401 (2015); **19**(1), 1–5 (2016); **19**(4), 365–377 (2016); **20**(2), 115–124, (2017).
G.P. Cherepanov, Philosophical Transactions A **373**, the Royal Society, pp 1–13. 20140119.
G.P. Cherepanov, J. Appl. Math. Mech. **79**(1), 112–143 (2015).
G.P. Cherepanov, A neoclassic approach to cosmology based on the invariant integral, Chapter 1, in *Horizons in World Physics*, vol 88 (Nova, New York, 2017), pp. 1–30.
Gena Cher (G.P. Cherepanov's pseudonym), *Poems of Ancient Egypt* (in Russian and English), (Ridero, 2017), pp. 1–440.

G. P. Cherepanov, *Invariant Integrals in Physics*,
https://doi.org/10.1007/978-3-030-28337-7

Printed in the United States
By Bookmasters